Laurence Johnson

A Medical Formulary

Based on the United States and British Pharmacopias

Laurence Johnson

A Medical Formulary
Based on the United States and British Pharmacopias

ISBN/EAN: 9783337187170

Printed in Europe, USA, Canada, Australia, Japan

Cover: Foto ©berggeist007 / pixelio.de

More available books at **www.hansebooks.com**

A

MEDICAL FORMULARY

BASED ON THE

UNITED STATES AND BRITISH PHARMACOPŒIAS

TOGETHER WITH NUMEROUS

FRENCH, GERMAN, AND UNOFFICINAL PREPARATIONS

BY

LAURENCE JOHNSON, A.M., M.D.,

LECTURER ON MEDICAL BOTANY, MEDICAL DEPARTMENT OF THE UNIVERSITY OF THE CITY OF NEW YORK; FELLOW OF THE NEW YORK ACADEMY OF MEDICINE, ETC.

NEW YORK
WILLIAM WOOD & COMPANY
27 GREAT JONES STREET
1881

Copyright
WILLIAM WOOD & COMPANY
1881

Trow's
Printing and Bookbinding Company
201–213 *East 12th Street*
New York

PREFACE.

The design of this work is to present, in a manner convenient for ready reference, the drugs and preparations in common use, together with formulæ illustrating the manner in which they are combined by good practitioners of the present day.

Although confining himself as closely as possible to the pharmacopœias, the author has found it necessary to include some unofficinal drugs and preparations, which have come into general use since the last editions of those works were issued.

In selecting illustrative formulæ, the author has drawn freely upon current medical literature, and has endeavored to give due credit for all material used. While it is impossible for him to enumerate, in this place, all the works which he has employed in this part of his task, he feels it incumbent upon him to make special mention of the Hospital Formulary and Posological Table, by Charles Rice, Ph.D.; The Pharmacopœias of the London Hospitals, by Peter Squire, F.L.S.; and The Pharmacopœia of the Hospital for Diseases of the Throat and Chest (London), by Morell Mackenzie, M.D.

He takes great pleasure also in acknowledging his indebtedness to the following named gentlemen, all of whom kindly furnished him formulæ which they have found useful:

Fordyce Barker, M.D., LL.D., James R. Leaming, M.D., Andrew H. Smith, M.D., A. A. Smith, M.D., F. A. Burrall, M.D., George H. Fox,

M.D., F. R. Sturgis, M.D., S. H. Dessau, M.D., James Knight, M.D., V. P. Gibney, M.D., Daniel Lewis, M.D., W. M. Chamberlain, M.D., G. M. Lefferts, M.D., F. H. Bosworth, M.D., S. Sexton, M.D., and R. Tauszky, M.D.

To Charles Rice, Ph. D., and to Prof. J. U. Lloyd, of Cincinnati, Ohio, the author is deeply indebted for many pharmaceutical processes, relating especially to unofficinal drugs and preparations; and to his friend H. G. Piffard, M.D., for much valuable assistance during the progress of the work.

L. J.

323 West Twenty-seventh Street,
New York, May, 1881.

INTRODUCTION.

This work, though based on the United States and British Pharmacopœias, does not include all the drugs and preparations contained in them, a number of the less important having been omitted. Since the last revision of the United States Pharmacopœia in 1870, and of the British in 1867, with additions made in 1874, a number of important therapeutic agents have come into use, and most of these are treated of here.

The preparations introduced from the French Codex and German Pharmacopœia are from the last editions of those works, while the unofficinal formulæ are derived mainly from recent sources, many of them having been furnished by their authors expressly for this work.

TITLES.—The leading titles are in capitals, first, the Latin title of the drug, followed, if it be officinal, by abbreviations in parentheses (U. S., Br.), or (U. S. et al. Ph.)—United States, British—United States and other Pharmacopœias—and after this the English name, and, in some instances, synonyms also.

Preparations.—The preparations of the United States Pharmacopœia have the Latin title in black-faced type, followed by abbreviations in parentheses, and the officinal English title. In this connection the abbreviations Br., Fr., Ger., are not to be considered as signifying that those pharmacopœias have preparations of exactly the same name and character as the one under consideration, but rather that their preparations are similar, or closely correspond with it.

PREPARATIONS (BR., FR., GER.) AND UNOFFICINAL.—These are distinguished by being printed in small capitals, the abbreviation standing first in the parenthesis indicating from which pharmacopœia the preparation is taken. Those taken from the British Pharmacopœia have the officina English name, while in the case of the French and German, the name has been translated into English; and the unofficinal formulæ have been named in such a manner as has seemed to the author most convenient for ready reference.

Weights.—The weights used in the preparations of the United States Pharmacopœia are derived from the troy pound, while in those from the British, the terms ounce and pound are to be understood as referring to avoirdupois weights. Their relations are exhibited in the following table:

Troy.		Avoirdupois.
1 pound..................	5,760 grains.	7,000 grains.
1 ounce..................	480 "	437.5 grains.
1 drachm..................	60 "	
1 scruple..................	20 "	

Wherever, then, in this work the word ounce is applied to a solid substance, it is to be understood as meaning a troy ounce, excepting in British preparations, and in formulæ derived from British sources, as those of the London Hospitals, etc.

In a very few instances French and German preparations are introduced with metrical weights, though in nearly all of them, quantities have been expressed in parts by weight.

The approximate relation of troy weights to metrical weights is exhibited in the following table:

Grain.		Gramme.	Grains.		Grammes.
1	=	0.065	15	=	1
½	=	0.032	30	=	2
¼	=	0.016	60	=	4
⅙	=	0.01	4 drachms	=	16
⅛	=	0.008	1 ounce	=	32

Measures.—The terms drachm and ounce, applied to liquids, are to be understood as meaning fluid drachm and fluid ounce, unless otherwise stated, as in certain preparations where quantities of liquids are expressed in weights. The term pint is to be understood as meaning the wine pint of sixteen fluid ounces, unless otherwise stated, as in British preparations, where the imperial pint of twenty fluid ounces is employed. To avoid confusion, in most of the British preparations the quantities of liquids have been expressed in ounces instead of pints. It is necessary, however, to bear in mind that the fluidounce derived from the wine pint, and that derived from the imperial pint, are not identical, the former, in distilled water at 60°, weighing 455.7 grains, while the latter weighs but 437.5 grains.

The relation of the weights and measures used in the United States Pharmacopœia, in distilled water at 60°, is exhibited in the following table:

Troy.	Measure.	Measure.	Troy.
1 pound	= 0.790 pint.	1 pint	= 1.265 pound.
1 ounce	= 1.053 fluidounce.	1 fluidounce	= 0.949 ounce.
1 drachm	= 1.053 fluidrachm.	1 fluidrachm	= 0.949 drachm.
1 grain	= 1.053 minim.	1 minim	= 0.949 grain.

Pharmaceutical Processes.—The language of the pharmacopœias describing the various processes has been much condensed and abbreviated, without, however, rendering it obscure.

The reader is presumed to be familiar with ordinary pharmaceutical manipulations, as the making of pills, powders, tinctures, etc., and hence specific directions are omitted in such cases. While in the case of extracts, fluid extracts, suppositories, etc., it is believed the directions will be found sufficiently full and explicit.

It may be necessary to observe that in all preparations, unless otherwise stated, the crude drugs are to be powdered as a preliminary step; that all temperatures are measured by Fahrenheit's thermometer, and that specific gravities are taken at the temperature of 60°.

Doses.—Unless otherwise stated, the doses, as usual, are to be considered as average doses for an adult. The proportionate dose for a child may be computed by any of the rules in common use, as, for instance, by taking the age of the child, in years, as the numerator, and the age plus 12 as the denominator of a fraction, thus: $\frac{4}{4+12}=\frac{1}{4}$—the dose for a child four years old.

As is well known, the doses of medicines must be regulated by the effects produced or desired, hence, only the most general directions can be given.

MEDICAL FORMULARY.

ABSINTHIUM (U. S., Fr., Ger.)—WORMWOOD.

The leaves and tops of Artemisia Absinthium, L. (*Nat. ord. Compositæ*), an herbaceous perennial, indigenous to the Old World, but cultivated and sparingly naturalized here.

Wormwood is a slightly aromatic bitter tonic, and in large doses acts as a vermifuge.

Its volatile oil possesses narcotic properties, and in very large doses may produce death.

Externally, fomentations of wormwood are employed in bruises, sprains, etc.

Dose: 10 to 60 grains.

PREPARATIONS.

Extract of Wormwood (Fr., Ger.).

Take of Dried tops of wormwood.................... 1 part.
Boiling distilled water...................... 8 parts.

Infuse the wormwood for twelve hours in 6 parts of water, express and strain. Treat the residue with the remainder of the water in like manner. Evaporate the infusions separately to a syrupy consistence, then mix, and evaporate to the consistence of a soft extract.

Dose: 2 to 30 grains. Made into pills with the powdered leaves, it is more active.

The German extract is an alcoholic one.

Oil of Wormwood (Fr.).

The oil obtained by distilling fresh wormwood with water.

Dose: 2 to 10 drops, in syrup, mucilage, or sweet spirit of nitre.

It enters into the composition of the French *liqueur, absinthe.*

Infused Oil of Wormwood (Fr.).

Take of Dried tops of wormwood.................... 1 part.
Olive oil................................ 10 parts.

Infuse two hours, express and filter.

Used as an embrocation in bruises, sprains, rheumatism, etc.

Tincture of Wormwood (Fr., Ger.).

Take of Leaves of wormwood	1 part.
Alcohol (60%)	sufficient.

Moisten, pack and percolate to 5 parts.

Dose: ½ to 2 drachms.

Compound Tincture of Wormwood (Fr.).

Take of Dried tops of wormwood	25 parts.
Dried tops of germander	25 parts.
Gentian	25 parts.
Bitter-orange peel	25 parts.
Rhubarb	15 parts.
Aloes	5 parts.
Cascarilla	5 parts.
Alcohol (60%)	1,000 parts.

Macerate ten days, express and filter.

Dose: 1 to 2 drachms.

Wine of Wormwood (Fr.).

Take of Dried leaves of wormwood	30 parts.
Alcohol (60%)	60 parts.
White wine	1,000 parts.

Macerate the wormwood in the alcohol for twenty-four hours, add the wine, continue the maceration for ten days, express and filter.

Dose: 1 to 2 ounces.

Distilled Water of Wormwood (Fr.).

Take of Fresh tops of wormwood	10 parts.
Water	sufficient.
Distil	10 parts.

Dose: 1 to 4 ounces.

ACACIA (U. S. et al. Ph.)—GUM ARABIC.

A gummy exudation from the bark of Acacia Senegal, Willd., and other species of acacia *(Nat. ord. Leguminosæ)* shrubs or trees indigenous to Africa and Asia.

Gum arabic occurs in tears or fragments of various sizes, colorless, or of a yellowish or brownish tint, odorless, and having a sweetish, mucilaginous taste. It is entirely soluble in water, forming a mucilage which is demulcent and nutritive. Used in catarrhal and febrile affections, and as a vehicle.

PREPARATIONS.

Mucilago Acaciæ (U. S. et al. Ph.)—Mucilage of Gum Arabic.

Gum arabic	4 ounces.
Water	½ pint.

Agitate until dissolved, and strain.

Used as a demulcent and vehicle.

Syrupus Acaciæ (U. S., Fr., Ger.)—Syrup of Gum Arabic.

Gum arabic	2 ounces.
Sugar	14 ounces.
Water	8 ounces.

Dissolve the gum in the water, add the sugar, heat gently until dissolved, and strain. Used as a vehicle, especially in cough mixtures.

MIXTURE OF GUM ARABIC (Ger.).

Take of Gum arabic	15 parts.
White sugar	15 parts.
Distilled water	170 parts.

Dissolve.

Used as a vehicle.

TROCHES OF GUM ARABIC (Fr.).

Take of Gum arabic	100 parts.
White sugar	900 parts.
Distilled orange-flower water	75 parts.

Make a mucilage with the water and an equal weight of the gum, then add the remainder and the sugar, and form troches of 15 grains each.

Used to allay irritation of the throat.

ACETUM (U. S. et al. Ph.)—VINEGAR.

Impure and very dilute acetic acid, the product of the acetous fermentation of any liquid susceptible of the vinous fermentation.

In this country it is generally obtained from cider.

Vinegar is diuretic and refrigerant. It is used internally in febrile and inflammatory affections, and topically as a fomentation in bruises and sprains, as a gargle in sore throat, etc.

For pharmaceutical purposes, it should be distilled, or what is better, the pure acetic acid may be employed.

Dose: 1 to 2 drachms.

PREPARATIONS.

Acetum Destillatum (U. S., Fr.)—Distilled Vinegar.

Take of Vinegar	8 pints.
Distil	7 pints.

Dose: 1 to 2 drachms. It may be used instead of diluted acetic acid in the preparation of the officinal vinegars.

Aromatic Vinegar (Fr., Ger.).

Take of Balm	25 parts.
Peppermint	25 parts.
Rosemary	25 parts.
Sage	25 parts.
Lavender	50 parts.
Garlic	10 parts.
White vinegar	2,000 parts.

Macerate ten days, express and filter. The German Pharmacopœia employs volatile oils and diluted acetic acid.

Camphorated Vinegar (Fr.).

Take of Camphor	10 parts.
Glacial acetic acid	10 parts.
White vinegar	400 parts.

Pulverize the camphor by rubbing it with some of the acid, add the remainder and the vinegar gradually, and after several days filter.

Syrup of Vinegar (Fr.).

Take of Vinegar	100 parts.
White sugar	175 parts.

Dissolve with a gentle heat and strain. Largely diluted, this forms a pleasant drink in febrile affections.

Lotion of Vinegar.

Vinegar	5 drachms.
Water	10 ounces.

Mix. *St. Bartholomew's Hospital.*

Lotion of Vinegar and Cantharides.

Distilled vinegar	3½ ounces.
Tincture of cantharides	6 drachms.
Rose-water	3½ ounces.

Mix. A stimulant to the scalp. *Tilbury Fox.*

ACHILLEA (U. S., Fr., Ger.)—YARROW—MILFOIL.

The leaves and flowering tops of Achillea Millefolium, L. (*Nat. ord. Compositæ*), an herbaceous perennial, common in all temperate regions.

Yarrow is a mild bitter tonic, and possesses also some astringent properties. Used in atonic dyspepsia, chronic catarrhal affections, passive hemorrhages, etc.

Dose: ½ to 1 drachm.

PREPARATIONS.

Extract of Yarrow.

Take of Yarrow .... a convenient quantity.

Exhaust by percolation with diluted alcohol, and evaporate to a proper consistence.

Dose: 10 to 30 grains.

Infusion of Yarrow.

Take of Yarrow .... 1 part.
Boiling water .... 10 parts.

Infuse an hour, express and strain.

Dose: 1 to 4 ounces. Used successfully in hemorrhoids and metrorrhagia. *Cazin.*

ACIDUM ACETICUM (U. S. et al. Ph.)—ACETIC ACID.

Obtained as one of the products of the distillation of wood. The officinal acid is a colorless liquid, having a pungent odor free from empyreuma, and of the specific gravity 1.047.

The medicinal properties of acetic acid are identical with those of vinegar, which see.

PREPARATIONS.

Acidum Aceticum Dilutum (U. S., Br.)—Diluted Acetic Acid.

Take of Acetic acid .... 1 pint.
Distilled water .... 7 pints.

Mix. It has the specific gravity 1.006. The diluted acetic of the German Pharmacopœia has sp. gr. 1.040, and hence nearly corresponds with the preceding.

Oxymel (Br., Fr., Ger.).

Take of Clarified honey (by weight) .... 40 ounces.
Acetic acid .... 5 ounces.
Distilled water .... 5 ounces.

Liquefy the honey by heat, then add the acid and water.

Dose: 1 to 2 drachms.

The French Codex employs 1 part of vinegar with 4 of honey.

Gargles of Acetic Acid.

Take of Acetic acid .... 2½ drachms.
Glycerin.............................. 3 drachms.
Water ....to 10 ounces.

Mix. Stimulant and antiseptic. *G. M. Lefferts.*

Take of Acetic acid ....15 minims.
Oxymel .... 30 minims.
Distilled water..............................to 1 ounce.

Mix. *St. Bartholomew's Hospital.*

Inhalation of Acetic Acid.

Take of Glacial acetic acid.............................. 1 part.
Acetic acid.............................. 1 part.

Mix. Add 2 drachms to a pint of water at 140° for each inhalation. Antiseptic ; used for the inflammatory sore throat of scarlet fever.

London Throat Hospital.

Lotion of Acetic Acid.

Take of Diluted acetic acid .... 2 ounces.
Water.............................. 20 ounces.

Mix. *London Chest Hospital.*

ACIDUM ACETICUM GLACIALE (Br., Fr., Ger.)—GLACIAL ACETIC ACID.

Prepared by distilling dried acetate of sodium with sulphuric acid.

It is a colorless liquid at the mean temperature of the air, but crystallizes at 34° and remains crystalline until the temperature rises to, or above 48°. Its specific gravity is 1.065.

Used as a caustic, and in preparations.

PREPARATIONS.

Aromatic Acetic Acid (Ger., Fr.).

Take of Oil of cloves.............................. 9 parts.
Oil of lemon .... 6 parts.
Oil of lavender .... 6 parts.
Oil of bergamot .... 3 parts.
Oil of thyme .... 3 parts.
Oil of cinnamon .... 1 part.
Glacial acetic acid.............................. 25 parts.

Dissolve by agitation.

The French preparation contains camphor and the oils of cinnamon, cloves, and lavender.

ACIDUM ARSENIOSUM (U. S. et al. Ph.)—ARSENIOUS ACID.—ARSENIC.—WHITE ARSENIC.

An anhydrous acid obtained by roasting arsenical ores, and purified by sublimation. It occurs as a heavy white powder, or in opaque or semi-transparent lumps, having a conchoidal fracture. It is entirely volatilized by heat, is without odor, has little taste, and is but sparingly soluble in water.

In medicinal doses, arsenic is tonic and alterative, in overdoses a violent corrosive poison. It is used in intermittent fevers, especially those which have resisted quinia, in chronic skin diseases, neuralgia, chorea, etc. Locally it is employed as an escharotic for the destruction of malignant growths.

Dose: $\frac{1}{60}$ to $\frac{1}{12}$ grain.

PREPARATIONS.

GRANULES OF ARSENIOUS ACID (Fr.).

Take of Arsenious acid	1 part.
Sugar of milk	40 parts.
Gum arabic	9 parts.
Syrup of honey	sufficient.

Triturate the arsenic first with the sugar, then with the gum, and with the syrup form a mass to be divided into granules containing $\frac{1}{60}$ of a grain each of arsenious acid.

PILLS OF ARSENIOUS ACID (Fr.).

Take of Arsenious acid	1 part.
Black pepper	10 parts.
Gum arabic	2 parts.
Distilled water	sufficient.

Triturate the arsenic with the pepper and the gum, then, with the water, form a mass to be divided into pills containing $\frac{1}{12}$ of a grain each of arsenious acid.

Take of Arsenious acid	5 grains.
Gum arabic	$\frac{1}{2}$ drachm.
Cinnamon powder	3 drachms.
Glycerin	sufficient.

Mix, and divide into 100 pills. Dose: 1 pill two or three times a day. In skin diseases. *Tilbury Fox.*

Take of Arsenious acid	1 grain.
Black pepper	6 grains.
Extract of gentian	24 grains.

Beat together into a pilular mass and divide into 12 pills.

British Skin Hospital.

Caustic of Arsenious Acid.

Take of Arsenious acid	20 grains.
Vermilion	1 drachm.
Lard	1 ounce.

Mix. *British Skin Hospital.*

Paste of Arsenious Acid.

Take of Arsenious acid	2 drachms.
Mucilage of gum arabic	1 drachm.

Mix. Used as a caustic for cancers. *Marsden.*

Arsenical Powder of Côme.

Take of Arsenious acid	20 parts.
Vermilion	60 parts.
Dragon's blood	6 parts.
Animal charcoal	4 parts.

Mix. Used in lepra, cancer, etc.

Arsenical Powder.

Take of Arsenious acid	1 drachm.
Vermilion	2 scruples.
Calomel	2½ ounces.

Mix. Used in lupus, syphilis, and scrofulous ulcers. *Startin.*

ACIDUM BENZOICUM (U. S. et al. Ph.)—BENZOIC ACID.

Take of benzoin, 12 troy ounces. Spread the benzoin over the bottom of an iron dish 8 inches in diameter, and 2 inches deep, cover with a piece of filtering paper, and by means of paste attach it to the rim. Then cover all with a conical receiver of thick, sized paper, and apply heat to the dish until vapors of benzoic acid cease to rise. Lastly, remove the acid from the receiver and diaphragm.

Benzoic acid, thus obtained, is in white, feathery crystals of a peculiar, agreeable odor, and a warm, acidulous taste. It is sparingly soluble in water, freely soluble in alcohol, and is dissolved by alkaline solutions, forming combinations from which it is precipitated by hydrochloric acid.

Benzoic acid is a local irritant, and also possesses antiseptic properties. Taken internally, it acts as a general stimulant, but affects the mucous membranes chiefly. Used in chronic bronchitis, and to prevent the formation of phosphatic calculi. It is also employed as an antiseptic surgical dressing.

Dose : 10 to 15 grains.

LOTIONS OF BENZOIC ACID.

Take of Benzoic acid.............................. 1 grain.
Rectified spirit.............................. 24 minims.
Water.............................. 1 ounce.

Mix. *Middlesex Hospital.*

Take of Benzoic acid.............................. 2 grains.
Water.............................. 1 ounce.

Mix. *British Skin Hospital.*

OINTMENT OF BENZOIC ACID.

Take of Benzoic acid.............................. 40 grains.
Acetate of morphia.............................. 6 grains.
Cerate.............................. 1 drachm.
Glycerin.............................. sufficient.

Mix. Apply several times daily. For rectal fistula and ulcer.
R. Tauszky.

ACIDUM BORICUM (Fr., Br., Ger.)—BORACIC or BORIC ACID

Take of Borax.............................. 300 grammes.
Distilled water..............................1,200 grammes.
Sulphuric acid.............................. 100 grammes.
The white of 1 egg.

Dissolve the borax in half the water by the aid of heat. Divide the remainder of the water into two equal parts, with one of which dilute the sulphuric acid, with the other mix the albumen. Mix the albuminous solution with the solution of borax, heat to the boiling point, add the diluted acid, filter, and set aside to crystallize. Lastly, drain the crystals, and dry them on bibulous paper.

Boracic acid, thus obtained, is in white, shining scales, soluble in 26 parts of cold, and in 3 parts of warm water, and freely soluble in alcohol. Used externally as an antiseptic and deodorant.

LOTIONS OF BORACIC ACID.

A saturated solution. *Lister.*

Take of Boracic acid.............................. 15 grains.
Water.............................. 1 ounce.

Mix. *British Skin Hospital.*

OINTMENTS OF BORACIC ACID.

Take of Boracic acid.............................. 1 part.
White wax.............................. 1 part.
Expressed oil of almonds.............................. 2 parts.
Paraffin.............................. 2 parts.

Rub the oil and the acid together in a warm mortar, then add the melted wax and paraffin, and triturate until cold. *Lister.*

Take of Boracic acid.............................. 1 drachm.
Lard.............................. 1 ounce.

Mix. *British Skin Hospital.*

ACIDUM CARBOLICUM (U. S. et al. Ph.)—CARBOLIC ACID.

Obtained from coal tar by distillation. When pure, it crystallizes in minute flakes or rhomboidal needles, clear and colorless, of an empyreumatic odor resembling that of creasote, and of a caustic taste. It is, however, generally of a slightly reddish tinge, due to impurity. Upon exposure to air it deliquesces, and assumes the liquid state in the presence of water without being dissolved by it. It is soluble in 20 parts of water, and very soluble in alcohol, ether, acetic acid, glycerin, and the fixed and volatile oils. It does not act like an acid upon vegetable colors, though it readily combines with bases.

Applied locally it is an escharotic. Taken internally in large doses it is a corrosive poison. It exerts a very destructive influence upon the lower forms of vegetable and animal life, arrests fermentation, and is a valuable disinfectant and antiseptic. It is used, like creasote, to arrest obstinate vomiting, and in zymotic diseases, sarcina ventriculi, etc. It is, however, more generally employed externally than internally. In the pure state it is sometimes used as a caustic, and, largely diluted with water, oil, glycerin, etc., it is very frequently employed as a dressing for wounds, burns, and scalds, and as a gargle in diphtheria, scarlatina, etc. It is also largely employed for disinfecting foul rooms, closets, etc. For this purpose the impure acid (Acidum Carbolicum Impurum, U. S.) may be used.

Dose: 1 to 2 grains.

PREPARATIONS.

Aqua Acidi Carbolici (U. S.)—Carbolic Acid Water.

Take of Glycerite of carbolic acid.................... 10 drachms.
Distilled water............................ sufficient.

Mix the glycerite with sufficient distilled water to make 1 pint.

Each drachm contains about 1 grain of carbolic acid.

Dose: 1 to 2 drachms.

Glyceritum Acidi Carbolici (U. S., Br.)—Glycerite of Carbolic Acid.

Take of Carbolic acid............................ 2 ounces.
Glycerin............................ ½ pint.

Rub together until the acid is dissolved.

Dose: 5 to 10 minims.

The British preparation, termed *glycerine of carbolic acid*, is almost identical with this.

Suppositoria Acidi Carbolici (U. S., Br.)—Suppositories of Carbolic Acid.

Take of Carbolic acid.............................. 12 grains.
Oil of theobroma.......................... 348 grains.
Water.............................. sufficient.

Mix the acid, previously dissolved in a few drops of water, with 1 drachm of the oil, and then, having melted the remainder and cooled it to 95°, mix the whole together and pour into suitable moulds, making 12 suppositories.

The British Pharmacopœia directs: carbolic acid, 12 grains; curd soap, 180 grains; starch, sufficient. Mix and divide into 12 equal parts, each of which is to be made into a proper form.

Unguentum Acidi Carbolici (U. S.)—Ointment of Carbolic Acid.

Take of Carbolic acid.............................. 60 grains.
Ointment.............................. 420 grains.

Mix them thoroughly.

Caustic of Carbolic Acid.

Take of Carbolic acid.............................. 1 ounce.
Water.............................. 30 minims.

Mix. *British Skin Hospital.*

Carbolized Collodion.

Take of Collodion.............................. 1 ounce.
Castor oil.............................. ½ drachm.
Carbolic acid.............................. ½ drachm.

Mix. *Hospital Formulary.*

Gargles of Carbolic Acid.

Take of Carbolic acid.............................. 20 grains.
Glycerin.............................. 1 ounce.
Common salt.............................. 1 drachm.
Warm water.............................. ½ pint.

Dissolve the salt in the water, then add the acid and glycerin.
Use every half-hour at the first intimation of sore throat.

F. A. Burrall.

Take of Glycerite of carbolic acid.................. 1 to 2 drachms.
Water.............................. to 10 ounces.

Mix. *G. M. Lefferts.*

Take of Carbolic acid.............................. 20 grains.
Glycerin.............................. ½ ounce.
Water.............................. to 10 ounces.

Mix. *London Throat Hospital.*

Lotions of Carbolic Acid.

Take of Carbolic acid.............................. 20 parts.
Glycerin.............................. 20 parts.
Water.............................. 20 parts.

Mix. To be applied frequently in ringworm of the beard.

G. H. Fox.

Take of Carbolic acid.............................. 2 drachms.
Glycerin.............................. 1 ounce.
Rose-water..............................to 8 ounces.

Mix. Used in tinea circinata. *Tilbury Fox.*

Take of Carbolic acid.............................. 1 grain.
Chloral.............................. 1 grain.
Iodide of potassium.............................. 1 grain.
Water.............................. 1 ounce.

Mix. Apply to the urethra on absorbent cotton, once or twice a week, and inject the bladder with a warm solution of salicylic acid (2 grains to 1 ounce), administering also tincture of belladonna, 3 minims three times a day. For irritable bladder. *R. Tauszky.*

Take of Carbolic acid.............................. 8 minims.
Solution of subsulphate of iron..............2 to 3 drachms.
Glycerin.............................. 1 ounce.

Mix. Apply to the throat with a camel's-hair pencil, two or three times a day. In diphtheria. *J. Lewis Smith.*

Take of Carbolic acid.............................. 40 grains.
Borate of sodium.............................. 1 drachm.
Bicarbonate of sodium.............................. 1 drachm.
Glycerin.............................. 1 ounce.
Rose-water.............................. 1 ounce.
Water..............................to 1 pint.

Mix. Used by means of nasal douche, nasal syringe, or atomizer, for cleansing purposes. *G. M. Lefferts.*

Take of Glycerite of carbolic acid.............................. $1\frac{1}{2}$ drachm.
Borate of sodium.............................. 1 drachm.
Water.............................. 1 pint.

Mix. Used in the same manner as the preceding. *G. M. Lefferts.*

Take of Glycerite of carbolic acid.............................. $2\frac{1}{2}$ drachms.
Water..............................to 10 ounces.

Mix. *St. Bartholomew's Hospital.*

Take of Carbolic acid.............................. 10 grains.
Water.............................. 1 ounce.

Mix. *British Skin Hospital.*

ACIDUM CARBONICUM—CARBONIC ACID.

Carbonic acid gas is readily absorbed by water, a property which admits of its being administered internally. It is obtained from marble-dust (carbonate of lime), by the action of sulphuric acid, and water is charged with it by means of machinery in general use.

Carbonic acid water is a refreshing drink, and is often of great service in controlling vomiting.

ACIDUM CHROMICUM (U. S., Fr., Ger.)—CHROMIC ACID.

Obtained by decomposing bichromate of potassium with sulphuric acid. It crystallizes in brilliant crimson-colored needles, which are deliquescent and very soluble in water. It is an energetic caustic, one of the best for the destruction of venereal and other warts, condylomata, etc.

Caustics of Chromic Acid.

Take of Chromic acid .... 100 grains.
Water .... 1 ounce.

Mix. Used in vegetations about the genital organs. *Bumstead.*

Take of Chromic acid .... 1 ounce.
Water .... 1 ounce.

Mix. *British Skin Hospital.*

ACIDUM CHRYSOPHANICUM—CHRYSOPHANIC ACID.

Obtained chiefly from Goa powder (Araroba), which is composed mainly of it. It crystallizes in bright yellow needles, but, as generally met with, it is a granular yellow powder, odorless, and with little taste. It is not, strictly speaking, an acid.

Used in the form of an ointment in psoriasis and other diseases of the skin.

Ointments of Chrysophanic Acid.

Take of Chrysophanic acid .... 20 grains.
Vaseline (or a similar preparation of petroleum) 190 grains.

Melt the vaseline in a water-bath, add the acid, stir and heat for about ten minutes, then quickly strain through muslin into a capsule standing on ice, and stir briskly until cold. *Hospital Formulary.*

Take of Chrysophanic acid	10 grains.
Lard	1 ounce.

Mix. *St. Mary's Hospital.*

Take of Chrysophanic acid	2 drachms.
Lard	1 ounce.

Heat together in a water-bath for half an hour; when set, mix with pestle and mortar. *British Skin Hospital.*

ACIDUM CITRICUM (U. S. et al. Ph.)—CITRIC ACID.

Citric acid exists in a large number of plants, but is obtained chiefly from the juice of lemons and limes. It occurs in colorless crystals, of an agreeable acid taste, and freely soluble in water. Used as a refrigerant in febrile diseases, especially when fresh lemon-juice cannot be obtained.

Dose: 5 to 30 grains.

PREPARATIONS.

Syrupus Acidi Citrici (U. S., Fr.)—Syrup of Citric Acid.

Take of Citric acid	120 grains.
Oil of lemon	4 minims.
Syrup	2 pints.

Rub the acid and oil with 1 ounce of the syrup, then add the remainder, and dissolve with a gentle heat.

An agreeable vehicle for the administration of certain salines.

Lemonade Powder (Ger.).

Take of Citric acid	2½ drachms.
White sugar	4 ounces.
Oil of lemon	1 drop.

Mix.

A tablespoonful to a glass of water makes a good substitute for lemonade when fresh lemons are not obtainable.

ACIDUM GALLICUM (U. S. et al. Ph.)—GALLIC ACID.

Obtained chiefly from galls, though it exists in many plants. It occurs in small, silky, nearly colorless crystals, without odor, of a slightly acid and astringent taste, soluble in 100 parts of cold, and in 3 parts of boiling water.

It is used as an astringent in passive hemorrhages from the lungs, stomach, kidneys, and uterus, in night-sweats, etc.

Dose: 5 to 10 grains.

PREPARATIONS.

Glyceritum Acidi Gallici (U. S., Br.)—Glycerite of Gallic Acid.

Take of Gallic acid	2 ounces.
Glycerin	½ pint.

Rub them together, then heat gently until the acid is dissolved.

Dose: 20 to 60 minims.

Glycerine of Gallic Acid, Br.

MIXTURES OF GALLIC ACID.

Take of Gallic acid	½ drachm.
Diluted sulphuric acid	1 drachm.
Deodorized tincture of opium	1 drachm.
Compound infusion of rose	4 ounces.

Mix. Dose: ½ ounce every four hours or oftener.

In menorrhagia, hæmaturia, purpura hæmorrhagica, and the hemorrhagic diathesis. *Bartholow.*

Take of Gallic acid	12 grains.
Compound tincture of cinnamon	1½ drachm.
Tincture of opium	8 minims.
Caraway water	to 2 ounces.

Mix. Dose: 2 drachms for a child two years old. In chronic diarrhœa. *Hillier.*

Take of Gallic acid	10 grains.
Tincture of opium	3 minims.
Diluted sulphuric acid	15 minims.
Water	to 1 ounce.

Mix. One dose. *University College Hospital.*

Take of Gallic acid	10 grains.
Camphorated tincture of opium	20 minims.
Diluted sulphuric acid	15 minims.
Water	to 1 ounce.

Mix. One dose. *Brompton Consumption Hospital.*

PILLS OF GALLIC ACID.

Take of Gallic acid	½ drachm.
Extract of belladonna	2 grains.
Mix and divide into	10 pills.

Dose: 2 pills at bed-hour. For the sweating of phthisis.

Bartholow.

Take of Gallic acid	1 drachm.
Aqueous extract of ergot (ergotin)	20 grains.
Digitalis	20 grains.
Mix and divide into	20 pills.

Dose: 1 pill every four hours. In passive hemorrhages.

Bartholow.

Take of Gallic acid .... 2½ grains.
Extract of henbane .... 1 grain.

Make 1 pill. *Royal Chest Hospital.*

Take of Gallic acid .... 3 grains.
Extract of rhatany .... 2 grains.
Glycerin .... sufficient.

Make 1 pill. *Samaritan Hospital.*

Take of Gallic acid .... 3½ grains.
Hydrochlorate of morphia .... ⅛ grain.
Extract of gentian .... sufficient.

Make 1 pill. *Brompton Consumption Hospital.*

Take of Gallic acid .... 4 grains.
Extract of opium .... ¼ grain.
Confection of roses .... sufficient.

Make 1 pill. *London Chest Hospital.*

PYROGALLIC ACID.

When gallic acid is heated to about 400° it is decomposed, yielding carbonic acid and a sublimate of pyrogallic acid.

LOTION OF PYROGALLIC ACID.

Take of Pyrogallic acid .... 15 parts.
Glycerin .... 5 parts.
Water .... 80 parts.

Mix. For alopecia areata. *G. H. Fox.*

OINTMENT OF PYROGALLIC ACID.

Take of Pyrogallic acid .... 10 grains.
Lard .... 1 ounce.

Mix. *British Skin Hospital.*

ACIDUM HYDROBROMICUM—HYDROBROMIC ACID.

Prepared in various ways, by the action of bromine upon phosphorus, sulphuric or tartaric acid upon bromide of potassium, etc. When pure, it is a colorless gas of a pungent and irritating odor, but it is readily absorbed by water, a property which permits of its being administered internally. The concentrated aqueous solution is colorless and has a pungent, acid taste. It is chiefly employed to prevent the disagreeable cephalic symptoms occasioned by quinia and iron. It has, however, been employed, like the bromides, in a variety of nervous affections.

Dose: Of the diluted acid, 10 to 60 minims.

MIXTURE OF HYDROBROMIC ACID.

Take of Diluted hydrobromic acid (34%) .... ½ drachm.
Spirit of chloroform .... 20 minims.
Syrup of squill .... 1 drachm.
Water ....to 1 ounce.

Mix. One dose: to be taken twice or thrice daily, for cough.

Hospital Formulary.

Take of Hydrobromic acid .... 20 minims.
Syrup .... 20 minims.
Water .... 1 ounce.

Mix. One dose. *London Chest Hospital.*

ACIDUM HYDROCHLORICUM (U. S. et al. Ph.)—HYDROCHLORIC ACID—MURIATIC ACID.

An aqueous solution of hydrochloric acid gas, of the specific gravity 1.160, obtained by the action of sulphuric acid upon common salt.

It is a colorless liquid of a suffocating odor, and possessing energetic caustic properties. The commercial acid commonly contains chlorine, iron, and other impurities, and is of a light yellowish color.

The concentrated acid is employed externally as a caustic.

PREPARATIONS.

Acidum Hydrochloricum Dilutum (U. S., Br., Ger.)—Diluted Hydrochloric Acid.

Take of Hydrochloric acid (by weight) .... 4 ounces.
Distilled water .... sufficient.

Mix the acid with sufficient distilled water to make 1 pint.

Dose: 10 to 30 minims. Used in dyspepsia, chronic diarrhœa and dysentery, typhus, typhoid and scarlet fevers, etc.; topically as a gargle in various forms of sore throat, and as a bath in skin diseases.

BATH OF HYDROCHLORIC ACID.

Take of Hydrochloric acid .... 1 ounce.
Water .... 30 gallons.

Mix. Used in chronic lichen and prurigo. *Tilbury Fox.*

GARGLE OF HYDROCHLORIC ACID.

Take of Diluted hydrochloric acid .... 12 minims.
Glycerin .... 24 minims.
Water ....to 1 ounce.

Mix. Stimulant. *London Throat Hospital.*

Mixtures of Hydrochloric Acid.

Take of Hydrochloric acid .... 3 drachms.
Compound tincture of gentian .... 8 ounces.
Water .... 8 ounces.

Mix. Dose: 1 drachm. *Hospital Formulary.*

Take of Diluted hydrochloric acid .... 2½ minims.
Sugar .... 12½ grains.
Water .... 1 ounce.

Mix. One dose. *Guy's Hospital.*

Take of Diluted hydrochloric acid .... 10 minims.
Tincture of chiretta .... 15 minims.
Water ....to 1 ounce.

Mix. One dose. *London Ophthalmic Hospital.*

Take of Diluted hydrochloric acid .... 1 drachm.
Spirit of chloroform .... 1 drachm.
Syrup .... 1 drachm.
Camphor water ....to 3 ounces.

Mix Dose: 1 to 2 drachms. For children. *Middlesex Hospital.*

ACIDUM HYDROCYANICUM DILUTUM (U. S., Br., Fr.)—DILUTED HYDROCYANIC ACID—DILUTED PRUSSIC ACID.

Take of Ferrocyanide of potassium .... 2 ounces.
Sulphuric acid (by weight) .... 1½ ounce.
Distilled water .... sufficient.

Mix the acid with 4 ounces of distilled water, and, when cool, pour into a glass retort, and add the ferrocyanide, dissolved in 10 ounces of distilled water, connect with a cooled receiver containing 8 ounces of distilled water, and distil 6 ounces. Lastly, add to the product 5 ounces of distilled water, or sufficient to render it of such a strength that $12\frac{7}{10}$ grains of nitrate of silver may be exactly neutralized by 100 grains of the acid.

When required for immediate use, prepare it thus:

Take of Cyanide of silver ....50½ grains.
Hydrochloric acid .... 41 grains.
Distilled water .... 1 ounce.

Mix the acid and water, add the cyanide, shake well, let the precipitate subside, then decant and preserve the clear liquid.

Diluted hydrocyanic acid is a clear, colorless liquid of a peculiar odor and a slightly irritating taste. It is a powerful antispasmodic and sedative. Used in whooping-cough, phthisis, obstinate vomiting, etc., and topically in skin diseases, chiefly to allay itching. In overdoses, it is one of the most deadly poisons known.

Dose: 1 to 4 minims.

Inhalation of Hydrocyanic Acid.

Take of Diluted hydrocyanic acid.................... 1 drachm.
Water.................................... 1 ounce.

Mix. Use a drachm in a pint of water at 80°, for each inhalation. A very useful sedative in the cough of laryngeal phthisis, and in some spasmodic affections. *London Throat Hospital.*

An inhalation containing 10 to 15 minims of the diluted acid with 1 drachm of water, is officinal in the British Pharmacopœia.

Lotions of Hydrocyanic Acid.

Take of Diluted hydrocyanic acid .... ½ drachm.
Infusion of marshmallow .... 5 ounces.

Mix. Used in pruritus. *Tilbury Fox.*

Take of Diluted hydrocyanic acid.................... 2 drachms.
Borax .... 1 drachm.
Rose-water.................................. 8 ounces.

Mix. Used in the pruritus of old people. *Neligan.*

Take of Diluted hydrocyanic acid.................... 10 minims.
Water.................................to 1 ounce.

Mix. *St. Bartholomew's Hospital.*

Take of Diluted hydrocyanic acid.................... 30 minims.
Water.................................to 1 ounce.

Mix. *St. George's Hospital.*

Take of Diluted hydrocyanic acid.................... 5 minims.
Bicarbonate of sodium...................... 5 grains.
Borax .... 5 grains.
Water.................................... 1 ounce.

Mix. *London Ophthalmic Hospital.*

MIXTURES OF HYDROCYANIC ACID.

Take of Diluted hydrocyanic acid .... 1 drachm.
Tincture of sanguinaria .... 4 drachms.
Syrup of seneka .... ½ ounce.
Syrup of tolu .... 2 ounces.
Cherry-laurel water .... 7 drachms.

Mix. Dose: 1 to 2 drachms, according to age, every three or four hours. For irritable cough. *Bartholow.*

Take of Diluted hydrocyanic acid .... 3 minims.
Camphor-water .... 1 ounce.

Mix. One dose. *London Throat Hospital.*

Take of Diluted hydrocyanic acid .... 3 minims.
Bicarbonate of sodium .... 10 grains.
Tincture of belladonna .... 10 minims.
Infusion of gentian .... 1 ounce.

Mix. One dose. *Samaritan Hospital.*

Take of Diluted hydrocyanic acid .... 10 minims.
Tincture of stramonium .... 20 minims.
Bromide of ammonium .... 1 drachm.
Syrup of tolu .... 2 ounces.
Mucilage of gum arabic .... 2 ounces.

Mix. Dose: 1 drachm three or four times a day for a child two years old. In whooping-cough. *Hospital for Ruptured and Crippled.*

ACIDUM LACTICUM (U. S., Fr., Ger.)—LACTIC ACID.

The product of the fermentation of the sugar of milk. It is a syrupy, nearly transparent liquid, of a slight, bland odor, a very sour taste, and having the specific gravity 1.212.

It is occasionally used in dyspepsia, and to prevent phosphatic deposits in the urine. Topically it has been employed, chiefly by inhalation, to dissolve the membrane of croup and diphtheria.

Dose: ½ to 1 drachm.

INHALATION OF LACTIC ACID.

Take of Lactic acid .... 30 minims.
Distilled water .... to 1 ounce.

Mix. Use by means of a spray apparatus. Of great service in diphtheria; it appears to have the effect of dissolving the membranous exudation. *London Throat Hospital.*

MIXTURE OF LACTIC ACID.

Take of Lactic acid .... 4 drachms.
Glycerite of pepsin .... 12 drachms.

Mix. Dose: 1 drachm after meals. In dyspepsia. *Bartholow.*

ACIDUM NITRICUM (U. S. et al. Ph.)—NITRIC ACID.

Obtained by the action of sulphuric acid upon nitrate of potassium.

It is a colorless liquid, having the specific gravity 1.420, and possessing energetic caustic properties. The commercial acid is of a pale yellow color, owing to impurities.

In this concentrated form, nitric acid is only employed as a caustic.

PREPARATIONS.

Acidum Mitricum Dilutum (U. S., Br., Ger.)—Diluted Nitric Acid.

Take of Nitric acid (by weight) .... 3 ounces.
Distilled water .... sufficient.

Mix the acid with sufficient distilled water to make 1 pint. Used as a tonic in dyspepsia, fevers, dysentery, etc., and topically as a stimulating gargle and lotion.

Dose: 10 to 30 minims.

Nitric Acid Lemonade (Fr.).

Take of Nitric acid (sp. gr. 1.420) .... 2 parts.
Water ....900 parts.
Syrup ....100 parts.

Mix. Used as a refrigerant drink in febrile affections.

Bath of Nitric Acid.

Take of Nitric acid.............................. 1 ounce.
Water.............................. 30 gallons.

Mix. Used in chronic lichen and prurigo. *Tilbury Fox.*

Gargle of Nitric Acid.

Take of Nitric acid.............................. 1 drachm.
Syrup............................. 1 ounce.
Water..............................to 12 ounces.

Mix. Stimulant. *St. Mary's Hospital.*

Lotions of Nitric Acid.

Take of Diluted nitric acid .... ½ drachm.
Acetate of lead .... 5 grains.
Water............................. 6 ounces.

Mix. Used in eczematous and lichenous affections. *Tilbury Fox.*

Take of Diluted nitric acid .... 30 minims.
Liquid extract of opium .... 45 minims.
Water............................ to 2 ounces.

Mix. *Middlesex Hospital.*

Take of Diluted nitric acid .... 1 ounce.
Water .... 1 pint.

Mix. In mucous patches, condylomata, torpid and ill-conditioned ulcers. *Bartholow.*

MIXTURES OF NITRIC ACID.

Take of Diluted nitric acid .... 20 minims.
Compound tincture of cardamom .... 1 drachm.
Compound infusion of gentian .... to 1½ ounce.

Mix. One dose. *St. Thomas's Hospital.*

Take of Diluted nitric acid .... 10 minims.
Tincture of columbo .... ½ drachm.
Water .... to 1 ounce.

Mix. One dose. *Brompton Consumption Hospital.*

Take of Nitric acid .... 8 minims.
Tincture of opium .... 40 minims.
Camphor water .... 8 ounces.

Mix. Dose ½ ounce. Known as *Hope's mixture.* Much used in chronic dysentery.

ACIDUM NITRO-HYDROCHLORICUM (U. S., Fr., Ger.)—NITRO-HYDROCHLORIC ACID—NITRO-MURIATIC ACID.

Take of Nitric acid (by weight) .... 3 ounces.
Hydrochloric acid (by weight) .... 5 ounces.

Mix the acids in a glass vessel, and when effervescence has ceased, put the product in a well-stopped bottle, and keep it in a cool, dark place.

Nitro-hydrochloric acid is a very corrosive liquid, of a deep golden-yellow color, and having the odor of chlorine. The concentrated acid is occasionally used as a caustic, but is chiefly employed, largely diluted, as a tonic in dyspepsia, fevers, oxaluria, etc. The diluted acid is rather more convenient for dispensing, and hence is more generally employed.

PREPARATIONS.

Acidum Nitro-Hydrochloricum Dilutum (U. S., Br.)—Diluted Nitro-Hydrochloric Acid.

Take of Nitric acid (by weight) .... 1½ ounce.
Hydrochloric acid (by weight) .... 2½ ounces.
Distilled water .... sufficient.

Mix the acids in a well-stopped bottle, and shake occasionally during twenty-four hours, then add sufficient distilled water to make 1 pint. Keep in a cool, dark place.

Dose: 10 to 30 minims.

Diluted nitro-hydrochloric acid is used for the same purposes, and in about the same manner as diluted nitric, and diluted hydrochloric acids.

When administered as an aid to digestion in dyspepsia, it is best combined with a bitter infusion, as infusion of quassia or gentian.

ACIDUM OLEICUM—OLEIC ACID.

One of the constituents of fats, and commercially obtained as a secondary product in the manufacture of stearin candles. It is, when pure, a clear, colorless liquid at ordinary temperatures, but crystallizes at about 40° F. The commercial article is of a yellow color, owing to impurities.

Oleic acid is used in making a class of preparations termed oleates, which are often employed by inunction instead of the officinal ointments.

ACIDUM PHOSPHORICUM GLACIALE (U. S.)—GLACIAL PHOSPHORIC ACID.

Obtained by digesting calcined bones in sulphuric acid, filtering, neutralizing with ammonia, filtering again, evaporating to dryness, and heating to redness.

It occurs in colorless, transparent, glass-like masses, slowly deliquescent, soluble in water, and in alcohol. Used in preparations.

PREPARATIONS.

Acidum Phosphoricum Dilutum (U. S. et al. Ph.)—Diluted Phosphoric Acid.

Take of Phosphorus........................360 grains.
Nitric acid (by weight) .... 5 ounces, or sufficient.
Distilled water...................... sufficient.

Mix the acid with ½ pint of distilled water, in a porcelain capsule of the capacity of 2 pints. Add the phosphorus, and invert over it a glass funnel with its rim resting on the inside of the capsule, near the liquid, Heat till the phosphorus is dissolved, adding a little distilled water if the reaction becomes too violent, and, if red vapors cease to be evolved before solution is effected, adding more nitric acid, diluted as before. Then remove the funnel, evaporate the solution until it weighs 2 ounces, and mix this, when cold, with sufficient distilled water to make the filtered liquid measure 20 ounces.

It may also be prepared thus:

Take of	Glacial phosphoric acid	1 ounce.
	Nitric acid	40 grains.
	Distilled water	sufficient.

Dissolve the glacial phosphoric acid in 3 ounces of the water, add the nitric acid, boil to a syrupy consistence, and add sufficient distilled water to make it measure 12½ ounces.

Diluted phosphoric acid is a colorless liquid of the specific gravity 1.056, and having a strongly acid taste.

Used as a tonic and refrigerant, like other mineral acids, but it is believed to be especially useful in cases of nervous depression.

Dose: 20 to 60 minims.

Mixtures of Phosphoric Acid and Strychnia.

Take of	Diluted phosphoric acid	½ ounce.
	Tincture of chloride of iron	1 ounce.
	Strychnia	1 grain.
	Mix and add of glycerin	1½ ounce.
	Syrup of orange peel	1 ounce.

Mix. Dose: 1 drachm in a wineglass of sweetened water directly after eating, taken through a tube. For nervous irritability, depression and anæmia. *Fordyce Barker.*

Take of	Diluted phosphoric acid	15 minims.
	Solution of strychnia (4 grs. to 1 oz.)	3 minims.
	Spirit of chloroform	15 minims.
	Infusion of quassia	1 ounce.

Mix. One dose. *London Chest Hospital.*

Mixture of Phosphoric Acid and Quassia.

Take of	Diluted phosphoric acid	15 minims.
	Tincture of quassia	30 minims.
	Syrup of orange	20 minims.
	Water	to 1 ounce.

Mix. One dose. *Brompton Consumption Hospital.*

ACIDUM SALICYLICUM.—SALICYLIC ACID.

Salicylic acid may be prepared from oil of wintergreen, salicin, and other vegetable products, but not in sufficient quantities to be of commercial importance. For medical use it is prepared from carbolic acid.

It occurs in small, white, acicular crystals, odorless, and of a sweetish, astringent, and slightly acrid taste. It is very sparingly soluble in cold water, but freely soluble in boiling water, alcohol, ether, and glycerin. It also dissolves freely in solutions of sulphite or phosphate of sodium, etc.

Salicylic acid is used with great success in acute rheumatism, and with variable results in a number of other diseases. Salicylate of sodium is, however, generally employed instead of the acid, on account of its greater solubility.

Externally salicylic acid is employed as an antiseptic and disinfectant.

Dose: 10 to 40 grains.

MIXTURES OF SALICYLIC ACID.

Take of Salicylic acid .... 160 grains.
Acetate of potassium .... 320 grains.
Glycerin .... 1 ounce.
Water .... to 4 ounces.

Mix. Dose: 1 drachm. In rheumatism. *Hospital Formulary.*

Take of Salicylic acid .... 20 grains.
Bicarbonate of sodium .... 20 grains.
Water .... 1 ounce.

Mix. One dose. *St. Mary's Hospital.*

Take of Salicylic acid .... 2 drachms.
Solution of acetate of ammonium .... 2 ounces.
Water .... 6 ounces.

Mix. Dose: 1 ounce. *National Dispensatory.*

ACIDUM SULPHURICUM (U. S. et al. Ph.).—SULPHURIC ACID—OIL OF VITRIOL.

Sulphuric acid is commonly prepared by burning sulphur and nitrate of potassium, or sodium, in a furnace so constructed that the current of air which supports combustion, conveys the gaseous product into a leaden chamber whose bottom is covered with water.

It is a colorless, oily liquid, of the specific gravity 1.843, without odor, and intensely corrosive. It mixes with water in all proportions, with the evolution of heat. Occasionally used as a caustic. For internal administration, the diluted, or the aromatic acid, is generally employed.

PREPARATIONS.

Acidum Sulphuricum Dilutum (U. S. et al. Ph.)—Diluted Sulphuric Acid.

Take of Sulphuric acid (by weight) .... 2 ounces.
Distilled water .... sufficient.

Add the acid gradually to 14 ounces of distilled water, and, after twenty-four hours, filter, adding sufficient distilled water through the filter to make 1 pint.

Used as a tonic and refrigerant in febrile affections, etc.

Dose : 5 to 30 minims.

Acidum Sulphuricum Aromaticum (U. S., Br.)—Aromatic Sulphuric Acid—Elixir of Vitriol.

Take of Sulphuric acid (by weight)	6 ounces.
Ginger	1 ounce.
Cinnamon	1½ ounce.
Alcohol	sufficient.

Add the acid gradually to 1 pint of alcohol. Mix the ginger and cinnamon, and with alcohol obtain 1 pint of tincture, by percolation. Mix this with the diluted acid.

This is the form in which sulphuric acid is most commonly prescribed. It is often used in colliquative sweats, and in colliquative and other diarrhœas.

Dose : 5 to 30 minims.

MIXTURES OF SULPHURIC ACID.

Take of Diluted sulphuric acid	10 minims.
Glycerin	30 minims.
Peppermint water	to 1 ounce.

Mix. One dose. *London Ophthalmic Hospital.*

Take of Diluted sulphuric acid	12 minims.
Tincture of catechu	15 minims.
Water	to 1 ounce.

Mix. One dose. *University College Hospital.*

Take of Diluted sulphuric acid	15 minims.
Sulphate of magnesium	20 grains.
Alum	10 grains.
Water	to 1 ounce.

Mix. One dose. *Brompton Consumption Hospital.*

Take of Diluted sulphuric acid	10 minims.
Sulphate of magnesium	40 grains.
Treacle	40 minims.
Peppermint water	7 drachms.

Mix. One dose. *Westminster Ophthalmic Hospital.*

Take of Diluted sulphuric acid	20 minims.
Tincture of opium	10 minims.
Spirit of chloroform	20 minims.
Camphor water	to 1 ounce.

Mix. One dose. *St. Mary's Hospital.*

ACIDUM SULPHUROSUM (U. S., Br.)—SULPHUROUS ACID.

Take of Sulphuric acid (by weight) .... 8 ounces.
Charcoal, in coarse powder .... 1 ounce.
Distilled water .... 36 ounces.

Pour the acid upon the charcoal, previously introduced into a matrass; apply heat, and, by means of proper apparatus, pass the evolved gas into the water, contained in a bottle and kept cool.

It is a colorless liquid, having the odor of burning sulphur, and a sour, sulphurous taste. Used internally in dyspepsia characterized by fermentation of the food—though the sulphites are preferable—and topically in skin diseases due to vegetable parasites, by inhalation in diphtheria, etc.

Dose: 20 to 60 minims.

Inhalation of Sulphurous Acid.

Take of Sulphurous acid .... 1 drachm.
Water (60° to 100°) .... 20 ounces.

For one inhalation. Stimulant. The pure acid may be used by means of a spray apparatus in diphtheria, etc., 40 to 60 minims being employed for each inhalation. *London Throat Hospital.*

Lotion of Sulphurous Acid.

Take of Sulphurous acid .... 5 ounces.
Glycerin .... to 10 ounces.

Mix. *St. Bartholomew's Hospital.*

This lotion will occasionally afford great relief in pruritus ani.

ACIDUM TANNICUM (U. S. et al. Ph.)—TANNIC ACID—TANNIN.

Take of Nutgall in fine powder .... sufficient.
Ether .... sufficient.

Expose the nutgall to a damp atmosphere for twenty-four hours, then mix it with sufficient ether, previously washed with water, to form a soft paste. After six hours, enclose in canvas, and express strongly and quickly so as to obtain the liquid. Reduce the cake to powder, form a paste with washed ether, and express as before. Mix the liquids, evaporate spontaneously to a syrupy consistence, then spread upon plates and dry quickly.

It is a light, non-crystalline powder, of a yellowish color and a very astringent taste. It is soluble in water, glycerin, and alcohol. It is powerfully astringent, and is used in all cases where vegetable astringents are required, as diarrhœa and dysentery, passive hemorrhages, colliquative sweats, etc. Externally it is employed in hemorrhages, catarrhal affections, etc.

Dose: 2 to 10 grains.

PREPARATIONS.

Glyceritum Acidi Tannici (U. S., Br.)—Glycerite of Tannic Acid.

Take of Tannic acid	2 ounces.
Glycerin	½ pint.

Rub them together, then heat gently until the acid is dissolved.

Dose: 10 to 40 minims.

Termed *Glycerine of Tannic Acid* by the British Pharmacopœia.

Suppositoria Acidi Tannici (U. S., Br.)—Suppositories of Tannic Acid.

Take of Tannic acid	60 grains.
Oil of theobroma	300 grains.

Rub the acid with 60 grains of the oil, melt the remainder, and, having cooled it to 95°, mix all together and pour into suitable moulds, making 12 suppositories.

The British Pharmacopœia directs: tannic acid, 36 grains; benzoated lard, 44 grains; white wax, 10 grains; oil of theobroma, 90 grains.

It also prepares suppositories with tannic acid, 36 grains; glycerin of starch, 50 grains; curd soap, 100 grains; starch, sufficient.

Trochisci Acidi Tannici (U. S., Br.)—Troches of Tannic Acid.

Take of Tannic acid	1 ounce.
Sugar	10 ounces.
Tragacanth	120 grains.
Orange-flower water	sufficient.

Rub the powders thoroughly together, then with orange-flower water form a mass, to be divided into 480 troches.

Dose: 1 or 2 troches.

Unguentum Acidi Tannici (U. S.)—Ointment of Tannic Acid.

Take of Tannic acid	30 grains.
Lard	1 ounce.

Rub them thoroughly together, avoiding the use of an iron spatula.

Gargles of Tannic Acid.

Take of Tannic acid	12 grains.
Rectified spirit	6 minims.
Camphor mixture	to 1 ounce.

Dissolve. Astringent. *London Throat Hospital.*

Take of Tannic acid	360 grains.
Gallic acid	120 grains.
Water	1 ounce.

Rub the acids to a fine powder, and mix with the water. Slowly sipped, or held passively in the mouth, it is most useful for arresting hemorrhage from the uvula or tonsils after excision. *London Throat Hospital.*

Inhalation of Tannic Acid.

Take of Tannic acid	40 grains.
Water	8 ounces.

Mix. Astringent. Used by means of a steam atomizing apparatus.
G. M. Lefferts.

Injections of Tannic Acid.

Take of Tannic acid	1 drachm.
Alum	2 drachms.
Water	1 pint.

Mix. For gonorrhœa in women. *Bumstead.*

Take of Tannic acid	2 drachms.
Glycerin	1 ounce.
Water	3 ounces.

Mix. *London Fever Hospital.*

Mixture of Tannic Acid.

Take of Tannic acid	7 grains.
Diluted sulphuric acid	10 minims.
Cinnamon water	1 ounce.

Mix. One dose. *Charing Cross Hospital.*

Ointment of Tannic Acid with Opium.

Take of Tannic acid	½ drachm.
Powdered opium	½ drachm.
Lard	1 ounce.

Mix. *British Skin Hospital.*

Pills of Tannic Acid with Opium.

Take of Tannic acid	20 grains.
Opium	1 grain.
Mucilage	sufficient.

Mix and divide into 4 pills. *St. Thomas's Hospital.*

Powders of Tannic Acid.

Take of Tannic acid	½ ounce.
Powdered gum arabic	2 drachms.
White sugar	2 drachms.

Mix. For insufflation into the larynx or nares. Useful in hemorrhage.
G. M. Lefferts.

Take of Tannic acid .... 5 grains.
Iodoform .... 2 grains.
Gum arabic .... 3 grains.

Mix and form a powder. For insufflation, as an astringent and alterative in post-nasal catarrh. *London Throat Hospital.*

Take of Tannic acid .... 2 drachms.
Powdered opium .... 6 grains.
Sugar .... sufficient.

Mix and divide into 6 powders.

Dose: 1 powder every two hours. In profuse diarrhœa. *Oppolzer.*

ACIDUM TARTARICUM (U. S. et al. Ph.)—TARTARIC ACID.

This is the acid of grapes, and is prepared from tartar, or crude bitartrate of potassium, which is deposited in wine-casks after fermentation.

Tartaric acid occurs as a white, crystalline powder, of a very sour taste, and freely soluble in water. It is diuretic and refrigerant. Used chiefly in preparing effervescing draughts.

Dose: 30 to 60 grains.

PREPARATIONS.

See *Pulveres Effervescentes* and *Pulveres Effervescentes Aperientes.*

Lemonade of Tartaric Acid (Fr.).

Take of Syrup of tartaric acid .... 10 parts.
Water .... 90 parts.

Mix. A refrigerant drink in fevers.

Syrup of Tartaric Acid (Fr.).

Take of Tartaric acid .... 1 part.
Distilled water .... 2 parts.
Syrup .... 47 parts.

Dissolve the acid in the water, and add the solution to the syrup.

ACIDUM VALERIANICUM (U. S., Fr., Ger.)—VALERIANIC ACID.

Prepared by distilling valerianate of sodium with water and sulphuric acid. It is a colorless, oily liquid, having the disagreeable odor of valerian. Not used in medicine except in combination with bases, as valerianate of zinc, valerianate of ammonium, etc.

ACONITIA (U. S. et al. Ph.)—ACONITIA.

Take of Aconite root	48 ounces.
Diluted sulphuric acid	1½ ounce.
Alcohol	sufficient.
Stronger water of ammonia	sufficient.
Stronger ether	sufficient.
Distilled water	sufficient.

Digest the aconite in 8 pints of alcohol at 120° for twenty-four hours, then transfer to a percolator and, with alcohol, percolate to 24 pints. Distil off the alcohol until it is reduced to 1 pint. Add to this, 1 pint of distilled water mixed with the sulphuric acid, remove the oil and resin which separate on standing, and evaporate to 4 ounces. When cool, wash it with ether, and add ammonia-water in slight excess. Next, shake the mixture with 6 ounces of the ether, allow it to settle, and decant the upper ethereal layer of liquid. Repeat the process twice, mix the ethereal solutions, and allow the mixture to evaporate spontaneously. Reduce the dry residue to powder, and keep in a well-stopped bottle.

It is a yellowish-white powder, without odor, and having a bitter, acrid taste, followed by a sensation of numbness, soluble in 150 parts of cold water, and freely soluble in alcohol, ether, and chloroform. Seldom employed internally.

Dose: $\frac{1}{160}$ to $\frac{1}{130}$ grain.

PREPARATIONS.

Ointment of Aconitia (Br.).

Take of Aconitia	8 grains.
Rectified spirit	½ drachm.
Prepared lard	1 ounce.

Dissolve the aconitia in the spirit, add the lard and mix.

Used in neuralgia. Should be used with great care.

ACONITUM—ACONITE.

ACONITI FOLIA (U. S., Br., Fr.)—ACONITE LEAVES.

ACONITI RADIX (U. S. et al. Ph.)—ACONITE ROOT.

The leaves and root of Aconitum Napellus, L. (*Nat. ord., Ranunculaceæ*), a large, herbaceous perennial indigenous to Northern Europe and Asia, and the western coast of North America.

Aconite owes its medicinal virtues to an alkaloid termed *aconitia*, which exists in much larger proportion in the root than in the leaves. The latter, as found in the shops, vary greatly in strength, and are therefore much less used than the former.

Aconite is a cardiac sedative of great power, and is much used in the early stages of acute inflammatory affections, as pneumonia, pleurisy, tonsillitis, rheumatism, erysipelas, etc.

Applied locally, it temporarily paralyzes the sensory nerves, and hence is of service in certain cases of neuralgia.

Dose: Of the leaves, 1 to 3 grains; of the root, ½ to 2 grains.

PREPARATIONS.

Extractum Aconiti (U. S. et al. Ph.)—Extract of Aconite.

Take of Aconite leaves, recently dried	12 ounces.
Alcohol	1 pint.
Diluted alcohol	sufficient.

Moisten the aconite with the alcohol, then percolate with diluted alcohol until 1 pint is obtained. Allow this to evaporate spontaneously to 3 ounces. Continue the percolation until 2 pints more are obtained, or the aconite is exhausted. Evaporate this on a water-bath, at or below 160°, to the consistence of syrup; add the reserved portion, and continue the evaporation at, or below 120°, until reduced to a proper consistence.

Dose: ⅙ to ¼ grain.

The British Pharmacopœia employs the fresh leaves; the German the root. The French Codex prepares also an extract from the juice.

Emplastrum Aconiti (U. S.)—Aconite Plaster.

Take of Aconite root	16 ounces.
Alcohol	sufficient.
Resin plaster	sufficient.

Macerate the root with 1 pint of alcohol for four days, then percolate until 2 pints are obtained. Distil off 1½ pint of alcohol, evaporate the residue to a soft extract, and add sufficient resin plaster, previously melted, to make the whole weigh 16 ounces.

Tinctura Aconiti Radicis (U. S., Br., Ger.)—Tincture of Aconite Root.

Take of Aconite root	12 ounces.
Alcohol	sufficient.

Moisten, pack, and percolate to 2 pints.

Dose: 1 to 3 minims.

The French Codex prepares a tincture from the fresh leaves, by macerating them for ten days in an equal weight of alcohol.

LINIMENT OF ACONITE (Br.).

Take of Aconite root	20 ounces.
Camphor	1 ounce.
Rectified spirit	sufficient.

Macerate the aconite with spirit for three days, then percolate into a receiver containing the camphor, until it measures 20 ounces.

An excellent preparation for the relief of some forms of neuralgia. It should be used with caution.

FLEMING'S TINCTURE OF ACONITE.

Take of Aconite root	8 ounces.
Alcohol	sufficient.

Moisten, pack, and percolate to 12 ounces.
Dose: 1 to 2 minims.

MIXTURES OF ACONITE.

Take of Tincture of aconite root	2 drachms.
Deodorized tincture of opium	6 drachms.

Mix. Dose: 8 drops in water every hour or two. In acute pleurisy, previous to effusion. *Bartholow.*

Take of Tincture of aconite	3 minims.
Colchicum wine	10 minims.
Bicarbonate of potassium	10 grains.
Water	1 ounce.

Mix. One dose. *Westminster Ophthalmic Hospital.*

ADEPS (U. S. et al. Ph.)—LARD.

The prepared fat of the hog (*Sus scrofa*, L). Leaf lard, that which is obtained from the mesentery, omentum, and kidneys, should be selected for medicinal use. At and below 90° F., lard is a soft solid, white, and of a peculiar odor, which should be free from rancidity. It is used in the preparation of ointments and cerates.

ÆTHER (U. S. et al. Ph.)—ETHER—SULPHURIC ETHER.

Prepared by distilling a mixture of sulphuric acid and alcohol at a temperature between 260° and 280°, and subsequently purifying the product by redistillation.

It is a colorless, limpid, and very volatile and inflammable liquid, of the specific gravity 0.750. It has a peculiar odor, a sweetish taste, is slightly soluble in water, and dissolves iodine, iodoform, sulphur, oils, fats, resins, etc.

Ether is a diffusible stimulant suited to many cases of exhaustion. Administered by inhalation, it induces anæsthesia, and is very largely employed in this manner, being, all things considered, the best and safest anæsthetic yet discovered. It is also employed externally to produce local anæsthesia.

Dose: 20 to 30 minims.

PREPARATIONS.

Æther Fortior (U. S.)—Stronger Ether.

Take of Ether	3 pints.
Water	3 pints.
Chloride of calcium	1 ounce.
Lime	1 ounce.

Shake the ether and water together, and when the water has subsided decant the ether. Shake this with the chloride of calcium and lime, and after twenty-four hours decant the ether into a retort and distil 1½ pint. Its specific gravity should not exceed 0.728.

The ethers of the British Pharmacopœia have respectively the specific gravity 0.735 and 0.720, the latter being termed *pure ether;* of the French Codex, 0.723 and 0.720; of the German Pharmacopœia, only one recognized, 0.728.

Spiritus Ætheris Compositus (U. S.)—Compound Spirit of Ether—Hoffmann's Anodyne.

Take of Ether	½ pint.
Alcohol	1 pint.
Ethereal oil	6 drachms.

Mix. Dose: ½ to 2 drachms. Stimulant, antispasmodic, and anodyne. Used in hysteria, sleeplessness, nervousness, etc.

Spirit of Ether (Br., Fr., Ger.)—Hoffmann's Anodyne.

Take of Ether	10 ounces.
Rectified spirit	20 ounces.

Mix. Dose: ½ to 2 drachms. Properties and uses similar to those of the preceding preparation.

Syrup of Ether (Fr.).

Take of Syrup	800 parts.
Distilled water	100 parts.
Alcohol (90%)	50 parts.
Ether	50 parts.

Mix in a bottle having a stop-cock at the bottom, shake occasionally during five or six days, then draw off the clear syrup.

Dose: ½ to 1 ounce.

MIXTURES OF ETHER.

Take of Spirit of ether .... ½ drachm.
Aromatic spirit of ammonia .... ½ drachm.
Syrup of tolu .... 1 drachm.
Water .... to 1 ounce.

Mix. One dose. *Middlesex Hospital.*

Take of Ether .... ½ drachm.
Ammoniated tincture of valerian .... ½ drachm.
Infusion of quassia .... to 1 ounce.

Mix. One dose. *Royal Chest Hospital.*

Take of Spirit of ether .... 5 minims.
Compound tincture of camphor (paregoric) .... 15 minims.
Water .... to ½ ounce.

Mix. One dose. *Samaritan Hospital.*

Take of Spirit of ether .... 20 minims.
Bicarbonate of sodium .... 10 grains.
Peppermint water .... 1 ounce.

Mix. One dose. *London Chest Hospital.*

ÆTHER ACETICUS (Br., Fr., Ger.)—ACETIC ETHER.

May be obtained by distilling a mixture of 8 parts of dry acetate of sodium, 5 parts of rectified spirit, and 10 parts of sulphuric acid; adding the distilled product to half its weight of chloride of calcium in a stoppered bottle; letting them remain together for twenty-four hours, and then decanting and rectifying the ethereal liquid.

It is a colorless liquid, with an agreeable ethereal odor. Medicinal properties similar to those of sulphuric ether. It is, however, milder and more agreeable.

Dose: 20 to 60 minims.

ÆTHER NITROSUS—NITROUS ETHER.

PREPARATIONS.

Spiritus Ætheris Nitrosi (U. S. et al. Ph.)—Spirit of Nitrous Ether—Sweet Spirit of Nitre.

Prepared by distilling a mixture of sulphuric acid and alcohol, with nitric acid, in the presence of copper.

It is a volatile, inflammable liquid, of a pale yellow color, a fragrant, ethereal odor, and a sharp, burning taste. Used as a diaphoretic and diuretic, especially with children, and generally in combination with other, and more powerful agents.

Dose: ½ to 2 drachms.

MIXTURES OF NITROUS ETHER.

Take of		
Take of	Spirit of nitrous ether	20 minims.
	Carbonate of ammonium	3 grains.
	Tincture of tolu	10 minims.
	Compound infusion of gentian	to 1 ounce.

Mix. One dose. *King's College Hospital.*

Take of		
Take of	Spirit of nitrous ether	1 drachm.
	Syrup of ipecac	2 drachms.
	Castor-oil	2 drachms.
	Syrup of tolu	7 drachms.

Mix. Dose: 1 drachm every two to four hours, for infants one year old. In primary bronchitis. *J. Lewis Smith.*

OLEUM ÆTHEREUM (U. S.)—ETHEREAL OIL.

Prepared by distilling a mixture of sulphuric acid and alcohol, at a temperature between 302° and 315°.

It is a transparent, nearly colorless, volatile liquid, of a peculiar ethereal odor and taste, and of the specific gravity 0.910. Used solely in the preparation of compound spirit of ether.

ALCOHOL (U. S. et al. Ph.)—ALCOHOL—RECTIFIED SPIRIT (Br.).

Spirit of the specific gravity 0.835 U. S.; 0.838 Br.; 0.830 to 0.834 Ger.; 88 to 90 per cent. Fr.

Alcohol Fortius (U. S.)—Stronger Alcohol.

Spirit of the specific gravity 0.817.

Alcohol is the product of the vinous fermentation of saccharine liquids, and is commercially obtained by distillation from fermented corn, wheat, rye, potatoes, etc. It is a clear, colorless, volatile and inflammable liquid, of a peculiar, agreeable odor, and a very pungent taste.

Alcohol contains, by volume, about 90 per cent. of spirit; stronger alcohol, about 95 per cent. Pure, or absolute alcohol, has the specific gravity 0.817. These different grades of alcohol are obtained from the first product of distillation, by various processes of rectification which it is unnecessary to describe here.

Alcohol dissolves bromine, iodine, phosphorus, sulphur, the alkalies, most of the alkaloids, oils, resins, etc., and hence is of the greatest importance in the preparation of medicines. Under the various forms of brandy, whiskey, rum, gin, and wine, it is extensively employed as a diffusible stimulant.

PREPARATION.

Alcohol Dilutum (U. S. et al. Ph.)—Diluted Alcohol.

Alcohol mixed with an equal measure of distilled water. Its specific gravity is 0.941. Used in preparations.

Proof spirit, Br.: rectified spirit, 5 pints; distilled water, 3 pints. Diluted alcohol, Ger.: alcohol, 7 parts; distilled water, 3 parts. Fr.: alcohol of 60 per cent.

ALCOHOL AMYLICUM—AMYLIC ALCOHOL—FUSEL OIL.

A peculiar alcohol, obtained from fermented grain or potatoes, by continuing the process of distillation after the ordinary spirit has ceased to come over. It is an oily, colorless liquid, of a strong, disagreeable odor, an acrid taste, and has the specific gravity 0.818.

Amylic alcohol is occasionally used in small doses, as a nervous stimulant, in phthisis and in the nervous affections of habitual drunkards. Pharmaceutically, it is employed in preparing valerianic acid.

Dose: 2 to 8 minims.

ALLIUM (U. S., Fr.)—GARLIC.

The bulb of Allium sativum, L. (*Nat. ord., Liliaceæ*), a small perennial indigenous to Europe, but cultivated in all parts of the world. It has a peculiar, penetrating odor, and a bitter, acrid taste.

Applied locally, garlic irritates and reddens the skin; taken internally, it acts as a stimulant. It is chiefly employed internally in catarrhal affections of the respiratory organs, and externally as a rubefacient and revulsive.

Dose: ½ to 2 drachms.

Syrupus Allii (U. S.)—Syrup of Garlic.

Take of Garlic sliced and bruised.................... 6 ounces.
Diluted acetic acid........................ 1 pint.
Sugar.................................... 24 ounces.

Macerate the garlic in 10 ounces of the acid for four days, and express. Mix the residue with the remainder of the acid, and express until sufficient additional liquid has been obtained to make the whole measure 1 pint. Then add the sugar.

Dose: 1 to 2 drachms.

ALOE—ALOES.

ALOE BARBADENSIS (U. S., Br., Fr.)—BARBADOES ALOES.
ALOE CAPENSIS (U. S., Fr., Ger.)—CAPE ALOES.
ALOE SOCOTRINA (U. S., Br.)—SOCOTRINE ALOES.

The inspissated juice of Aloe vulgaris, Lamarck, Aloe spicata, Thunberg, and Aloe Socotrina, Lamarck, respectively, (*Nat. ord., Liliaceæ*), though other species also contribute to the production of Cape and Socotrine aloes. The first named is produced in the West Indies, the second is received from the Cape of Good Hope, and the third from the island of Socotra, off the eastern coast of Africa.

Socotrine aloes is most commonly employed in this country. It is in pieces of a yellowish or reddish color, growing darker with age, breaking with a conchoidal fracture, and of a lighter color inside. Its powder is yellow, and its taste exceedingly bitter and disagreeable.

Aloes is a purgative whose action is chiefly on the large intestine ; hence, it is contraindicated in inflammations of the pelvic viscera. It is also an emmenagogue, because of its effect upon the lower bowel, and, sympathetically, the pelvic organs generally. In small doses it is believed to be tonic, and is therefore of use in dyspepsia attended with constipation, and in hemorrhoids due to relaxation of the veins of the rectum.

Dose : 2 to 3 grains, laxative ; 10 to 20 grains, purgative.

PREPARATIONS.

Aloe Purificata (U. S.)—Purified Aloes.

Take of Socotrine aloes .... 24 ounces.
Stronger alcohol .... 4 ounces.

Melt the aloes in a water-bath, add the alcohol, strain, and evaporate until it solidifies and becomes brittle on cooling.

Dose : 2 to 3 grains, laxative ; 10 to 20 grains, purgative.

Pilulæ Aloes (U. S., Br., Fr.)—Pills of Aloes.

Take of Socotrine aloes .... 48 grains.
Soap .... 48 grains.

Beat them into a mass with water, and divide into 24 pills.

Dose : 1 to 5 pills, or more.

The British Pharmacopœia directs : Socotrine aloes, 2 ounces ; hard soap, 1 ounce ; volatile oil of nutmeg, 1 drachm ; confection of roses, 1 ounce, and leaves the mass undivided. It also prepares pills of Barbadoes aloes by the same formula, substituting oil of caraway for oil of nutmeg. The French Codex employs 2 parts of Cape aloes with 1 part of confection of roses.

Pilulæ Aloes et Asafœtidæ (U. S., Br.)—Pills of Aloes and Asafetida.

Take of Socotrine aloes	32 grains.
Asafetida	32 grains.
Soap	32 grains.

Beat them into a mass with water, and divide into 24 pills.

Dose: 2 to 5 pills.

Pilulæ Aloes et Mastiches (U. S.)—Pills of Aloes and Mastic.

Take of Socotrine aloes	48 grains.
Mastic	12 grains.
Red rose	12 grains.

Beat them into a mass with water, and divide into 24 pills.

Dose: 1 to 3 pills.

Pilulæ Aloes et Myrrhæ (U. S., Br.)—Pills of Aloes and Myrrh.

Take of Purified aloes	48 grains.
Myrrh	24 grains.
Aromatic powder	12 grains.
Syrup	sufficient.

Beat them together into a mass, and divide into 24 pills.

Dose: 3 to 6 pills.

Pulvis Aloes et Canellæ (U. S.)—Powder of Aloes and Canella.

Take of Socotrine aloes	12 ounces.
Canella	3 ounces.

Rub together until thoroughly mixed.

Dose: 10 to 20 grains.

Suppositoria Aloes (U. S.)—Suppositories of Aloes.

Take of Purified aloes	60 grains.
Oil of theobroma	300 grains.

Mix the aloes with 60 grains of the oil; then, having melted the remainder and cooled it to 95°, mix all together and pour into suitable moulds, making 12 suppositories.

Used as a remedy for seat-worms. A decoction of aloes in milk or water, administered by enema, answers the same purpose.

Tinctura Aloes (U. S. et al. Ph.)—Tincture of Aloes.

Take of Socotrine aloes	1 ounce.
Liquorice	3 ounces.
Alcohol	½ pint.
Distilled water	1½ pint.

Macerate seven days, and filter through paper.

Dose: 1 to 8 drachms.

The French and German preparations contain 1 part of aloes in 5 parts of alcohol.

Tinctura Aloes et Myrrhæ (U. S.)—Tincture of Aloes and Myrrh.

Take of Socotrine aloes	3 ounces.
Myrrh	3 ounces.
Alcohol	sufficient.

Moisten, pack, and percolate to 2 pints.

Dose: 1 to 2 drachms.

Vinum Aloes (U. S., Br.)—Wine of Aloes.

Take of Socotrine aloes	1 ounce.
Ginger	60 grains.
Cardamom	60 grains.
Sherry wine	1 pint.

Macerate for seven days, with occasional agitation, and filter.

Dose: 1 to 2 drachms, stomachic; 1 to 2 ounces, purgative.

Compound Decoction of Aloes (Br.).

Take of Socotrine aloes	2 drachms.
Myrrh	1½ drachm.
Saffron	1½ drachm.
Carbonate of potassium	1 drachm.
Extract of liquorice	1 ounce.
Compound tincture of cardamoms	8 ounces.
Distilled water	sufficient.

Put the extract of aloes and myrrh with the carbonate of potassium and extract of liquorice in a covered vessel with 20 ounces of distilled water; boil five minutes, then add the saffron, cool, add the tincture of cardamoms, cover closely, and macerate two hours; finally, strain, pouring on the strainer enough distilled water to make the product measure 30 ounces.

Dose: ½ to 2 ounces.

Extract of Aloes (Br., Ger.).

Take of Socotrine (or Barbadoes) aloes	1 pound.
Boiling distilled water	1 gallon.

Stir well together, and after twelve hours pour off the clear liquid, strain the remainder, and evaporate the mixed liquids to dryness.

Dose: 2 to 6 grains.

Vitriolated Extract of Aloes (Ger.).

Take of Extract of aloes	8 parts.
Distilled water	32 parts.
Sulphuric acid	1 part.

Mix the extract with the water, add the acid, drop by drop, then evaporate to dryness.

Pill of Aloes and Iron (Br., Ger.).

Take of	Sulphate of iron	1½ ounce.
	Barbadoes aloes	2 ounces.
	Compound powder of cinnamon	3 ounces.
	Confection of roses	4 ounces.

Rub the first three ingredients together, then add the confection gradually and make a uniform mass.

Dose: 5 to 10 grains.

The German Pharmacopœia directs: dried sulphate of iron, aloes, each equal parts, beat into a mass with alcohol, and divide into pills of 1½ grain each.

Compound Tincture of Aloes (Ger., Fr.).

Take of	Aloes	9 parts.
	Gentian	1 part.
	Rhubarb	1 part.
	Zedoary	1 part.
	Saffron	1 part.
	Larch agaric	1 part.
	Diluted alcohol	200 parts.

Prepare the tincture by digestion.

Dose: ½ to 1 drachm. A substitute for the *Elixir of Life*, much used in Europe. The French preparation is very similar.

Compound Mixture of Aloes.

Take of	Socotrine aloes	1 ounce.
	Bicarbonate of sodium	1½ ounce.
	Glycerin	2 ounces.
	Compound spirit of lavender	2 ounces.
	Oil of peppermint	25 minims.
	Water	1 pint.

Mix. Dose: 1 to 2 drachms, stomachic and tonic; ½ to 1 ounce purgative. *Hospital Formulary.*

Compound Pills of Aloes.

Take of	Aloes	1 ounce.
	Extract of gentian	½ ounce.
	Oil of caraway	40 minims.
	Sugar of milk	sufficient.

Mix and divide into 145 pills. *Hospital Formulary.*

Pills of Aloes and Ox-Bile.

Take of	Extract of aloes	30 grains.
	Purified ox-bile	20 grains.
	Resin of podophyllum	2½ grains.

Mix and divide into 10 pills.

Dose: 1 pill at night in chronic constipation; 1 pill night and morning in acute constipation. *Hospital Formulary.*

Pills of Aloes and Savin.

Take of Socotrine aloes .... 30 grains.
Myrrh .... 30 grains.
Extract of savin .... 30 grains.

Mix and divide into 30 pills.

Dose: 1 pill three times a day. In ordinary cases of dysmenorrhœa.
Hospital for Ruptured and Crippled.

Pills of Aloes and Hyoscyamus.

Take of Socotrine aloes .... 20 grains.
Extract of hyoscyamus .... 30 grains.
Ipecac .... 5 grains.
Soap .... 20 grains.

Mix and divide into 20 pills.

Dose: 1 pill morning and night. In hemorrhoids. *Fordyce Barker.*

Pills of Aloes and Opium.

Take of Socotrine aloes .... 10 grains.
Sulphate of iron .... 20 grains.
Extract of opium .... 10 grains.
Soap .... 10 grains.

Mix and divide into 20 pills.

Dose: 1 pill morning and night. In hemorrhoids associated with looseness of the bowels. *Fordyce Barker.*

Pill of Aloes and Belladonna.

Take of Socotrine aloes .... $1\frac{1}{2}$ grain.
Extract of belladonna .... $\frac{1}{4}$ grain.

Make 1 pill. *Brompton Consumption Hospital.*

Take of Extract of aloes .... 1 grain.
Extract of belladonna .... $\frac{1}{3}$ grain.

Make 1 pill. *St. Mary's Hospital.*

Pill of Aloes and Ipecacuanha.

Take of Extract of Socotrine aloes .... 1 grain.
Ipecacuanha .... $\frac{1}{2}$ grain.
Extract of gentian .... $1\frac{1}{2}$ grain.

Make 1 pill. *Brompton Consumption Hospital.*

Pill of Aloes and Nux Vomica.

Take of Extract of aloes .... 1 grain.
Extract of nux vomica .... $\frac{1}{4}$ grain.
Myrrh .... 2 grains.
Soap .... sufficient.

Make 1 pill. *London Chest Hospital.*

Take of Extract of Socotrine aloes .... 1 grain.
Extract of nux vomica .... ½ grain.
Extract of henbane.................... 1 grain.
Soap .... 1½ grain.
Make 1 pill. *St. Thomas's Hospital.*

Take of Extract of aloes .... 1½ grain.
Extract of nux vomica .... ½ grain.
Extract of belladonna.................... ⅓ grain.
Powdered liquorice.................... 2 grains.
Glycerin of tragacanth.................... sufficient.
Make 1 pill. *University College Hospital.*

ALTHÆA (U. S., Fr., Ger.)—MARSHMALLOW.

The root of Althæa officinalis, L. (*Nat. ord., Malvaceæ*), an herbaceous perennial indigenous to Europe, but naturalized in a few places along the coast of New England, New York, and New Jersey.

Marshmallow root abounds in mucilage and starch, which it yields to boiling water. It is employed internally as a demulcent, and externally as an emollient.

PREPARATIONS.

Syrup of Marshmallow (Fr.).

Take of Marshmallow root.................... 1 part.
Water.................... 6 parts.
Syrup .... 30 parts.

Macerate the root in the water for twelve hours, strain without expression, add the syrup, and evaporate to the specific gravity 1.260.

Troches of Marshmallow (Fr.).

Take of Marshmallow root .... 10 parts.
White sugar....................100 parts.
Tragacanth.................... 1 part.
Water .... sufficient.

Boil the root in 4 times its weight of water, strain, and evaporate to a proper consistence; add the tragacanth, then the sugar, and form into troches of 15 grains each.

Decoction of Marshmallow.

Take of Marshmallow root.................... 4 ounces.
Raisins, stoned .... 2 ounces.
Boiling water.................... 5 pints.

Boil down to 3 pints and strain. *Edinburgh Pharmacopœia.*

Used freely as a demulcent drink in febrile and inflammatory affections.

ALUMEN (U. S. et al. Ph.)—ALUM.

Two compound sulphates are known as alums, the sulphate of aluminium and ammonium, and the sulphate of aluminium and potassium. The former is the *Alumen* of the United States and British Pharmacopœias, while the latter has the title of *Aluminii et Potassii Sulphas*. The latter is, however, the only one generally found in market, and possesses all the medicinal virtues of the other, and is the officinal alum of the French Codex and German Pharmacopœia. Alum acts as a powerful astringent upon all organic fibres, and hence, when locally applied, has a wide range of usefulness in hemorrhages and catarrhal affections. In large doses, it is a prompt and efficient emetic, and is one of the best to employ when it is desired merely to empty the stomach, without producing subsequent nausea and prostration, as in narcotic poisoning, and especially in croup. In smaller doses, it is used with much success in the treatment of whooping-cough, especially when attended with bronchitis. It moderates the cough to some extent, doubtless from its astringent effect upon the bronchial mucous membrane ; but it has a further beneficial effect which cannot be so easily explained.

Dose : 5 to 20 grains.

PREPARATIONS.

Alumen Exsiccatum (U. S. et al. Ph.)—Dried Alum.

Take of Alum.................................... 4 ounces.

Subject it to a temperature not exceeding 400° until the residue weighs 2 ounces and 120 grains. Dried alum is astringent and mildly escharotic. Used to repress fungous granulations.

Collyrium of Alum.

Take of Alum .... 5 grains.
Rose-water............................ 1 ounce.

Dissolve. Used in catarrhal ophthalmia.

Collyrium of Alum and Belladonna.

Take of Alum.................................... 4 grains.
Extract of belladonna .... 7½ grains.
Water .... to 1 ounce.

Mix. *Middlesex Hospital.*

Electuary of Alum.

Take of Alum .... 2 drachms.
Molasses or syrup .... 1 ounce.

Mix. Dose : 1 to 2 drachms every ten minutes. An excellent emetic in spasmodic croup.

Gargles of Alum.

Take of Alum .... 8 grains.
Water .... 1 ounce.

Dissolve. Mildly astringent. *London Throat Hospital.*

Take of Alum .... 1 drachm.
Tannic acid .... 1 drachm.
Water .... 10 ounces.

Mix. Astringent. *G. M. Lefferts.*

Inhalation of Alum.

Take of Alum .... 8 grains.
Distilled water .... 1 ounce.

Dissolve. Used by means of an atomizer. *London Throat Hospital.*

Injection of Alum.

Take of Alum .... 12 to 30 grains.
Water .... 4 ounces.

Dissolve. Used in gonorrhœa. *Bumstead.*

Take of Alum .... 2 drachms.
Water .... 20 ounces.

Dissolve. *Women's Hospital, London.*

Take of Alum .... 1 drachm.
Sulphate of zinc .... ½ drachm.
Borax .... 4 grains.
Rose-water .... 8 ounces.

Mix. Use in chronic gonorrhœa and leucorrhœa. *Bartholow.*

Lotion of Alum.

Take of Alum .... 20 grains.
Sulphate of zinc .... 10 grains.
Glycerin .... 1 drachm.
Rose-water .... 4 ounces.

Mix. Used in erythema, intertrigo, and eczema. *Tilbury Fox.*

Mixtures of Alum.

Take of Alum .... 50 grains.
Syrup of ginger .... 1 ounce.
Syrup of gum arabic .... 1 ounce.
Water .... 1 ounce.

Mix. Dose: 1 drachm every four or six hours. In whooping-cough.
Meigs and Pepper.

Take of Alum .... 8 grains.
Diluted sulphuric acid .... 10 minims.
Tincture of hops .... 20 minims.
Infusion of roses .... 1 ounce.

Mix. One dose. *Guy's Hospital.*

Take of Alum .... 8 grains.
Compound tincture of camphor (paregoric) ... 20 minims.
Camphor-water .... 4 drachms.
Wine of ipecac .... 15 minims.
Dill-water .... to 1 ounce.

Mix. Dose: 1 to 1½ ounce. *Guy's Hospital.*

PILLS OF ALUM.

Take of Alum .... 2 drachms.
Extract of gentian .... ½ drachm.

Mix and divide into 30 pills.

Dose: 2 pills three times a day. In gastric catarrh. *Bartholow.*

Take of Alum .... 1 drachm.
Extract of opium .... 10 grains.
Catechu .... 1 drachm.

Mix and divide into 20 pills.

Dose: 2 pills every two, three, or four hours. In chronic diarrhœa and chronic dysentery. *Bartholow.*

POWDERS OF ALUM.

Take of Alum .... 3 grains.
Starch .... ½ grain.

Mix. Use by insufflation. A mild astringent in chronic tracheitis.
London Throat Hospital.

Take of Alum .... 1 part.
Subnitrate of bismuth .... 1 part.

Mix. Use by insufflation into the ear. Astringent.
London Throat Hospital.

Take of Dried alum .... 1 part.
Powdered yolk of egg .... 1 part.

Mix. A topical application in aphthous stomatitis. *V. P. Gibney.*

AMMONIA.

A gas which is generated during the decay of many organic substances, but obtained for use in medicine and the arts, by heating a mixture of chloride of ammonium and lime in the presence of water. It has a pungent, suffocating odor, and is freely soluble in water.

PREPARATIONS.

Aqua Ammoniæ (U. S. et al. Ph.)—Water of Ammonia—Solution of Ammonia.

Take of	Chloride of ammonium	12 ounces.
	Lime	12 ounces.
	Water	6 pints.
	Distilled water	sufficient.

Mix the lime with the water, stir well, decant the milky liquid into a retort, and add the chloride of ammonium. Then apply heat, and by means of proper apparatus, pass the gas which generates into a bottle containing the distilled water, kept ice-cold. Lastly, add sufficient distilled water to raise the specific gravity of the liquid to 0.960. Keep in small bottles, well-stopped.

It is a transparent, colorless liquid, having a very pungent odor, and a strongly alkaline taste. It is a powerful diffusible stimulant, and is used in low typhoid conditions. Applied externally, it acts as a rubefacient and vesicant.

Dose: 10 to 30 minims, diluted with water.

Aqua Ammoniæ Fortior (U. S., Br.)—Stronger Water of Ammonia.

An aqueous solution of ammonia, of the specific gravity 0.900. Used externally, and in preparations.

The British preparation has the specific gravity 0.891.

Linimentum Ammoniæ (U. S. et al. Ph.)—Liniment of Ammonia.

Take of	Water of ammonia	1 ounce.
	Olive oil (by weight)	2 ounces.

Mix.

The British Pharmacopœia directs the proportion of 1 to 3; the French Codex, 1 to 9; the German Pharmacopœia, 1 to 4.

Spiritus Ammoniæ (U. S., Ger.)—Spirit of Ammonia.

Take of	Chloride of ammonium	12 ounces.
	Lime	12 ounces.
	Water	6 pints.
	Alcohol	20 ounces.

Proceed as in the preparation of water of ammonia, passing the gas into the alcohol. Its properties are similar to those of water of ammonia, and it is used in like manner.

Dose: 10 to 30 minims, diluted with water.

Spiritus Ammoniæ Aromaticus (U. S., Br.)—Aromatic Spirit of Ammonia.

Take of Carbonate of ammonium	1 ounce.
Water of ammonia	3 ounces.
Oil of lemon	2½ drachms.
Oil of nutmeg	40 minims.
Oil of lavender	15 minims.
Alcohol	1½ pint.
Water	sufficient.

Dissolve the carbonate in the water of ammonia, previously mixed with 4 ounces of water. Dissolve the oils in the alcohol, mix the solutions, and add sufficient water to make the whole measure 2 pints.

Dose: ½ to 1 drachm, diluted with water. Often used as an antacid and stimulant in nervous and sick headache.

The British preparation, made by distillation, is very similar.

Anisated Spirit of Ammonia (Ger., Fr.).

Take of Oil of Anise	1 part.
Alcohol	24 parts.
Water of ammonia	5 parts.

Dissolve the oil in the alcohol, and mix with the water of ammonia. The proportions of the French preparation are 1, 32, 8.

Fetid Spirit of Ammonia (Br.).

Take of Asafetida	1½ ounce.
Strong solution of ammonia	2 ounces.
Rectified spirit	sufficient.

Macerate the asafetida in 15 ounces of the spirit for twenty-four hours, distil off the spirit, mix the product with the solution of ammonia, and add sufficient rectified spirit to make 20 ounces.

Dose: ½ to 1 drachm.

Inhalation of Ammonia.

Take of Solution of ammonia	4 drachms.
Water	4 drachms.

Mix. One drachm in a pint of water at 80° for each inhalation. Stimulant; useful in chronic laryngitis and functional aphonia.

The strong salts of ammonia are very useful in cases of obstinate sneezing, influenza, etc. The patient should smell the salts directly a disposition to sneeze is felt. *London Throat Hospital.*

LOTION OF AMMONIA.

Take of Aromatic spirit of ammonia.................. 1 part.
Water.................................. 3 parts.

Mix. Shampoo for the scalp. *G. H. Fox.*

MIXTURES OF AMMONIA.

Take of Aromatic spirit of ammonia .... 20 minims.
Spirit of ether.......................... 20 minims.
Camphor water..........................to 1 ounce.

Mix. One dose. *London Hospital.*

Take of Aromatic spirit of ammonia .... 20 minims.
Spirit of chloroform...................... 20 minims.
Water..................................to 1 ounce.

Mix. One dose. *Samaritan Hospital.*

Take of Aromatic spirit of ammonia................. 2 drachms.
Spirit of chloroform .... 2 drachms.
Tincture of camphor....................... 2 drachms.
Deodorized tincture of opium............... 2 drachms.
Tincture of capsicum....................... 1 drachm.

Mix. Dose: 20 to 30 minims, in a wineglass of water. A stimulating and anodyne diarrhœa mixture. *A. A. Smith.*

AMMONII ACETAS—ACETATE OF AMMONIUM.

Acetate of ammonium is used only in the following preparation:

Liquor Ammonii Acetatis (U. S. et al. Ph.)—Solution of Acetate of Ammonium—Spirit of Mindererus.

Take of Diluted acetic acid .... 2 pints.
Carbonate of ammonium .... sufficient.

Add the carbonate gradually to the acid until the latter is neutralized, and filter. It should be freshly made when dispensed.

It may also be made by mixing the following solutions:

Take of Carbonate of ammonium ....640 grains.
Distilled water .... sufficient.

Dissolve, filter, and add sufficient distilled water to make 1 pint.

Take of Acetic acid............................. 4 ounces.
Distilled water .... 12 ounces.

Mix. These solutions may be mixed in equal quantities when dispensed.

Solution of acetate of ammonium is a colorless liquid, of an acetous odor and a saline taste. It is a very valuable diaphoretic, diuretic, and refrigerant. Used in acute inflammatory and febrile affections.

Dose: 1 to 8 drachms.

Mixtures of Acetate of Ammonium.

Take of Solution of acetate of ammonium............ 4 drachms.
Spirit of nitrous ether .... ½ drachm.
Camphor water ....to 1 ounce.

Mix. One dose. *Royal Chest Hospital.*

Take of Tincture of aconite root.................... 16 minims.
Acetate of morphia .... 2 grains.
Solution of acetate of ammonium............ 4 ounces.

Mix. Dose: 2 drachms. In the early stages of inflammatory affections, as pneumonia, pleurisy, etc.

Take of Solution of acetate of ammonium............ ½ ounce.
Syrup of ipecacuanha...................... 1 drachm.
Solution of sulphate of morphia (*U. S. Ph.*) ... 40 drops.
Syrup of gum arabic....................... 1 ounce.
Water .... 1½ ounce.

Mix. Dose: 1 drachm for a child two years old, every two hours, in bronchitis. *Meigs and Pepper.*

Take of Solution of acetate of ammonium............ 1½ ounce.
Bromide of ammonium .... 1 drachm.
Syrup of ipecacuanha .... 3 drachms.
Spirit of nitrous ether.................... 3 drachms.
Syrup of tolu.............................to 3 ounces.

Mix. Dose: 1 drachm every two or three hours, in bronchitis.

Take of Solution of acetate of ammonium............ 2½ drachms.
Tincture of perchloride of iron.............. 10 minims.
Diluted acetic acid .... ½ drachm.
Water.................................to 1 ounce.

Mix. One dose. *Middlesex Hospital.*

AMMONII BENZOAS (U. S., Br., Fr.)—BENZOATE OF AMMONIUM.

Take of Benzoic acid............................ 2 ounces.
Water of ammonia..................3½ ounces, or sufficient.
Distilled water .... 4 ounces.

Dissolve the acid in the water of ammonia previously mixed with the distilled water, evaporate with a gentle heat, adding water of ammonia, if necessary, to keep the liquid alkaline; then set aside to crystallize. Dry the crystals without heat.

It is in minute white crystals, having a slight odor of benzoic acid, a bitterish, saline taste, soluble in water and alcohol. Its medicinal effects are similar to those of benzoic acid, than which, however, it is much more soluble, and less irritating to the stomach.

Dose: 10 to 20 grains.

AMMONII BROMIDUM (U. S., Br.)—BROMIDE OF AMMONIUM.

Take of	
Bromine	2 ounces.
Iron wine	1 ounce.
Water of ammonia	4½ ounces.
Distilled water	sufficient.

Add the iron, and then the bromine, to ½ pint of water in a flask, loosely cork the flask, and agitate until the odor of bromine can no longer be perceived. Then add the water of ammonia, previously mixed with ½ pint of distilled water, heat gently for half an hour, and filter, washing the filter with boiling distilled water. Lastly, evaporate in a capsule until a pellicle forms, then stir until it granulates.

It is a white, granular salt, becoming yellow on exposure, of a sharp, saline taste, and freely soluble in water.

Its medicinal properties are similar to those of bromide of potassium, though it is probably more active. Used in a great variety of nervous diseases, including epilepsy, whooping-cough, infantile convulsions, etc.

Dose: 5 to 20 grains.

Mixtures of Bromide of Ammonium.

Take of	
Bromide of ammonium	½ ounce.
Syrup of ginger	1 ounce.
Water	2 ounces.

Mix. Dose: 1 drachm.

Take of	
Bromide of ammonium	2 drachms.
Bromide of potassium	½ ounce.
Syrup of ginger	1 ounce.
Water	2 ounces.

Mix. Dose: 1 drachm.

AMMONII CARBONAS (U. S. et al. Ph.)—CARBONATE OF AMMONIUM.

Prepared by heating a mixture of chloride of ammonium and chalk in a retort, and passing the vapors into a large chamber, where they condense.

It occurs in white, translucent, crystalline masses, having a strong odor of ammonia. Exposed to the air it is converted into bicarbonate of ammonium, and falls into a dry powder.

It is a powerful, diffusible stimulant. Used in low fevers, and diseases of a marked asthenic character, as typhus, typhoid, and scarlet fevers, typhoid pneumonia, etc.

Dose : 3 to 10 grains.

PREPARATIONS.

SOLUTION OF CARBONATE OF AMMONIUM (Ger.).

Take of	Carbonate of ammonium	1 part.
	Distilled water	5 parts.

Dissolve.

MIXTURES OF CARBONATE OF AMMONIUM.

Take of	Carbonate of ammonium	5 grains.
	Spirit of chloroform	20 minims.
	Mucilage	2 drachms.
	Water	to 1 ounce.

Mix. One dose. *Royal Chest Hospital.*

Take of	Carbonate of ammonium	5 grains.
	Compound tincture of gentian	20 minims.
	Decoction of cinchona	1 ounce.

Mix. One dose. *Royal Free Hospital.*

Take of	Carbonate of ammonium	1½ drachm.
	White sugar	1½ drachm.
	Gum arabic	1½ drachm.
	Compound spirit of lavender	2 drachms.
	Mint water	4 ounces.

Mix. Dose : ½ ounce. *Ellis.*

Take of	Carbonate of ammonium	16 grains.
	Fluid extract of squill	2 drachms.
	Fluid extract of seneka	2 drachms.
	Tincture of tolu	2 drachms.
	Water	½ ounce.
	Syrup	to 2 ounces.

Mix. Dose : 1 drachm. Expectorant. *Hospital Formulary.*

Take of Carbonate of ammonium	8 grains.
Syrup of tolu	½ ounce.
Water	1½ ounce.

Mix. Dose: 1 drachm every two or three hours, for an infant of three months. In bronchitis. *J. Lewis Smith.*

Take of Carbonate of ammonium	½ drachm.
Citrate of iron and ammonium	½ drachm.
Syrup	4 ounces.

Mix. Dose: 1 to 2 drachms every second or third hour. In the declining stage of scarlatina. *J. Lewis Smith.*

AMMONII CHLORIDUM (U. S. et al. Ph.)—CHLORIDE OF AMMONIUM—MURIATE OF AMMONIA.

Obtained from the gas-liquor of gas-works. It is in translucent masses, entirely volatilized by heat, of a very sharp, saline taste, and freely soluble in water.

It is an alterative whose effects are exerted chiefly upon the mucous surfaces. Used mainly in catarrhal affections, as bronchitis, laryngitis, etc., though it is occasionally employed in rheumatism and neuralgia. Externally, lotions of chloride of ammonium are used in sprains, bruises, etc.

Dose: 5 to 30 grains.

PREPARATIONS.

Ammonii Chloridum Purificatum (U. S., Br.)—Purified Chloride of Ammonium.

Take of Chloride of ammonium	20 ounces.
Water of ammonia	5 drachms.
Water	2 pints.

Dissolve the chloride in the water, in a porcelain dish, with the aid of heat; add the water of ammonia, filter, and evaporate to dryness, stirring so as to produce a granular salt.

Dose: 5 to 30 grains.

Lotions of Chloride of Ammonium.

Take of Chloride of ammonium	30 grains.
Alcohol	1 drachm.
Distilled vinegar	1 drachm.
Water	to 1 ounce.

Mix. *St. George's Hospital.*

Take of Chloride of ammonium .... 6 grains.
Alum .... 6 grains.
Water .... 1 ounce.

Dissolve. For aural use. Should be introduced into the ear at about 100°. *London Throat Hospital.*

MIXTURES OF CHLORIDE OF AMMONIUM.

Take of Chloride of ammonium .... 1 ounce.
Wild-cherry bark .... 2 ounces.
Water .... 1 pint.

Make a cold infusion.

Dose: ½ ounce every hour. May be used in certain forms of fibroid phthisis. Liquorice root may be added to improve the taste of the mixture.

J. R. Leaming.

Take of Chloride of ammonium .... 2 drachms.
Chlorate of potassium .... 1 drachm.
Cinnamon (or camphor) water .... 3 ounces.
Syrup of seneka .... ½ ounce.
Spirit of nitrous ether .... ½ ounce.
Extract of liquorice .... 1½ drachm.

Mix. Dose: ½ ounce every two hours. May be used in different forms of sore throat, simple, croupal, or diphtheritic. Tartar emetic, belladonna, aconite, or tincture of iron, may be added, according to indications.

J. R. Leaming.

Take of Chloride of ammonium .... ½ ounce.
Nitrate of potassium .... ½ ounce.
Seneka root .... ½ ounce.
Liquorice root .... 1 ounce.
Water .... 1 pint.

Infuse. Dose: ½ ounce every half-hour. Useful in aborting an influenza or cold, if taken early. This combination of drugs, mixed and powdered, has for years been known as "White's red salts," in the vicinity of Hudson, N. Y. *J. R. Leaming.*

Take of Chloride of ammonium .... 2 drachms.
Camphorated tincture of opium .... ½ ounce.
Syrup of ipecacuanha .... ½ ounce.
Spirit of nitrous ether .... ½ ounce.
Syrup of tolu .... 1½ ounce.

Mix. Dose: 1 to 2 drachms every three hours.

Take of Chloride of ammonium .... 2 drachms.
Brown mixture .... 3 ounces.

Mix. Dose: 1 to 2 drachms.

Take of Chloride of ammonium .... 1 drachm.
Bromide of ammonium .... 1 drachm.
Spirit of nitrous ether .... ½ ounce.
Syrup of wild cherry .... 2½ ounces.

Mix. Dose: 1 drachm.

Take of Chloride of ammonium .... 1 drachm.
Tincture of cubeb.................... ½ ounce.
Compound tincture of gentian.................... ½ ounce.
Glycerin....................to 4 ounces.

Mix. Dose: 1 drachm in ½ ounce of water every two hours. In follicular pharyngitis with gastric dyspepsia. *A. A. Smith.*

Take of Chloride of ammonium .... 10 grains.
Carbonate of ammonium .... 5 grains.
Camphor water .... 1 ounce.

Mix. One dose. *Royal Chest Hospital.*

Take of Chloride of ammonium .... 1 ounce.
Fluid extract of cimicifuga.................... 2 ounces.
Syrup .... 1 ounce.
Cherry-laurel water .... 1 ounce.

Mix. Dose: 1 drachm three or four times a day. In muscular rheumatism. *Bartholow.*

PASTILLES OF CHLORIDE OF AMMONIUM.

Take of Chloride of ammonium ....1 or 2 grains.
Chlorate of potassium .... 1 grain.
Seneka root .... 1 grain.
Extract of liquorice .... 1 grain.
Cubeb.................... 1 grain.
Sugar .... sufficient.
Gum arabic.................... sufficient.

Make one pastille. Used in hoarseness and slight irritation of the throat. *J. R. Leaming.*

AMMONII IODIDUM (U. S.)—IODIDE OF AMMONIUM.

Take of Iodide of potassium .... 4 ounces.
Sulphate of ammonium .... 1 ounce.
Boiling distilled water .... 2 ounces.
Alcohol.................... sufficient.
Water .... sufficient.

Mix the salts, add to the boiling water, cool, add 1 ounce of alcohol, and reduce the temperature to about 40°. Then throw the mixture into a cooled glass funnel, stopped with moistened cotton, wash with 1 ounce of a mixture of 2 parts of water and 1 part of alcohol, and evaporate the solution rapidly to dryness.

It is a white, granular, very deliquescent salt. Occasionally used, both externally and internally, as a substitute for iodide of potassium in syphilis, cutaneous affections, enlarged glands, etc.

Dose: 3 to 10 grains.

AMMONII NITRAS (U. S.)—NITRATE OF AMMONIUM.

Prepared by neutralizing nitric acid with ammonia or carbonate of ammonium, evaporating and crystallizing.

It is a white, crystalline, deliquescent salt. Used in preparing nitrous oxide gas.

AMMONII SULPHAS (U. S.)—SULPHATE OF AMMONIUM.

A colorless, crystalline salt, obtained by neutralizing coal-gas liquor with sulphuric acid. Used in preparations.

AMMONII VALERIANAS (U. S., Fr.)—VALERIANATE OF AMMONIUM.

Take of		
Take of	Valerianic acid	4 ounces.
	Chloride of ammonium	sufficient.
	Lime	sufficient.

From the chloride of ammonium and lime obtain gaseous ammonia, as in preparing water of ammonia, and pass it into the acid until the latter is neutralized. Then set the solution aside to crystallize; and lastly, dry the crystals on bibulous paper.

It is a white salt, having the odor of valerianic acid, a sharp, sweetish taste, and is very soluble in water and alcohol. Used as an antispasmodic and nervous stimulant, in hysteria, nervous headache, etc.

Dose: 2 to 8 grains. It may be administered in simple elixir, or in aqueous solution, with the addition of any aromatic tincture as a flavoring agent.

AMMONIACUM (U. S. et al. Ph.)—AMMONIAC.

A gum-resin obtained from Dorema ammoniacum, Don. (*Nat. ord.*, *Umbelliferæ*), an herbaceous perennial indigenous to Persia.

It occurs in roundish grains or tears, from the size of a pea to that of a cherry, of a pale, creamy color externally, and white within, growing darker with age. It has a peculiar, characteristic odor, and a bitter, acrid taste.

Ammoniac is used internally as an antispasmodic and expectorant; externally, in the form of a plaster, as a discutient.

Dose: 6 to 12 grains.

PREPARATIONS.

Emplastrum Ammoniaci (U. S. et al. Ph.)—Ammoniac Plaster.

Take of Ammoniac	5 ounces.
Diluted acetic acid	½ pint.

Dissolve the ammoniac in the acid and strain. Then evaporate the solution on a water-bath, stirring constantly, until it acquires a proper consistence.

Emplastrum Ammoniaci cum Hydrargyro (U. S., Br.)—Plaster of Ammoniac with Mercury.

Take of Ammoniac	12 ounces.
Mercury	3 ounces.
Olive oil	60 grains.
Sublimed sulphur	8 grains.

Heat the oil, and add the sulphur to it gradually, stirring until they unite. Triturate the mercury with the mixture until globules are no longer visible, then add the ammoniac, previously liquefied, and mix the whole carefully.

Mistura Ammoniaci (U. S., Br.)—Ammoniac Mixture.

Take of Ammoniac	120 grains.
Water	½ pint.

Add the water gradually to the ammoniac, rubbing them thoroughly together, and strain.

Dose: ½ to 1 ounce.

TINCTURE OF AMMONIAC (Fr.).

Take of Ammoniac	1 part.
Alcohol (80%)	5 parts.

Macerate ten days, with occasional agitation, and filter.

Dose: ½ to 1 drachm.

MIXTURES OF AMMONIAC.

Take of Ammoniac mixture	½ an ounce.
Wine of ipecacuanha	10 minims.
Ethereal tincture of lobelia	10 minims.
Water	to 1 ounce.

Mix. One dose. *Royal Chest Hospital.*

Take of Ammoniac .... 50 grains.
Diluted nitric acid .... 2 drachms.
Water .... 8 ounces.

Add the acid to the water, then gradually rub the gum with the compound.

Dose: ½ ounce two or three times a day, in the chronic catarrh of elderly persons. *Ellis.*

Take of Compound tincture of camphor .... 30 minims.
Oxymel of squill .... 30 minims.
Ammoniac mixture .... to 1 ounce.

Mix. One dose. *Brompton Consumption Hospital.*

AMYGDALA AMARA (U. S. et al. Ph.)—BITTER ALMOND.
AMYGDALA DULCIS (U. S. et al. Ph.)--SWEET ALMOND.

The first named is the seed of Amygdalus communis, L., var. amara, DC., and the second of A. communis, L., var. dulcis, DC. (*Nat. ord., Rosaceæ*), trees indigenous to Asia, but extensively cultivated in Southern Europe and Northern Africa. These two varieties of the almond tree closely resemble each other, and their fruits are nearly alike in general appearance, but possess different constituents and different tastes.

Sweet almonds are used for the extraction of almond oil, to prepare an emulsion, and as a food for patients suffering with diabetes.

Bitter almonds are very poisonous, in large doses, owing to the hydrocyanic acid which they generate when mixed with water. The essential oil of bitter almonds also contains hydrocyanic acid, and is very poisonous.

PREPARATIONS.

Aqua Amygdalæ Amaræ (U. S., Ger.)—Bitter Almond Water.

Take of Oil of bitter almond .... 16 minims.
Carbonate of magnesium .... 60 grains.
Distilled water .... 2 pints.

Rub the oil, first with the carbonate of magnesium, and then with the water, added gradually, and filter.

Dose: 1 to 3 drachms.

Slightly sedative; used as a vehicle for narcotic medicines.

The concentrated bitter almond water of the German Pharmacopœia is prepared by distillation, and is a very dangerous preparation.

Mistura Amygdalæ (U. S., Br., Fr.)—Almond Mixture.

Take of Sweet almond .... ½ ounce.
Gum arabic .... 30 grains.
Sugar ....120 grains.
Distilled water .... 8 ounces.

Having blanched the almond, add the gum and sugar, and beat them until they are thoroughly mixed; then rub the mixture with the water, gradually added, and strain.

Demulcent and nutritive; may be used freely.

Oleum Amygdalæ Amaræ (U. S., Fr.)—Oil of Bitter Almond.

Obtained by distillation from the kernels of bitter almonds after their fixed oil has been removed by expression. Its effects are due to the hydrocyanic acid which it contains. It is about four times stronger than the officinal acid, though it varies greatly, and is therefore unreliable.

Dose: ¼ to 1 minim.

Oleum Amygdalæ Expressum (U. S. et al. Ph.)—Expressed Oil of Almond.

Obtained by expression from the kernels of both sweet and bitter almonds. It is emollient, demulcent, and nutritive, and is used for the same purposes as olive oil.

Syrupus Amygdalæ (U. S., Fr., Ger.)—Syrup of Almond.

Take of Sweet almond	12 ounces.
Bitter almond	4 ounces.
Sugar	72 ounces.
Water	3 pints.

Blanch the almonds, and rub them to a fine paste, adding, during the trituration, 3 ounces of water and 12 ounces of sugar. Mix the paste with the remainder of the water, strain, express, add the remainder of the sugar, and dissolve with a gentle heat. Lastly, strain through muslin, and keep in well-stopped bottles in a cool place.

Dose: 1 to 4 drachms.

Slightly sedative, from the small amount of hydrocyanic acid which it contains. Used chiefly in cough mixtures.

Compound Powder of Almond (Br.).

Take of Sweet almond	8 ounces.
Refined sugar	4 ounces.
Gum arabic	1 ounce.

Blanch the almonds and rub them to a smooth consistence. Mix the gum and sugar, and, adding them to the pulp gradually, rub the whole to a coarse powder.

Two and one-half ounces of the powder, triturated with 20 ounces of water, forms the almond mixture of the British Pharmacopœia.

AMYL NITRIS (Br.)—NITRITE OF AMYL.

An ethereal liquid produced by the action of nitric or nitrous acid upon amylic alcohol.

Nitrite of amyl is administered by inhalation in a variety of nervous and spasmodic affections; but it has attracted most attention lately as an agent for counteracting the poisonous effects of chloroform inhalation. Its use for this latter purpose was first suggested by Dr. F. A. Burrall, and a number of cases are already upon record in which it has evidently been the means of saving life. It may also be administered by the mouth.

Dose: 2 to 5 minims, internally, or by inhalation.

AMYLUM (U. S. et al. Ph.)—STARCH.

The fecula of common wheat, Triticum vulgare, Kunth. (*Nat. ord., Gramineæ*).

In the form of mucilage, starch is often used as a vehicle for the administration of opium, etc., by enema. Externally it is employed as a dusting-powder.

PREPARATIONS.

Glycerin of Starch (Br., Fr.).

Take of Starch .... 1 ounce.
Glycerin .... 8 ounces.

Rub them together until they are intimately mixed, then gradually heat to a temperature of 240° F., stirring constantly until a translucent jelly is formed.

Used as a substitute for simple ointments when grease is objectionable. It is, however, rather irritating, and should be diluted with one or two parts of water.

Mucilage of Starch (Br., Fr.).

Take of Starch .... 120 grains.
Distilled water .... 10 ounces.

Triturate the starch with the water, gradually added, then boil for a few minutes, constantly stirring.

AMYLI IODIDUM—IODIDE OF STARCH.

Prepared by triturating starch and iodine in the presence of a little water, and afterward carefully drying the product.

Used in cutaneous diseases.

ANGUSTURA (U. S., Br., Fr.)—ANGUSTURA.

The bark of Galipea Cusparia, St. Hilaire (*Nat. ord., Rutaceæ*), a small tree indigenous to Venezuela.

Angustura is an aromatic bitter tonic, suited to convalescence from acute diseases, especially in patients who are unable to bear cinchona or its derivatives.

Dose: 10 to 30 grains.

PREPARATIONS.

Infusum Angusturæ (U. S., Br.)—Infusion of Angustura.

Take of Angustura	½ ounce.
Water	sufficient.

Moisten, pack, and percolate to 1 pint.

Or, macerate the angustura in 1 pint of boiling water for two hours, and strain.

Dose: 1 to 2 ounces. *Infusion of Cusparia, Br.*

ANISUM (U. S. et al. Ph.)—ANISE—ANISEED.

The fruit of Pimpinella anisum, L. (*Nat. ord., Umbelliferæ*), an annual plant indigenous to Asia Minor, the Greek Islands, and Egypt, but cultivated in many subtropical countries.

Anise is an aromatic stimulant and carminative. It is employed to relieve the flatulent colic of infants and young children, and as an adjunct to other medicines.

Dose: 10 to 30 grains.

PREPARATIONS.

Aqua Anisi (U. S., Fr.)—Anise water.

Take of Oil of anise	½ drachm.
Carbonate of magnesium	60 grains.
Distilled water	2 pints.

Rub the oil, first with the carbonate, then with the water, gradually added, and filter.

It may also be prepared by mixing 10 ounces of powdered anise with 16 pints of water, and distilling 8 pints. The French Codex prepares it by distillation.

Used as a vehicle.

Oleum Anisi (U. S. et al. Ph.)—Oil of Anise.

The oil obtained from anise by distillation.

Dose : 5 to 15 minims.

Spiritus Anisi (U. S., Br., Fr.)—Spirit of Anise, Essence of Anise.

Take of Oil of anise.............................. 1 ounce.
Stronger alcohol.......................... 15 ounces.

Dissolve. Dose : 1 to 2 drachms.

The British Pharmacopœia directs the proportion of 1 to 4 ; the French Codex prepares it by distilling anise with alcohol.

OLEOSACCHARATE OF ANISE (Fr.).

Take of Oil of anise.............................. 1 part.
White sugar.............................. 80 parts.

Triturate them together in a mortar.

SYRUP OF ANISE (Fr.).

Take of Anise water..............................500 parts.
White sugar..............................950 parts.

Dissolve by agitation, and filter.

ANTHEMIS (U. S. et al. Ph.)—CHAMOMILE.

The flowers of Anthemis nobilis, L. (*Nat. ord.*, *Compositæ*), an herbaceous perennial indigenous to Europe.

Chamomile is an aromatic, bitter tonic and stimulant. In cold infusion it stimulates the appetite and aids digestion ; the warm infusion in large doses is diaphoretic and emetic.

Dose : 10 to 30 grains.

PREPARATIONS.

Infusum Anthemidis (U. S., Br.)—Infusion of Chamomile.

Take of Chamomile.............................. ½ ounce.
Boiling water.............................. 1 pint.

Macerate for ten minutes and strain.

Dose : 1 to 4 ounces.

EXTRACT OF CHAMOMILE (Br., Fr.).

Take of Chamomile flowers	1 pound.
Oil of chamomile	15 minims.
Distilled water	1 gallon.

Boil the chamomile with the water until its volume is reduced to one-half, strain, press, and filter. Then evaporate on a water-bath to the proper consistence, adding the oil at the end of the process.

Dose: 2 to 10 grains.

OIL OF CHAMOMILE (Br., Fr.).

The oil obtained from chamomile flowers by distillation. It possesses the purely stimulant properties of the plant.

Dose: 3 to 10 minims.

INFUSED OIL OF CHAMOMILE (Fr.).

Take of Chamomile flowers	1 part.
Olive oil	10 parts.

Digest for two hours in a covered vessel, strain with expression, and filter.

Used as an embrocation.

CAMPHORATED OIL OF CHAMOMILE (Fr.).

Take of Infused oil of chamomile	9 parts.
Camphor	1 part.

Dissolve and filter.

Used like the preceding.

ANTIMONIUM—ANTIMONY.

Metallic antimony is not used medicinally.

ANTIMONII ET POTASSII TARTRAS (U. S. et al. Ph.)—TARTRATE OF ANTIMONY AND POTASSIUM—TARTAR EMETIC.

Take of Oxide of antimony	2 ounces.
Bitartrate of potassium	2½ ounces.
Distilled water	18 ounces.

Heat the water to the boiling point in a glass vessel, add the powders, previously mixed, boil an hour, filter while hot, and set aside to crystallize. Lastly, dry the crystals and preserve them in a well-stopped bottle.

It is in white crystals, of a metallic taste, and soluble in 20 parts of water.

In small doses, tartar emetic is diaphoretic, expectorant, and nauseant; in large doses, emetic, cathartic, and sedative; in overdoses, an active poison.

It was formerly much used in the acute stage of inflammatory affections, as bronchitis, pneumonia, pleurisy, etc., and though it has been, in a measure, superseded by agents of a less dangerous character, it is doubtful if the substitution has always been wise. As an emetic, it is prompt and effective, but induces considerable prostration. Locally it is an irritant, producing an abundant pustular eruption. In the form of an ointment it is used as a counter-irritant.

Dose: $\frac{1}{30}$ to $\frac{1}{2}$ grain, diaphoretic and expectorant; 1 to 3 grains, emetic.

PREPARATIONS.

Emplastrum Antimonii (U. S.)—Antimonial Plaster.

Take of Tartrate of antimony and potassium.......... 1 ounce.
Burgundy pitch.......................... 4 ounces.

Melt and strain the pitch, then add the powder, and stir until cool.

Unguentum Antimonii (U. S. et al. Ph.)—Antimonial Ointment.

Take of Tartrate of antimony and potassium.........100 grains.
Lard.................................400 grains.

Rub them together.

Vinum Antimonii (U. S. et al. Ph.)—Antimonial Wine.

Take of Tartrate of antimony and potassium......... 32 grains.
Boiling distilled water...................... 1 ounce.
Sherry wine........................... sufficient.

Dissolve the antimony in the water, and, while the solution is hot, add sufficient wine to make 1 pint.

Dose: 5 to 60 minims.

Mixtures of Tartar Emetic.

Take of Tartar emetic............................ ½ grain.
Acetate of morphia....................... ½ grain.
Water............................ 2 ounces.

Mix. Dose: 1 drachm every hour or two. In acute catarrh, nasal, pharyngeal, and bronchial. *Bartholow.*

Take of Tartar emetic	1 grain.
Chloride of ammonium	80 grains.
Extract of liquorice	20 grains.
Hydrochlorate of morphia	1 grain.
Syrup of tolu	1 ounce.
Cherry-laurel water	1 ounce.

Mix. Dose: 1 drachm every two, three, or four hours. In acute inflammatory affections of the air-passages. *Bartholow.*

Take of Tartar emetic	$\frac{1}{6}$ grain.
Solution of acetate of ammonium	2 drachms.
Spirit of nitrous ether	1 drachm.
Camphor water	to 1 ounce.

Mix. One dose. *Women's Hospital, London.*

Take of Tartar emetic	$\frac{1}{8}$ grain.
Sulphate of magnesium	1 drachm.
Water	to 1 ounce.

Mix. One dose. *St. Bartholomew's Hospital.*

Take of Tartar emetic	$\frac{1}{6}$ grain.
Nitrate of potassium	10 grains.
Water	1 ounce.

Mix. One dose. *Brompton Consumption Hospital.*

Pill of Tartar Emetic and Opium.

Take of Tartar emetic	$\frac{1}{4}$ grain.
Opium	$\frac{1}{2}$ to 1 grain.
Treacle	sufficient.

Make 1 pill. *Guy's Hospital.*

Powder of Tartar Emetic and Ipecacuanha.

Take of Tartar emetic	1 to 3 grains.
Ipecacuanha	$\frac{1}{2}$ drachm.

Mix and divide into 3 powders.

Dose: 1 powder every ten or fifteen minutes until vomiting is produced.

ANTIMONII OXIDUM (U. S., Br., Fr.)—OXIDE OF ANTIMONY.

Prepared from sulphuret of antimony. It is a grayish-white powder, insoluble in water, and producing the general effects of tartar emetic. It is, however, seldom employed, except in preparing tartar emetic.

Dose: 2 to 3 grains.

PREPARATION.

ANTIMONIAL POWDER (Br.).

Take of Oxide of antimony........................ 1 ounce.
Phosphate of lime......................... 2 ounces.

Mix them thoroughly.

Dose: 3 to 10 grains.

ANTIMONII OXYSULPHURETUM (U. S., Fr., Ger.)—OXYSULPHURET OF ANTIMONY—KERMES' MINERAL.

Prepared by boiling sulphuret of antimony in a solution of carbonate of sodium. It is a purplish-brown, tasteless and insoluble powder, producing the general effects of tartar emetic, but is less efficient and less reliable. It is not much used.

Dose: 1 to 3 grains.

ANTIMONII SULPHURETUM (U. S. et al. Ph.)—SULPHURET OF ANTIMONY.

Native sulphuret of antimony, purified by fusion. Used in preparations.

ANTIMONIUM SULPHURATUM (U. S. et al. Ph.)—SULPHURATED ANTIMONY—GOLDEN SULPHURET OF ANTIMONY.

Prepared by boiling sulphuret of antimony in a solution of potassa, and adding the filtered liquid to sulphuric acid as long as it produces a precipitate. It is a reddish-brown, tasteless and insoluble powder, producing the general effects of other antimonials. It is seldom used, except in the following preparation.

PREPARATION.

Pilulæ Antimonii Compositæ (U. S., Br.)—Compound Pills of Antimony—Plummer's Pill.

Take of Sulphurated antimony .... 12 grains.
Mild chloride of mercury.................... 12 grains.
Guaiac.............................. 24 grains.
Molasses .... 24 grains.

Rub the antimony, first with the calomel, then with the guaiac and molasses, and divide into 24 pills.

Dose: 1 to 3 pills.

APOCYNUM CANNABINUM (U. S.)—INDIAN HEMP.

The root of Apocynum cannabinum, L. (*Nat. ord., Apocynaceæ*), a perennial herb indigenous to North America.

In large doses apocynum is a powerful emeto-cathartic; in smaller doses its action is exerted mainly upon the skin and kidneys, producing diaphoresis and diuresis. It has been most frequently employed in dropsy, especially when dependent upon cardiac or hepatic disease.

Apocynum androsæmifolium, L. (*Dogs' Bane*), quite as common as the above named species, possesses similar properties, but is believed to be less active.

Dose: of the dried root, 5 to 10 grains, diaphoretic and diuretic; 15 to 30 grains, emetic.

PREPARATIONS.

Decoction of Apocynum.

Take of Apocynum	½ ounce.
Water	1½ pint.

Boil to 1 pint. Dose: 1 to 2 ounces.

Oleoresin of Apocynum—Apocynin.

Prepared in the same manner as oleoresin of iris, which see.

Dose: ½ to 3 grains. *J. U. Lloyd.*

APORMORPHIÆ HYDROCHLORAS—HYDROCHLORATE OF APOMORPHIA.

Obtained by the action of hydrochloric acid upon morphia at a temperature of about 300°, in a sealed tube. It is a crystalline substance, readily soluble in water, and is a prompt and violent emetic. It is chiefly administered hypodermically in cases of great urgency, as in poisoning.

Dose: $\frac{1}{60}$ to $\frac{1}{10}$ grain, hypodermically.

AQUA—WATER.

Aqua Destillata (U. S. et al. Ph.)—Distilled Water.

Take of water	80 pints.

Distil 2 pints, using a tin or glass condenser, and throw them away; then distil 64 pints, and keep in well-stopped bottles.

Though distilled water is directed for most of the preparations in which water is employed, in many of them pure river or rain water, after boiling, may be substituted.

The medicated waters are treated of under the various drugs used in their preparation, except the following:

Aqua Chlorini (U. S. et al. Ph.)—Chlorine Water.

Take of		
Take of	Black oxide of manganese	½ ounce.
	Hydrochloric acid (by weight)	3 ounces.
	Water	4 ounces.
	Distilled water	20 ounces.

Introduce the oxide into a flask, add the acid previously diluted with 2 ounces of the water, heat gently, and, by means of proper apparatus, pass the gas which generates through the remainder of the water and into a four-pint bottle containing the distilled water.

Chlorine water is used as an antiseptic and disinfectant.

Dose: 1 to 4 drachms.

CHERRY-LAUREL WATER (Br., Fr., Ger.).

Take of		
Take of	Fresh leaves of cherry-laurel	1 pound.
	Water	50 ounces.

Chop, crush and bruise the leaves, macerate them in the water for twenty-four hours, then distil 20 ounces.

Dose: 5 to 30 minims.

It owes its virtues to the hydrocyanic acid which it contains; but, as the percentage of acid is very variable, the preparation is uncertain. It is better to use the officinal acid.

ARGENTUM (U. S. et al. Ph.)—SILVER.

Metallic silver is without medicinal properties.

The following compounds of silver are employed in medicine and pharmacy.

ARGENTI CYANIDUM (U. S.)—CYANIDE OF SILVER.

A white, tasteless and insoluble powder. Used in preparing diluted hydrocyanic acid.

ARGENTI IODIDUM—IODIDE OF SILVER.

Obtained by mixing solutions of iodide of potassium and nitrate of silver. It is a greenish yellow powder, producing, when administered internally, the general effect of nitrate of silver, with the asserted advantage of not discoloring the skin.

Dose : 1 to 2 grains.

ARGENTI NITRAS (U. S. et al. Ph.)—NITRATE OF SILVER.

Take of Silver, in small pieces	2 ounces.
Nitric acid by (weight)	2½ ounces.
Distilled water	sufficient.

Add the silver to the acid previously mixed with 1 ounce of distilled water in a capsule, cover with an inverted funnel, heat gently until the silver is dissolved, then remove the funnel, evaporate to dryness and melt the mass, stirring until all free nitric acid is expelled. Dissolve, when cold, in 6 ounces of distilled water, decant the clear solution, mix the residue with 1 ounce of distilled water, filter, add the filtrate to the decanted solution, evaporate, and crystallize. Drain the crystals, and preserve them in a well-stopped bottle.

It is a heavy, colorless salt, soluble in water, and possessing caustic properties.

Applied locally, nitrate of silver acts as a caustic, the cauterized surface turning brown or black when exposed to light.

Taken internally it is tonic, astringent, and antispasmodic. It is used in dysentery, diarrhœa, gastric ulcer, epilepsy, chorea, etc. When administered for too great a length of time, it produces a blue discoloration of the skin, which is indelible. Topically it is employed as a caustic in chancres and other specific sores, and, in solution, to produce healthy granulations upon wounds and ulcers, as an injection in gonorrhœa, etc.

Dose : ¼ to 2 grains.

PREPARATIONS.

Argenti Nitras Fusa (U. S., Fr., Ger.)—Fused Nitrate of Silver.

Take of Nitrate of silver	sufficient.

Melt in a porcelain capsule, continuing the heat until frothing ceases, then pour into suitable moulds.

Used as a caustic.

Nitrated Lunar Caustic (Ger.).

Take of Crystallized nitrate of silver .... 1 part.
Nitrate of potassium .... 2 parts.

Rub them together, melt in a porcelain vessel, and pour into suitable moulds. Much milder in action than the preceding.

Collyrium of Nitrate of Silver.

Take of Nitrate of silver .... 1 to 5 grains.
Water .... 1 ounce.

Dissolve. In common use.

Injection of Nitrate of Silver.

Take of Nitrate of silver .... 1 to 1½ grain.
Water .... 6 ounces.

Dissolve. Use every three hours in the first stage of gonorrhœa, to abort the disease. *Bumstead.*

Lotion of Nitrate of Silver.

Take of Nitrate of silver .... 2 to 10 grains.
Water .... 1 ounce.

Mix. Used in eczema and erythema. *Tilbury Fox.*

Mixture of Nitrate of Silver.

Take of Nitrate of silver .... 1 grain.
Diluted nitric acid .... 8 minims.
Deodorized tincture of opium .... 8 minims.
Mucilage of gum arabic .... ½ ounce.
Syrup .... ½ ounce.
Cinnamon-water .... 1 ounce.

Mix. Dose: 1 drachm every three, four, or six hours, for a child one year old. In cholera infantum, after the acuter symptoms have subsided. *Bartholow.*

Pills of Nitrate of Silver.

Take of Nitrate of silver .... 3 grains.
Powdered opium .... 6 grains.
Powdered ipecac .... 6 grains.

Mix and make 12 pills.

Dose: 1 pill every four or six hours. In the diarrhœa of typhoid fever. *Bartholow.*

Take of Nitrate of silver .... 15 grains.
Distilled water .... sufficient.
Extract of belladonna .... 10 grains.
Oil of cloves .... 10 minims.
Powdered gentian .... sufficient.
Extract of gentian .... sufficient.

Mix and divide into 60 pills.

Dose: 1 pill three times a day. In chronic gastric catarrh. *Frerichs.*

Take of Nitrate of silver........................... $\frac{1}{6}$ grain.
Dover's powder........................... 2 grains.

Mucilage sufficient to make 1 pill. *London Chest Hospital.*

Take of Nitrate of silver........................... $\frac{1}{2}$ grain.
Opium........................... $\frac{1}{4}$ grain.
Extract of henbane........................... 1 grain.

Make 1 pill. *London Fever Hospital.*

Powder of Nitrate of Silver.

Take of Nitrate of silver........................... 5 to 40 grains.
Sulphate of potassium........................... $1\frac{1}{2}$ drachm.
Subnitrate of bismuth...........................to 1 ounce.

Mix and make a powder. To be blown with a powder-blower into the anterior and posterior nares, daily, or on alternate days, in naso-pharyngeal catarrh with muco-purulent discharge. The surfaces should first be cleansed with a solution of sodium nitrate. *Andrew H. Smith.*

ARGENTI OXIDUM (U. S., Br.)—OXIDE OF SILVER.

Take of Nitrate of silver........................... 4 ounces.
Distilled water........................... $\frac{1}{2}$ pint.
Solution of potassa...........................$1\frac{1}{2}$ pint or sufficient.

Dissolve the nitrate in the water, and to the solution add solution of potassa as long as it produces a precipitate. Collect, wash, and dry the precipitate. It is an olive-brown powder, slightly soluble in water. Taken internally it produces the general effects of nitrate of silver.

Dose: $\frac{1}{2}$ to 2 grains.

Pills of Oxide of Silver.

Take of Oxide of silver........................... 5 grains.
Extract of hyoscyamus........................... 5 grains.

Mix and divide into 10 pills.

Dose: 1 pill three times a day, before meals. In nervous dyspepsia, and chronic gastric catarrh. *Bartholow.*

Take of Oxide of silver........................... $\frac{1}{2}$ grain.
Extract of hops........................... 2 grains.

Make 1 pill. *London Chest Hospital.*

ARNICÆ FLORES (U. S., Fr., Ger.)—ARNICA FLOWERS.
ARNICÆ RADIX (Br., Fr., Ger.)—ARNICA ROOT

The flowers and root of Arnica montana, L. (*Nat. ord., Compositæ*), a small, perennial herb indigenous to Central and Northern Europe.

Both flowers and root of arnica act as irritants when taken internally or applied externally, and hence the plant has been found most useful in cases requiring stimulation. Arnica has been employed internally in a great variety of affections, but is at present little used in this country, except as an external application to bruises, sprains, etc.

Dose: 5 to 20 grains.

PREPARATIONS.

Emplastrum Arnicæ (U. S.)—Arnica Plaster.

Take of Extract of arnica	1½ ounce.
Resin plaster	3 ounces.

Melt the plaster, then add the extract and mix thoroughly.

Extractum Arnicæ (U. S.)—Extract of Arnica.

Take of Arnica flowers	24 ounces.
Alcohol	4 pints.
Water	2 pints.
Diluted alcohol	sufficient.

Mix the alcohol and water, and make a tincture by percolation, continuing with diluted alcohol until 6 pints are obtained. Evaporate to the proper consistence.

Dose: 5 to 10 grains.

Tinctura Arnicæ (U. S. et al. Ph.)—Tincture of Arnica.

Take of Arnica flowers	6 ounces.
Alcohol	1½ pint.
Water	½ pint.
Diluted alcohol	sufficient.

Mix the alcohol and water, moisten, pack and percolate the arnica with the mixture, continuing with diluted alcohol until 2 pints are obtained.

Dose: 5 to 30 minims.

The British preparation is made with the root, in the proportion of 1 to 20. The French Codex prepares also a tincture from the fresh plant.

INFUSION OF ARNICA.

Take of Arnica flowers or root	½ ounce.
Boiling water	1 pint.

Infuse for half an hour in a covered vessel, and strain.

Dose: ½ to 1 ounce, chiefly externally.

ARSENICUM—ARSENIC.

The most important of the medicinal compounds of arsenic are treated of elsewhere. See *Arsenious Acid* and *Solution of Arsenite of Potassium*.

ARSENICI CHLORIDUM—CHLORIDE OF ARSENIC.

PREPARATION.

Liquor Arsenici Chloridi (U. S., Br.)—Solution of Chloride of Arsenic.

Arsenious acid	64 grains.
Hydrochloric acid	2 drachms.
Distilled water	sufficient.

Boil the arsenious acid with the hydrochloric acid and 4 ounces of the distilled water, until it is dissolved, and, when cold, add sufficient distilled water to make 1 pint.

Dose: 2 to eight minims.

It has the same strength as Fowler's solution—4 grains to an ounce—and is used for the same purposes, though it is said to be less reliable.

ARSENICI IODIDUM (U. S.)—IODIDE OF ARSENIC.

Take of Arsenic	60 grains.
Iodine	300 grains.

Rub them together thoroughly, then heat in a flask until liquefaction occurs, cool, break in pieces, and keep in a well-stopped bottle. It is an orange-red, crystalline compound, completely soluble in water. Used as an alterative in skin diseases.

Dose: $\frac{1}{60}$ to $\frac{1}{10}$ grain.

Pills of Iodide of Arsenic.

Take of Iodide of arsenic	2 grains.
Manna	4 grains.
Mucilage	sufficient.

Mix, and divide into 20 pills.

Dose: 1 pill three times a day, in psoriasis. *Tilbury Fox.*

Liquor Arsenici et Hydrargyri Iodidi (U. S.)—Solution of Iodide of Arsenic and Mercury.

Iodide of arsenic	35 grains.
Red iodide of mercury	35 grains.
Distilled water	½ pint.

Rub the iodides with ½ ounce of the water until dissolved, then add the remainder of the water, and filter.

Dose: 2 to 10 minims.

Commonly known as *Donovan's Solution.* Used in skin diseases, especially those of a chronic or syphilitic character.

ASAFŒTIDA (U. S. et al. Ph.)—ASAFETIDA.

A gum-resin obtained by incisions made into the living roots of Ferula Narthex, Boiss., and Ferula Scorodesma, B. et H. (*Nat. ord., Umbelliferæ*), large herbaceous perennials indigenous to Asia.

It occurs in masses of tears varying in size, consistence, and color, but of an odor which is characteristic and unmistakable. Its taste is acrid, bitter, and disagreeable.

Asafetida is an antispasmodic and nervine. Used in hysteria, asthma, whooping-cough, and a variety of other nervous diseases.

Dose: 5 to 10 grains.

PREPARATIONS.

Emplastrum Asafœtidæ (U. S., Ger.)—Asafetida Plaster.

Take of Asafetida	12 ounces.
Lead plaster	12 ounces.
Galbanum	6 ounces.
Yellow wax	6 ounces.
Alcohol	3 pints.

Dissolve the gums in the alcohol, on a water-bath, strain, and evaporate to the consistence of honey; then add the plaster and wax, previously melted, stir well, and evaporate to the proper consistence.

Mistura Asafœtidæ (U. S., Br.)—Asafetida Mixture.

Take of Asafetida	120 grains.
Water	½ pint.

Rub together until thoroughly mixed.

Dose: ½ to 1 ounce. Often used by enema in hysterical and infantile convulsions. *Enema of Asafetida, Br.*

Pilulæ Asafœtidæ (U. S.)—Pills of Asafetida.

Take of Asafetida	72 grains.
Soap	24 grains.

Beat them together into a pilular mass, and divide into 24 pills.

Dose: 2 to 4 pills. See also *Pilulæ Galbani Compositæ.*

Suppositoria Asafœtidæ (U. S.)—Suppositories of Asafetida.

Take of Tincture of asafetida	1 ounce.
Oil of theobroma	320 grains.

Evaporate the tincture to the consistence of a thick syrup, mix it with 1 drachm of the oil; then, having melted the remainder, mix thoroughly, and pour into suitable moulds, making 12 suppositories.

Tinctura Asafœtidæ (U. S. et al. Ph.)—Tincture of Asafetida.

Take of Asafetida	4 ounces.
Alcohol	2 pints.

Macerate for seven days and filter.

Dose : ½ to 1 drachm.

The French Codex prepares also an ethereal tincture.

MIXTURES OF ASAFETIDA.

Take of Asafetida mixture	4 ounces.
Chloride of ammonium	1 drachm.

Mix. Dose: ½ ounce, as necessary. In the cough maintained by habit, which may succeed whooping-cough, and the sympathetic cough of mothers whose children have whooping-cough. *Bartholow.*

Take of Tincture of asafetida	½ drachm.
Tincture of valerian	1 drachm.
Carbonate of ammonium mixture	½ ounce.
Mucilage of tragacanth	½ ounce.

Mix. One dose. *Charing Cross Hospital.*

PILL OF ASAFETIDA AND IRON.

Take of Asafetida	2 grains.
Sulphate of iron	1 grain.
Extract of gentian	1 grain.

Make 1 pill. *Brompton Consumption Hospital.*

PILL OF ASAFETIDA AND ZINC.

Take of Compound asafetida pill	4 grains.
Sulphate of zinc	1 grain.

Make 1 pill. *London Chest Hospital.*

ASCLEPIAS TUBEROSA (U. S.)—PLEURISY ROOT.

The root of Asclepias tuberosa, L. (*Nat. ord., Asclepiadaceæ*), an herbaceous perennial indigenous to North America.

The root of A. incarnata, L., and of A. Syriaca, L., two other common species, possess the same properties as the above, though perhaps to a less degree.

Asclepias, in moderate doses, acts as a diuretic and diaphoretic ; in large doses as an emetic. It is used in the earlier stages of bronchitis, pleurisy, pneumonia, rheumatism, etc.

Dose : of the powdered root, 20 to 60 grains.

PREPARATIONS.

Decoction of Pleurisy Root.

Take of Pleurisy root.............................. 1 ounce.
Boiling water............................ 1 pint.

Boil for half an hour, strain and add sufficient water to make 1 pint.

Dose: 1 to 4 ounces.

Oleoresin of Asclepias—Asclepidin.

Prepared in the same manner as oleoresin of iris, which see.

Dose: 1 to 5 grains. *J. U. Lloyd.*

ATROPIA (U. S.' et al. Ph.)—ATROPIA.

Take of Belladonna root.......................... 48 ounces.
Purified chloroform (by weight)............. $4\frac{1}{2}$ ounces.
Diluted sulphuric acid...................... sufficient.
Solution of potassa......................... sufficient.
Alcohol.................................. sufficient.
Water.................................. sufficient.

Percolate the belladonna with the alcohol until 16 pints are obtained; reduce to 4 pints by distilling off the alcohol. Acidulate this with diluted sulphuric acid, evaporate to $\frac{1}{2}$ pint, add an equal bulk of water, and filter. To the filtered liquid add, first, $1\frac{1}{2}$ ounce of chloroform, then solution of potassa in slight excess, and shake occasionally for half an hour. When the heavier liquid has subsided, separate it, add $1\frac{1}{2}$ ounce of chloroform to the lighter liquid, shake, and separate as before. Repeat with the remainder of the chloroform, mix the heavier liquids in a capsule, and set aside until, by evaporation, the atropia is left dry.

Atropia, thus prepared, is in yellowish-white crystals, odorless, of an acrid taste, soluble in 300 parts of water, 25 of ether, and in a smaller proportion of alcohol.

It produces essentially the same effects as belladonna.

Dose: $\frac{1}{100}$ to $\frac{1}{25}$ grain.

PREPARATIONS.

Atropiæ Sulphas (U. S. et al. Ph.)—Sulphate of Atropia.

Take of Atropia.................................. 60 grains.
Stronger ether........................... $4\frac{1}{2}$ ounces.
Sulphuric acid........................... 6 grains.
Alcohol.................................. 1 drachm.

Dissolve the atropia in the ether, mix the alcohol and acid, and carefully drop the mixture into the ethereal solution. When the sulphate has been deposited, decant the ether and dry the salt.

Dose: $\frac{1}{100}$ to $\frac{1}{25}$ grain. It is very soluble, and on this account is generally used instead of the alkaloid.

SOLUTION OF ATROPIA (Br.).

Take of Atropia	4 grains.
Rectified spirit	1 drachm.
Distilled water	7 drachms.

Dissolve the atropia in the spirit, and add this gradually to the water, shaking them together.

The British Pharmacopœia prepares also a solution of sulphate of atropia, 4 grains to 1 ounce. Used hypodermically, and instilled into the eyes to dilate the pupil.

OINTMENT OF ATROPIA (Br.).

Take of Atropia	8 grains.
Rectified spirit	½ drachm.
Lard	1 ounce.

Dissolve the atropia in the spirit, add the lard, and mix thoroughly.

MIXTURES OF ATROPIA.

Take of Sulphate of atropia	1 grain.
Sulphate of zinc	½ drachm
Distilled water	1 ounce.

Mix. Dose: 3 to 5 drops, twice or thrice a day. In gastralgia and gastric ulcer. *Bartholow.*

Take of Atropia	½ grain.
Diluted hydrochloric acid	1 ounce.

Mix. Dose: 5 drops in water, before meals. In heart-burn, water-brash, etc. *Bartholow.*

AURANTIUM—ORANGE.

AURANTII AMARI CORTEX (U. S. et al. Ph.)—BITTER ORANGE PEEL.

The rind of the fruit of Citrus vulgaris, Risso (*Nat. ord., Aurantiaceæ*), a tree indigenous to India, but cultivated in most warm countries.

AURANTII DULCIS CORTEX (U. S., Fr.)—SWEET ORANGE PEEL.

The rind of the fruit of Citrus Aurantium, Risso, a tree of the same origin, and having the same distribution as the bitter orange.

AURANTII FLORES (U. S. et al. Ph.)—ORANGE FLOWERS.

The flowers of either of the above named species of orange.

The British Pharmacopœia recognizes also the fruit of the bitter orange, and the French Codex that of the sweet orange.

Orange peel is an aromatic bitter. Used as a flavoring agent and as an adjunct to other medicines.

Dose: ½ to 1 drachm.

Orange flowers yield a fragrant, volatile oil, and are used, chiefly in the form of orange flower water, as a vehicle.

PREPARATIONS.

Aqua Aurantii Florum (U. S. et al. Ph.)—Orange Flower Water.

Take of Recent orange flowers .... 48 ounces.
Water .... 16 pints.

Mix them, and, by means of steam, distil 8 pints. Used as a vehicle.

Confectio Aurantii Corticis (U. S.)—Confection of Orange Peel.

Take of Recent sweet orange peel, grated .... 12 ounces.
Sugar .... 36 ounces.

Beat them together.

Dose: 1 to 4 drachms.

Syrupus Aurantii Corticis (U. S. et al. Ph.)—Syrup of Orange Peel.

Take of Sweet orange peel, recently dried .... 2 ounces.
Carbonate of magnesium .... ½ ounce.
Sugar .... 28 ounces.
Alcohol .... sufficient.
Water .... sufficient.

Percolate the orange peel with alcohol until 6 ounces of tincture are obtained. Evaporate this, at or below 120°, to 2 ounces, add the carbonate and 1 ounce of sugar, and rub them together, adding gradually ½ pint of water. Then filter, add water to 1 pint, and dissolve in it the remainder of the sugar with a gentle heat, and strain.

Used as a vehicle.

The British, French, and German preparations are made with bitter orange peel.

Syrupus Aurantii Florum (U. S. et al. Ph.)—Syrup of Orange Flowers.

Take of Orange flower water .... 20 ounces.
Sugar .... 36 ounces.

Dissolve with a gentle heat. Used as a vehicle.

Tinctura Aurantii (U. S. et al. Ph.)—Tincture of Orange.

Take of Bitter orange .... 4 ounces.
Diluted alcohol .... sufficient.

Moisten, pack, and percolate to 2 pints.

Dose: 1 to 2 drachms.

The British Pharmacopœia prepares also a tincture from the fresh peel, and the French Codex one from fresh sweet orange peel.

INFUSION OF ORANGE PEEL (Br.).

Take of	Bitter orange peel	½ ounce.
	Boiling distilled water	10 ounces.

Infuse in a covered vessel for fifteen minutes, and strain.

Dose: 1 to 2 ounces.

COMPOUND INFUSION OF ORANGE PEEL (Br.).

Take of	Bitter orange peel	¼ ounce.
	Fresh lemon peel	60 grains.
	Cloves	30 grains.
	Boiling distilled water	10 ounces.

Infuse in a covered vessel for fifteen minutes, and strain.

Dose: 1 to 2 ounces.

AURUM (Fr., Ger.)—GOLD.

Metallic gold, in a very finely divided state, is occasionally used in syphilis, and in some diseases of the skin. It may be prepared by triturating gold leaf with sulphate of potassium or sugar of milk, and then washing out the triturant with water.

Dose: $\frac{1}{6}$ to 1 grain. Applied by frictions to the sides of the tongue.

AURI CHLORIDUM (Fr.)—CHLORIDE OF GOLD.

Prepared by dissolving gold leaf in nitro-hydrochloric acid, evaporating, and crystallizing.

Dose: $\frac{1}{50}$ to $\frac{1}{15}$ grain.

AURI ET SODII CHLORIDUM (Fr., Ger.)—CHLORIDE OF GOLD AND SODIUM.

Prepared by dissolving 10 parts of gold leaf in nitro-hydrochloric acid, evaporating to a syrupy consistence, adding an equal volume of water, and then 3 parts of chloride of sodium. The mixture is then evaporated to dryness, on a sand-bath.

Dose: $\frac{1}{50}$ to $\frac{1}{15}$ grain.

AZEDARACH (U. S.)—AZEDARACH.

The bark of the root of Melia Azedarach, L. *(Nat. ord., Meliaceæ)*, a tree indigenous to India and China, but cultivated for ornament in the Southern United States.

Azedarach is emetic and cathartic, and in large doses produces narcotic effects similar to those of spigelia. It is used in the Southern States as a vermifuge for lumbricoid worms, and is considered nearly as efficient as spigelia.

Dose: 1 to 2 drachms.

PREPARATION.

Decoction of Azedarach.

Take of Azedarach	....	4 ounces.
Water	....	1½ pint.

Boil to 1 pint and strain.

Dose: 1 ounce every two or three hours.

BALSAMUM PERUVIANUM (U. S. et al. Ph.)—BALSAM OF PERU.

A semi-liquid balsam obtained from Myroxylon Pereiræ, Klotzsch (*Nat. ord., Leguminosæ*), a tree indigenous to Central America.

Balsam of Peru is of a light brown color, in thin layers perfectly transparent, and has a very agreeable odor, and a warm, pungent taste. Taken internally it acts as a stimulant to the mucous membranes, and is occasionally used in catarrhal affections. Externally it is employed as a stimulating dressing for indolent ulcers, wounds, etc.

PREPARATIONS.

Syrup of Balsam of Peru (Ger.).

Take of Balsam of Peru	....	1 part.
Distilled water	....	11 parts.

Digest for several hours with frequent agitation, decant, and filter. To 10 parts of the filtrate, add 18 parts of sugar, and dissolve.

Dose: 1 to 3 drachms.

Mixture of Balsam of Peru.

Take of Balsam of Peru	....	25 minims.
Honey	....	40 minims.
Water	....	to 1 ounce.

Mix. One dose. *Guy's Hospital.*

BALSAMUM TOLUTANUM (U. S. et al. Ph.)—BALSAM OF TOLU.

A semi-liquid balsam obtained from Myroxylon Toluifera, H. B. K. (*Nat. ord., Leguminosæ*), a tree indigenous to the northern part of South America.

When first imported it has a soft consistence, but by age it becomes hard and brittle. It is of a reddish-brown color, an agreeable odor, and a sweetish, pungent taste. Like balsam of Peru, it acts as a stimulant to the mucous membranes, and is used in catarrhal affections. Owing to its agreeable odor and taste, it is often employed in cough syrups and mixtures.

Dose: 10 to 30 grains.

PREPARATIONS.

Syrupus Tolutanus (U. S. et al. Ph.)—Syrup of Tolu.

Take of Tincture of tolu	2 ounces.
Carbonate of magnesium	120 grains.
Sugar	26 ounces.
Water	1 pint.

Rub the tolu with the carbonate of magnesium and 2 ounces of sugar, then with the water, gradually added, and filter. Dissolve the remainder of the sugar in the filtered liquid with a gentle heat, and strain.

Used as a vehicle, chiefly in cough mixtures.

Tinctura Tolutana (U. S., Br., Fr.)—Tincture of Tolu.

Take of Balsam of tolu	3 ounces.
Alcohol	2 pints.

Macerate until dissolved, and filter.

Dose: 15 to 40 minims.

BAPTISIA—WILD INDIGO.

The root of Baptisia tinctoria, R. Br. (*Nat. ord., Leguminosæ*), an herbaceous perennial indigenous to the United States and Canada.

In the early part of the present century, baptisia was considered useful in low fevers and as a topical application to unhealthy ulcers. It is largely employed at the present day, by homœopathists, in the treatment of typhoid fever, and careful experiments seem to justify their esteem of it.

PREPARATIONS.

Decoction of Baptisia.

Take of Fresh root of baptisia	1 ounce.
Boiling water	1½ pint.

Boil to 1 pint and strain.

Dose: ½ to 1 drachm.

Tincture of Baptisia.

Take of Fresh root of baptisia	1 part.
Diluted alcohol	2 parts.

Macerate fourteen days, and filter.

Dose: 2 to 5 drops hourly in the early stages of typhoid fever. The best effects have been obtained from small doses, frequently repeated.

BELÆ FRUCTUS (Br.)—BAEL FRUIT.

The dried, half-ripe fruit of Ægle Marmelos, DC. (*Nat. ord., Aurantiaceæ*), a medium-sized tree indigenous to India.

Bael fruit is an astringent, which is very useful in chronic relaxation of the bowels. It is not administered in substance.

PREPARATIONS.

LIQUID EXTRACT OF BAEL (Br.).

Take of Bael fruit	1 pound.
Distilled water	12 pints (imp.).
Rectified spirit	2 ounces.

Macerate the bael for twelve hours in one-third of the water; pour off the clear liquor; repeat a second and third time for one hour in the remainder of the water, express, filter the mixed liquids, evaporate to 14 ounces, and, when cold, add the spirit.

Dose: 1 to 2 ounces.

DECOCTION OF BAEL FRUIT.

Take of Bael fruit	1 ounce.
Boiling water	1 pint.

Make a decoction. Dose: a small wineglassful three times a day.

James Knight.

Dr. Knight writes: "I have used bael fruit for eighteen years, having imported it for my own use, and have found it one of the most efficient remedies for the cure of chronic diarrhœa."

BELLADONNA—BELLADONNA.

BELLADONNÆ FOLIA (U. S. et al. Ph.)—BELLADONNA LEAVES.

BELLADONNÆ RADIX (U. S. et al. Ph.)—BELLADONNA ROOT.

The leaves and root of Atropa Belladonna, L. (*Nat. ord., Solanaceæ*), an herbaceous perennial, indigenous to Europe, but occasionally cultivated in this country.

Belladonna is an acro-narcotic, and may produce its characteristic effects when applied externally as well as when taken internally. Among these effects, are dilation of the pupil, with dimness of vision, redness, dryness, and heat of the fauces, flushing of the face, an efflorescence upon the skin, and delirium. In overdoses it is an active poison. It owes its activity to an alkaloid, named *atropia*, which is capable of producing all the effects of the crude drug.

Belladonna is administered internally for its narcotic effects in a great variety of painful affections, and in some of a convulsive character. It is also useful in nocturnal incontinence of urine, constipation, etc. Externally it is applied to painful tumors, inflamed breasts, to check the secretion of milk, and is instilled into the eye to dilate the pupil in iritis, and to facilitate ophthalmoscopic examinations.

It is used by many as an antidote to opium.

Dose: Of the leaves, 1 to 10 grains; of the root, 1 to 5 grains.

PREPARATIONS.

Emplastrum Belladonnæ (U. S. et al. Ph.)—Belladonna Plaster.

Take of Belladonna root	16 ounces.
Alcohol	sufficient.
Resin plaster	sufficient.

Macerate the belladonna in 1 pint of alcohol for four days, then percolate to 2 pints, evaporate to the consistence of a soft extract, and add sufficient resin plaster, previously melted, to make the whole weigh 16 ounces.

The British and French preparations are made with the extract.

Extractum Belladonnæ (U. S. et al. Ph.)—Extract of Belladonna.

Take of Belladonna leaves, fresh .... 12 ounces.

Bruise the leaves, sprinkling on them a little water, and express the juice; heat this to the boiling point, strain, and evaporate to the proper consistence.

Dose: $\frac{1}{4}$ to 1 grain. As the preparation found in the shops is of uncertain strength, caution is necessary in increasing the dose.

The British, French, and German extracts are similar to the above, and of about the same strength.

Extractum Belladonnæ Alcoholicum (U. S.)—Alcoholic Extract of Belladonna.

Take of Belladonna leaves	24 ounces.
Alcohol	4 pints.
Water	2 pints.
Diluted alcohol	sufficient.

Mix the alcohol and water, and prepare a tincture by percolation, continuing the process with diluted alcohol until 6 pints are obtained. Evaporate this, on a water-bath, to the proper consistence.

Dose: $\frac{1}{8}$ to $\frac{1}{2}$ grain. Rather more reliable than the preceding, but still of uncertain strength, since it is prepared from dry leaves, which may have undergone deterioration by long keeping.

Extractum Belladonnæ Radicis Fluidum (U. S.)—Fluid Extract of Belladonna Root.

Take of Belladonna root .... 16 ounces.
Glycerin .... 4 ounces.
Alcohol .... sufficient.
Water .... sufficient.

Mix 12 ounces of alcohol, 3 of glycerin, and 1 of water; moisten the belladonna with 4 ounces of the mixture, and proceed according to the general formula, page 161.

Dose: 1 to 5 minims.

Suppositoria Belladonnæ (U. S.)—Suppositories of Belladonna.

Take of Alcoholic extract of belladonna.............. 6 grains.
Oil of theobroma..........................354 grains.
Water.......................... sufficient.

Rub the extract with a drop or two of water, then mix it thoroughly with 60 grains of the oil, add the remainder, previously melted and cooled to 95°, and pour into suitable molds, making 12 suppositories.

Tinctura Belladonnæ (U. S. et al. Ph.)—Tincture of Belladonna.

Take of Belladonna leaves, recently dried............ 4 ounces.
Diluted alcohol .... sufficient.

Moisten, pack, and percolate to 2 pints.

Dose: 15 to 30 minims.

The British Pharmacopœia directs the proportion of 1 to 20; the French Codex, 1 to 5; while the German Pharmacopœia macerates 5 parts of the fresh leaves in 6 parts of alcohol. The French Codex also prepares a tincture from the fresh leaves.

Unguentum Belladonnæ (U. S., Br., Ger.)—Ointment of Belladonna.

Take of Extract of belladonna...................... 60 grains.
Water .... $\frac{1}{2}$ drachm.
Lard ....420 grains.

Rub the extract first with the water, then with the lard, gradually added.

LINIMENT OF BELLADONNA (Br.).

Take of Belladonna root.......................... 20 ounces.
Camphor .... 1 ounce.
Rectified spirit.......................... sufficient.

Macerate the belladonna with some of the spirit for three days, then percolate it with enough more, into a receiver containing the camphor, to make 20 ounces.

This is one of the best of all belladonna preparations, for external use.

A number of other preparations of belladonna are officinal with the French, but as they are of no special importance they are omitted.

Mixtures of Belladonna.

Take of Extract of belladonna	4 grains.
Syrup of opium	1 ounce.
Syrup of orange flowers	1 ounce.

Mix. Dose: 1 drachm several times a day, in whooping-cough.

Trousseau and Pidoux.

Take of Extract of belladonna	1 grain.
Alum	$\frac{1}{2}$ drachm.
Syrup of ginger	1 ounce.
Syrup of gum arabic	1 ounce.
Water	1 ounce.

Mix. Dose: 1 drachm four times a day, in whooping-cough.

Meigs and Pepper.

Take of Tincture of belladonna	2 drachms.
Tincture of aconite root	1 drachm.

Mix. Dose: 4 drops in water every hour or two. In ordinary sore throat. *Bartholow.*

Pills of Belladonna.

Take of Extract of belladonna	4 grains.
Extract of stramonium	5 grains.
Extract of hyoscyamus	5 grains.
Sulphate of quinia	40 grains.

Mix and divide into 20 pills.

Dose: 1 pill three times a day. In dysmenorrhœa of a neuralgic character, and in ovarian neuralgia. *Bartholow.*

Take of Extract of belladonna	$\frac{1}{4}$ grain.
Extract of gentian	4 grains.

Make 1 pill. *Royal Chest Hospital.*

Take of Extract of belladonna	$\frac{1}{4}$ grain.
Ipecacuanha	$\frac{1}{2}$ grain.
Extract of taraxacum	3 grains.

Make 1 pill. *Brompton Consumption Hospital.*

Take of Extract of belladonna	$\frac{1}{8}$ grain.
Sulphate of zinc	1 grain.
Sugar of milk	1 grain.
Treacle	sufficient.

Make 1 pill. *London Ophthalmic Hospital.*

BENZOINUM (U. S. et al. Ph.)—BENZOIN.

A resin obtained from Styrax Benzoin, Dryander (*Nat. ord., Styraceæ*), a tree indigenous to Sumatra and Java.

The best quality of benzoin occurs in whitish tears, loosely agglutinated into a mass, though generally it is quite compact, the tears being imbedded in a reddish-brown connecting medium. It has a fragrant odor, and an aromatic taste. Its most important constituents are benzoic acid and resin, and to these it owes its medicinal effects.

Benzoin acts as a stimulant to the mucous membranes, and is occasionally employed in catarrhal affections. Externally it is a stimulant and irritant, and is applied to cracked nipples, fissure of the anus, etc.

Dose: 10 to 30 grains; seldom used in substance.

PREPARATIONS.

Tinctura Benzoini (U. S., Fr., Ger.)—Tincture of Benzoin.

Take of Benzoin	6 ounces.
Alcohol	2 pints.

Macerate for seven days, and filter.

Dose: 20 to 30 minims. Used chiefly in the preparation of ointment of benzoin.

Tinctura Benzoini Composita (U. S., Br.)—Compound Tincture of Benzoin.

Take of Benzoin	3 ounces.
Socotrine aloes	½ ounce.
Storax	2 ounces.
Balsam of tolu	1 ounce.
Alcohol	2 pints.

Macerate for seven days, and filter.

Dose: ½ to 2 drachms. Often used externally.

Unguentum Benzoini (U. S., Br., Fr.)—Ointment of Benzoin.

Take of Tincture of benzoin	2 ounces.
Lard	16 ounces.

Melt the lard on a water-bath, add the tincture, constantly stirring, and, when the alcohol has evaporated, remove from the water-bath, and stir until cold. *Benzoated Lard* (Br., Fr.).

The addition of benzoin to lard prevents its becoming rancid, hence ointment of benzoin is used as the basis of many other ointments.

INHALATION OF BENZOIN.

Take of Compound tincture of benzoin.............. 1 drachm.
Water, at 140° F......................... 20 ounces.

Mix. The vapor to be inhaled in acute inflammation of the pharynx and larynx. *London Throat Hospital.*

MIXTURE OF BENZOIN.

Take of Compound tincture of benzoin .... 20 minims.
Oxymel of squill .... 30 minims.
Wine of ipecacuanha .... 5 minims.
Tincture of tolu.......................... 5 minims.
Water ....to 1 ounce.

Mix. One dose. *Brompton Consumption Hospital.*

BISMUTHUM (U. S., Br., Fr.)—BISMUTH.

Metallic bismuth is not used medicinally. Its compounds are employed chiefly in painful disorders of the stomach and bowels, their effect being to allay irritation, quiet pain, and neutralize acidity.

As met with in commerce, bismuth contains a small proportion of arsenic, copper, and silver, to the former of which, existing in its compounds as an impurity, are attributed, by some, a part of their medicinal effects. For the removal of these impurities, the British Pharmacopœia and French Codex direct the following

PREPARATION.

PURIFIED BISMUTH (Br., Fr.).

Take of Bismuth .... 10 ounces.
Nitrate of potassium, in powder............. 2 ounces.

Melt the bismuth with 1 ounce of the nitrate in a crucible, stirring until the salt has solidified over the metal. Then remove the salt, add the remainder of the nitrate, and repeat the process. Finally, pour the fused bismuth into a suitable mould.

BISMUTHI SUBCARBONAS (U. S., Br.)—SUBCARBONATE OF BISMUTH.

Take of Bismuth................................ 2 ounces.
Nitric acid (by weight) .... 8½ ounces.
Water of ammonia .... 5 ounces.
Carbonate of sodium .... 10 ounces.
Distilled water .... sufficient.

Mix 4¼ ounces of the nitric acid with 4 ounces of distilled water, add the bismuth, and set aside for twenty-four hours. Then dilute the solution with 10 ounces of distilled water and, after twenty-four hours, filter. Dilute the filtrate with 4 pints of distilled water, add the water of ammonia, previously diluted with an equal measure of distilled water, strain, wash the precipitate with 2 pints of distilled water, and place it in a capacious vessel; then add the remainder of the nitric acid, and afterward 4 ounces of distilled water. After twenty-four hours, filter.

Dissolve the carbonate of sodium in 12 ounces of distilled water, with the aid of heat, and filter. To the filtrate, when cold, add the solution of bismuth, collect, wash, and dry the precipitate.

It is a white, tasteless, and insoluble powder.

Dose: 10 to 60 grains.

Powders of Subcarbonate of Bismuth.

Take of Subcarbonate of bismuth.................... 10 grains.
Wood charcoal............................ 10 grains.
Bicarbonate of sodium.................... 5 grains.

Mix. One dose. *University College Hospital.*

Take of Subcarbonate of bismuth.................... 5 grains.
Carbonate of magnesium.................. 3 grains.
Powdered gum arabic .... 2 grains.

Mix. One dose. *Brompton Consumption Hospital.*

Take of Subcarbonate of bismuth.................... 10 grains.
Dover's powder........................... 10 grains.

Mix. One dose. *University College Hospital.*

BISMUTHI SUBNITRAS (U. S. et al. Ph.)—SUBNITRATE OF BISMUTH.

Take of Bismuth.................................. 2 ounces.
Nitric acid (by weight)..................... 10 ounces.
Carbonate of sodium........................ 10 ounces.
Water of ammonia .... 6 ounces.
Distilled water............................ sufficient.

Dissolve the bismuth in 4½ ounces of the acid, as in the preceding process, dilute the solution in like manner, and precipitate it with solution of carbonate of sodium, made in the same way; then dissolve the moist precipitate in the remainder of the acid mixed with 4 ounces of water, dilute the solution, precipitate it with the water of ammonia, collect, wash, and dry the precipitate.

This process differs from the preceding in that the water of ammonia is used for the final precipitation.

It is a heavy, white, tasteless, and insoluble powder.

Dose: 5 to 10 grains.

PREPARATIONS.

Bismuth Lozenges (Br., Fr.).

Take of		
	Subnitrate of bismuth	1,440 grains.
	Carbonate of magnesium	4 ounces.
	Precipitated carbonate of lime	6 ounces.
	Refined sugar	29 ounces.
	Gum arabic	1 ounce.
	Mucilage of gum arabic	2 ounces.
	Rose-water	sufficient.

Mix the dry ingredients, add the mucilage, form a proper mass with rose water, and divide into 720 lozenges.

Dose: 1 to 6 lozenges.

Mixtures of Subnitrate of Bismuth.

Take of		
	Subnitrate of bismuth	2 drachms.
	Diluted hydrocyanic acid	½ drachm.
	Mucilage of gum arabic	2 ounces.
	Peppermint water	2 ounces.

Mix. Dose: ½ ounce three times a day. In gastric ulcer, and scirrhus of the stomach. *Bartholow.*

Take of		
	Subnitrate of bismuth	3 drachms.
	Carbolic acid	2 to 4 grains.
	Mucilage of gum arabic	1 ounce.
	Peppermint water	3 ounces.

Mix. Dose: ½ ounce three or four times a day. In vomiting of pregnancy, acidity, pyrosis, and in the vomiting of teething children. In the case of children the dose must, of course, be reduced proportionately.

Bartholow.

Take of		
	Subnitrate of bismuth	20 grains.
	Mucilage of tragacanth	½ ounce.
	Tincture of cinnamon	10 minims.
	Water	to 1 ounce.

Mix. One dose. *Brompton Consumption Hospital.*

Take of		
	Subnitrate of bismuth	10 grains.
	Bicarbonate of sodium	10 grains.
	Compound tragacanth powder	10 grains.
	Water	1 ounce.

Mix. One dose. *Charing Cross Hospital.*

Take of		
	Subnitrate of bismuth	80 grains.
	Camphorated tincture of opium	2 drachms.
	Chalk mixture	to 2 ounces.

Mix. Dose: 1 drachm, in diarrhœas of children.

Powders of Subnitrate of Bismuth.

Take of Subnitrate of bismuth .... 2 drachms.
Sulphate of morphia...................... 1 grain.

Mix and divide into 6 powders.

Dose: One powder three times a day, in milk. In gastric ulcer, and scirrhus of the stomach. *Bartholow.*

Take of Subnitrate of bismuth .... 5 grains.
Compound powder of chalk and opium....... 5 grains.

Mix. One dose. *Westminster Hospital.*

Take of Subnitrate of bismuth .... 8 grains.
Dried carbonate of sodium .... 8 grains.
Hydrochlorate of morphia..................... $\frac{1}{10}$ grain.

Mix. One dose. *Guy's Hospital.*

Take of Subnitrate of bismuth...................... 6 drachms.
Chromate of lead........................... 10 grains.
Carmine .... 5 grains.
Vermilion.................................. 5 grains.

Mix. Used as a dusting powder in skin diseases. *British Skin Hospital.*

BISMUTHI ET AMMONII CITRAS—CITRATE OF BISMUTH AND AMMONIUM.

PREPARATION.

Liquor Bismuthi et Ammonii Citratis (Br.)—Solution of Citrate of Bismuth and Ammonium.

Take of Purified bismuth.........................430 grains.
Nitric acid.............................. 2 ounces.
Citric acid .... 2 ounces.
Solution of ammonia .... sufficient.
Distilled water .... sufficient.

Dissolve the bismuth in the nitric acid diluted with one ounce of distilled water, heat nearly to the boiling point for ten minutes, decant, and evaporate to 2 ounces. Then add the citric acid dissolved in 4 ounces of distilled water, and afterward the solution of ammonia, in small quantities, until the precipitate is redissolved, and the solution is neutral or slightly alkaline. Dilute with distilled water to 20 ounces.

Dose: $\frac{1}{2}$ to 1 drachm.

This, and other soluble compounds of bismuth, are of doubtful utility, since the best effects of bismuth are produced by its local, and, probably, mechanical action upon the mucous surfaces.

BRAYERA (U. S. et al. Ph.)—KOUSSO.

The flowers and unripe fruit of Brayera anthelmintica, Kunth (*Nat. ord., Rosaceæ*), a tree indigenous to Abyssinia.

The dried flowers occur in clusters of a light brown or yellowish color, of a fragrant odor, and a taste which is at first faint, but afterward acrid and unpleasant.

Kousso is one of the best remedies for tape-worm.

Dose: 2 to 4 drachms.

PREPARATION.

INFUSION OF KOUSSO (Br.).

Take of	Kousso in coarse powder	½ ounce.
	Boiling water	8 ounces.

Infuse in a covered vessel for fifteen minutes, without straining.

Dose: 4 to 8 ounces, taken with the dregs.

BROMINIUM (U. S. et al. Ph.)—BROMINE.

A non-metallic element which exists in sea water and the waters of some saline springs. It is a volatile liquid of a dark red color, a caustic taste, and a very disagreeable odor.

Bromine, in its medicinal effects, bears some analogy to iodine, and is used as an alterative in scrofula, bronchocele, cutaneous diseases, etc. In combination with potassium, sodium, etc., in the form of bromides, it has a powerful, quieting effect upon the nervous system. Externally, in solution, it is one of the best applications for foul or gangrenous ulcers and wounds, and especially hospital gangrene.

Dose: ½ to 2 grains, in a large quantity of water.

SOLUTION OF BROMINE.

Take of	Bromine (by weight)	1 ounce.
	Bromide of potassium	160 grains.
	Distilled water	4 ounces.

Dissolve the bromide in 2 ounces of the water, add the bromine, agitate, and finally add the remainder of the water.

Dose: 1 to 2 minims. Chiefly externally. *J. Lawrence Smith.*

BRYONIA (Fr.)—BRYONY.

The root of Bryonia alba, L., and of B. dioica, L. (*Nat. ord., Cucurbitaceæ*), an herbaceous perennial indigenous to Europe.

Bryony is an active hydragogue cathartic and diuretic, and has long been used in dropsies. It has also been employed with good effect in pleurisy, pericarditis, rheumatism, etc.

PREPARATION.

Tincture of Bryony.

Take of Bryony root	1 ounce.
Diluted alcohol	9 ounces.

Prepare a tincture by maceration.

Dose: 3 to 10 minims.

BUCHU (U. S., Br., Fr.)—BUCHU LEAVES.

The leaves of Barosma betulina, Bartling, B. crenulata, Hooker, and B. serratifolia, Willd. (*Nat. ord., Rutaceæ*), erect shrubs indigenous to South Africa.

Buchu leaves have a strong, aromatic odor, a bitterish, mint-like taste, and act as a stimulant to the mucous membranes, especially of the genito-urinary tract. Used in chronic nephritis, cystitis, urethritis, incontinence and retention of urine, etc.

Dose: 20 to 30 grains.

PREPARATIONS.

Extractum Buchu Fluidum (U. S.)—Fluid Extract of Buchu.

Take of Buchu	16 ounces.
Alcohol	sufficient.

Moisten the buchu with 6 ounces of alcohol, and proceed according to the general formula, page 161.

Dose: 20 to 30 minims. This preparation is generally less efficient than the infusion.

Infusum Buchu (U. S., Br.)—Infusion of Buchu.

Take of Buchu	1 ounce.
Boiling water	1 pint.

Macerate for two hours in a covered vessel, and strain.

Dose: 1 to 2 ounces.

TINCTURE OF BUCHU (Br.).

Take of Buchu leaves	2½ ounces.
Proof spirit	20 ounces.

Macerate the buchu for forty-eight hours in 15 ounces of the spirit, then percolate with the remainder and enough more to make 20 ounces.

Dose: 1 to 2 drachms.

MIXTURES OF BUCHU.

Take of Infusion of buchu	1 ounce.
Tincture of henbane	12 minims.
Mucilage	2 drachms.

Mix. One dose. *Royal Free Hospital.*

Take of Infusion of buchu	1½ ounce.
Bicarbonate of potassium	10 grains.
Tincture of henbane	15 minims.

Mix. One dose. *St. Thomas's Hospital.*

CAFFEA (U. S., Fr.)—COFFEE.

The seeds of Caffea Arabica, L. (*Nat. ord., Rubiaceæ*), a small tree indigenous to Africa, but widely cultivated in tropical countries.

Coffee is a nervous stimulant, its effect being due to a neutral principle termed *caffeinum.* It is used in certain forms of nervous headache, and to counteract the effects of poisonous doses of opium.

Dose: 20 to 40 grains.

CAFFEINUM—CAFFEIN.

Obtained by precipitating a decoction of coffee with acetate of lead, filtering, removing the excess of lead with sulphuretted hydrogen, neutralizing with ammonia, evaporating, and recrystallizing.

It is in white, silky crystals of a slightly bitter taste, and soluble in 58 parts of water.

Dose: 1 to 5 grains.

CITRATE OF CAFFEIN.

Obtained by saturating a solution of citric acid with caffein, evaporating, and crystallizing. It is much more soluble than caffein.

Dose: 1 to 5 grains.

CALAMUS (U. S., Fr., Ger.)—CALAMUS—SWEET FLAG.

The rhizome of Acorus calamus, L. (*Nat. ord.*, *Araceæ*), an herbaceous perennial indigenous to North America.

Sweet flag has a fragrant odor, and a warm, aromatic taste. It is an aromatic stimulant, and is used in dyspepsia, and as an adjunct to other more active remedies.

Dose: 20 to 60 grains.

PREPARATIONS.

EXTRACT OF CALAMUS (Ger.).

Take of Calamus	2 parts.
Alcohol	9 parts.
Water	9 parts.

Digest the calamus twenty-four hours in 6 parts each of alcohol and water, and express. Treat the residue in like manner with the remainder of the alcohol and water, and evaporate the mixed and filtered liquids to a thick extract.

TINCTURE OF CALAMUS (Ger.).

Take of Calamus	1 part.
Diluted alcohol	5 parts.

Macerate eight days and filter.

CALCIUM—CALCIUM.

CALCII BROMIDUM—BROMIDE OF CALCIUM.

Prepared by saturating hydrobromic acid with pure carbonate of lime, and evaporating the solution to dryness.

It is a whitish, granular salt, of a sharp, bitter taste, and freely soluble in water.

It produces essentially the same effects as the other bromides, and is used, like them, in insomnia, epilepsy, hysteria, infantile convulsions, etc.

Dose: 5 to 30 grains.

MIXTURE OF BROMIDE OF CALCIUM.

Take of Bromide of calcium	1 ounce.
Syrup of lacto-phosphate of lime	4 ounces.

Mix. Dose: 1 drachm three times a day in a little water, in epileptic cases.

National Dispensatory.

CALCII CARBONAS PRÆCIPITATA (U. S. et al. Ph.)—PRECIPITATED CARBONATE OF CALCIUM—PRECIPITATED CARBONATE OF LIME.

Take of	Solution of chloride of calcium	5½ pints.
	Carbonate of sodium	72 ounces.
	Distilled water	sufficient.

Dissolve the carbonate in 6 pints of distilled water, heat this and the solution of calcium to the boiling point, and mix them. Decant the clear liquid, wash the precipitate with boiling distilled water, and dry it on bibulous paper.

It is a fine white powder, tasteless, and insoluble.

In this connection will be considered :

Creta Præparata (U. S., Br.)—Prepared Chalk.

Take of chalk a convenient quantity. Add a little water to the chalk, and rub it into a fine powder. Throw this into a vessel of water, stir briskly, and after a short interval decant into another vessel the supernatant liquid, while yet turbid. Treat the coarser particles of chalk remaining in the first vessel in a similar manner, and add the turbid liquid to that previously decanted. Lastly, let the powder subside, and having poured off the water, dry it.

Either of the above preparations may be employed when the effects of chalk are desired.

Chalk is an absorbent, astringent, and antacid. It is one of the best astringents in nearly all forms of diarrhœa, but especially in those accompanied with acidity of the intestinal secretions. It is also employed in dyspepsia, gout, etc. Externally it is employed as an absorbent powder in cutaneous diseases.

Dose: 10 to 40 grains.

PREPARATIONS.

Mistura Cretæ (U. S., Br.)—Chalk Mixture.

Take of	Prepared chalk	½ ounce.
	Glycerin	½ ounce.
	Gum arabic	120 grains.
	Cinnamon water	4 ounces.
	Water	4 ounces.

Rub the chalk and gum with the water gradually added; then add the other ingredients and mix.

Dose: 1 to 2 ounces.

Trochisci Cretæ (U. S.)—Troches of Chalk.

Take of Prepared chalk	4 ounces.
Gum arabic	1 ounce.
Nutmeg	60 grains.
Sugar	6 ounces.

Mix thoroughly, then with water form a mass, and divide into 480 troches.

Aromatic Powder of Chalk (Br.).

Take of Cinnamon	4 ounces.
Nutmeg	3 ounces.
Saffron	3 ounces.
Cloves	1½ ounce.
Cardamom seeds	1 ounce.
Refined sugar	25 ounces.
Prepared chalk	11 ounces.

Mix thoroughly, pass the powder through a fine sieve, and finally rub in a mortar.

Dose: 10 to 60 grains.

Aromatic Powder of Chalk and Opium (Br.).

Take of Aromatic powder of chalk	9¾ ounces.
Opium	¼ ounce.

Mix thoroughly.

Dose: 10 to 40 grains.

Compound Mixtures of Chalk.

Take of Tincture of opium	1 drachm.
Tincture of catechu	3 drachms.
Chalk mixture	3½ ounces.

Mix. Dose: ½ ounce every three hours, in diarrhœa.

Take of Chalk mixture	½ ounce.
Tincture of catechu	30 minims.
Decoction of logwood	to 1 ounce.

Mix. One dose. *King's College Hospital.*

CALCII CHLORIDUM (U. S., Br., Fr.)—CHLORIDE OF CALCIUM.

Prepared by neutralizing hydrochloric acid with carbonate of lime, evaporating and fusing the product. It is in dry, white masses, which are very deliquescent. It is alterative and resolvent, and has been used with success in scrofulous swellings of the glands, ovarian and fibroid tumors, and in various skin diseases.

Dose: 10 to 20 grains.

PREPARATIONS.

Liquor Calcii Chloridi (U. S.)—Solution of Chloride of Calcium.

Take of Chloride of calcium.......................... sufficient.

Dissolve in one and a half times its weight of distilled water, and filter.

Dose: 30 to 60 minims.

CALCII HYPOPHOSPHIS (U. S., Br.)—HYPOPHOSPHITE OF CALCIUM.

Prepared by boiling phosphorus with milk of lime, filtering and evaporating the product. It is a white, crystalline salt, of a bitter and nauseous taste. Used in phthisis, caries of the bones, scrofula, and other wasting diseases.

Dose: 10 to 30 grains.

MIXTURE OF HYPOPHOSPHITE OF CALCIUM.

Take of Hypophosphite of calcium	3 grains.
Saccharated solution of lime	10 minims.
Glycerin	20 minims.
Camphor water	1 ounce.

Mix. One dose. *London Chest Hospital.*

CALCII PHOSPHAS PRÆCIPITATA (U. S. et al. Ph.)—PRECIPITATED PHOSPHATE OF CALCIUM.

Calcined bone is macerated in hydrochloric acid until dissolved, the solution filtered, and precipitated with water of ammonia. The precipitate is then washed with boiling distilled water, and dried. It is a white, amorphous powder, without odor or taste. It is used for the same purposes as the hypophosphite, in wasting diseases, rachitis and ununited fractures.

Dose: 10 to 30 grains.

SYRUP OF LACTO-PHOSPHATE OF LIME.

A solution of calcium phosphate in lactic acid and syrup, containing, in 1 ounce, 16 grains of calcium phosphate and 33 grains of lactic acid.

Hospital Formulary.

CALCII SULPHIDUM (Fr.)—SULPHIDE OF CALCIUM.

Take of Sulphur	10 parts.
Lime	30 parts.
Water	50 parts.

Mix thoroughly, and boil until a small portion placed upon a cold surface becomes solid in cooling. Then turn the mass upon a marble slab, and when cool, break into pieces, and preserve in well-stopped bottles.

Sulphide of calcium appears to prevent and arrest suppuration. Dr. Sidney Ringer extols it as a remedy for furuncular and glandular inflammations, etc.; and more recently, Dr. Samuel Sexton writes that he employs it with great satisfaction in nearly all cases where inflammation of the external meatus of the ear is a symptom, but especially in those where suppuration has occurred, or threatens to occur.

Dose: $\frac{1}{20}$ to 1 grain. It may be conveniently administered in the form of a trituration: 1 part of the sulphide with 9 parts of sugar of milk.

PILLS OF SULPHIDE OF CALCIUM.

Take of Sulphide of calcium	10 grains.
Compound tragacanth powder	30 grains.
Water	sufficient.

Mix and divide into 30 pills. *University College Hospital.*

These pills should be used while fresh, as the salt speedily undergoes change.

CALX (U. S. et al. Ph.)—LIME.

Prepared from carbonate of lime, by calcination. Chalk, marble, or even any common limestone may be used.

Lime is in white, or grayish-white masses, the color depending upon the relative purity of the stone employed, and of a caustic, alkaline taste. When exposed to the air, it absorbs moisture and falls into powder, forming a hydrate (*slaked lime*). The same effect is rapidly produced by pouring water upon it.

Unslaked lime acts as a caustic, and, with arsenic, forms the arsenical paste formerly much used for the destruction of cancerous and other tumors.

PREPARATIONS.

Linimentum Calcis (U. S., Br., Fr.)—Liniment of Lime.

Take of Solution of lime	8 ounces.
Flaxseed oil (by weight)	7 ounces.

Mix.

The British Pharmacopœia employs equal parts of lime water and olive oil; the French Codex, 9 parts of lime water with 1 part of oil of sweet almonds.

Used as an application to burns and scalds, the surface being thickly coated with it, and then covered with cotton wool. Commonly known as *carron oil.*

Liquor Calcis (U. S. et al. Ph.)—Solution of Lime—Lime Water.

Take of Lime	4 ounces.
Distilled water	8 pints.

Slake the lime with a little of the water, then pour on the remainder, and stir them together. Keep the solution, together with the undissolved lime, in well-stopped bottles.

Lime water is an antacid and astringent. Much used in acid indigestion, diarrhœa, vomiting, etc.

Dose: 2 to 4 ounces.

Saccharated Solution of Lime (Br.).

Take of Slaked lime	1 ounce.
Refined sugar	2 ounces.
Distilled water	20 ounces.

Triturate the lime and sugar together, transfer the mixture to a bottle containing the water, shake occasionally for a few hours, and then draw off the clear liquid with a siphon.

Dose: 15 to 60 minims.

Lotions of Lime.

Take of Lime water	1 ounce.
Oil of almonds	1 drachm.

Mix. *British Skin Hospital.*

Take of Bicarbonate of sodium	6 grains.
Glycerin	10 minims.
Lime water	1 ounce.

Mix. *Westminster Hospital.*

CALX CHLORINATA (U. S. et al. Ph.)—CHLORINATED LIME.

Prepared by exposing slaked lime to chlorine gas as long as it is absorbed.

It is a whitish powder, with the odor of chlorine. It is used chiefly as a disinfectant and deodorizer, being applied in solution to ulcers, burns, etc. It has been employed internally in typhus, typhoid, and scarlet fevers, and in other contagious and infectious diseases.

Dose: 3 to 6 grains.

PREPARATIONS.

Solution of Chlorinated Lime (Br., Fr.).

Take of Chlorinated lime	1 pound.
Distilled water	1 gallon.

Triturate well together, transfer to a bottle, shake occasionally for three hours, then filter.

Lotion of Chlorinated Lime.

Take of Solution of chlorinated lime	½ ounce.
Water	to 10 ounces.

Mix. *Middlesex Hospital.*

CALENDULA—MARIGOLD.

The leaves and flowers of Calendula officinalis, L. (*Nat. ord., Compositæ*), an annual herb indigenous to the Old World, but in common cultivation here for ornament.

Calendula was anciently held in high esteem, but has fallen into disrepute. It was considered stimulant, antispasmodic, sudorific, diuretic, and emmenagogue, and was used in hysterical and dropsical affections, and topically as a dressing for wounds, ulcers, etc.

Dose: ½ to 1 drachm.

PREPARATION.

Tincture of Calendula.

Take of Fresh calendula	1 part.
Alcohol	5 parts.

Macerate two weeks and filter.

Dose: ½ to 2 drachms. Diluted with twenty parts of water, it is used externally.

CALUMBA (U. S. et al. Ph.)—COLUMBO.

The root of Jateorrhiza palmata, Miers (*Nat. ord., Menispermaceæ*), an herbaceous climbing plant, with large, perennial roots, indigenous to Eastern Africa.

Columbo occurs in commerce in transverse slices, an inch or more in diameter, and a quarter or half inch thick, of a yellowish color, and an aromatic, persistent, bitter taste. It is a pure, bitter tonic, without astringency, and is used in atonic dyspepsia, diarrhœa, dysentery, and convalescence from acute diseases generally. It is usually acceptable to the stomach when nearly all other bitters disagree.

Dose: 15 to 30 grains.

PREPARATIONS.

Extractum Calumbæ Fluidum (U. S. et al. Ph.)—Fluid Extract of Columbo.

Take of Columbo	16 ounces.
Glycerin	2 ounces.
Alcohol	sufficient.
Water	sufficient.

Mix the glycerin with 14 ounces of alcohol, moisten the columbo with 4 ounces of the mixture, and proceed according to the general formula, page 161.

Dose: 15 to 30 minims.

Infusum Calumbæ (U. S., Br.)—Infusion of Columbo.

Take of Columbo .... ½ ounce.
Water .... sufficient.

Moisten, pack, and percolate to 1 pint, then heat to the boiling point, and strain. Or, macerate the columbo in a pint of boiling water for two hours, and strain.

Dose: 1 to 2 ounces. Often used as a vehicle for other tonics, etc.

The British Pharmacopœia directs to macerate ½ ounce of columbo, for one hour, in 10 ounces of cold water.

Tinctura Calumbæ (U. S., Br., Fr.)—Tincture of Columbo.

Take of Columbo .... 4 ounces.
Diluted alcohol .... sufficient.

Moisten, pack, and percolate to 2 pints.

Dose: 1 to 4 drachms.

EXTRACT OF COLUMBO (Br., Fr., Ger.).

Take of Columbo .... 1 pound.
Distilled water .... 4 pints (imp.).

Macerate the columbo with 2 pints of water, for twelve hours, strain and press; macerate again with the same quantity of water, strain and press as before; mix the liquors, filter, and evaporate on a water-bath to the proper consistence.

Dose: 2 to 10 grains.

The French and German preparations are alcoholic extracts.

MIXTURES OF COLUMBO.

Take of Tincture of columbo .... 15 drachms.
Deodorized tincture of opium .... 1 drachm.

Mix. Dose: 1 drachm in a wineglass of water before meals, in indigestion attended with diarrhœa. *Bartholow.*

Take of Bicarbonate of sodium .... 10 grains.
Tincture of orange .... 30 minims.
Infusion of columbo .... to 1 ounce.

Mix. One dose. *St. Bartholomew's Hospital.*

CAMPHORA (U. S. et al. Ph.)—CAMPHOR.

A concrete, volatile substance obtained by steaming the wood of Camphora officinarum, C. Bauhin (*Nat. ord., Lauraceæ*), a large tree indigenous to China and Japan, but cultivated elsewhere in tropical and subtropical regions.

Camphor occurs in crystalline, translucent masses, of a characteristic odor and taste, slightly soluble in water, but freely so in alcohol, ether, chloroform, volatile and fixed oils.

Camphor is stimulant, antispasmodic, diaphoretic, and anodyne. Used in typhus and typhoid fevers, and typhoid conditions generally, also in strangury, chordee, dysmenorrhœa, nervous and mental affections, etc. Applied externally, it is at first irritant, afterward anodyne. Few remedies, indeed, have a wider range of usefulness.

Dose: 1 to 10 grains. It may be reduced to powder for administration by triturating it with a few drops of alcohol.

PREPARATIONS.

Aqua Camphoræ (U. S., Br., Fr.)—Camphor Water.

Take of Camphor	120 grains.
Alcohol	40 minims.
Carbonate of magnesium	½ ounce.
Distilled water	2 pints.

Rub the camphor with the alcohol, then with the carbonate of magnesium, and lastly with the water, gradually added; then filter.

Dose: 2 to 8 drachms.

The processes of the British Pharmacopœia and the French Codex differ from the above, though they obtain substantially the same result—a saturated solution of camphor in water.

Linimentum Camphoræ (U. S. et al. Ph.)—Liniment of Camphor.

Take of Camphor	3 ounces.
Olive oil (by weight)	12 ounces.

Dissolve.

The French and German preparations, termed *Camphorated Oil*, are made in the proportion of 1 to 9.

Spiritus Camphoræ (U. S. et al. Ph.)—Spirit of Camphor.

Take of Camphor	4 ounces.
Alcohol	2 pints.

Dissolve and filter.

Dose: 5 to 60 minims.

AMMONIATED CAMPHOR LINIMENT (Ger., Fr.).

Take of Camphorated oil	4 parts.
Water of ammonia	1 part.

Mix thoroughly.

The French Codex directs the proportion of 9 to 1.

Compound Liniment of Camphor (Br.).

Take of Camphor	2½ ounces.
Oil of lavender	1 drachm.
Strong solution of ammonia	5 ounces.
Rectified spirit	15 ounces.

Dissolve the camphor and the oil in the spirit, then add the solution of ammonia gradually, shaking well together.

Ointment of Camphor (Fr.).

Take of Camphor	3 parts.
White wax	1 part.
Lard	9 parts.

Melt the lard and wax together, add the camphor, and stir while cooling.

Wine of Camphor (Ger.).

Take of Camphor	1 part.
Gum arabic	1 part.
White wine	48 parts.

Rub the camphor and gum together, and then gradually add the wine.

Dose: ½ to 1 drachm.

Liniments of Camphor with Cantharides.

Take of Liniment of camphor	14½ drachms.
Tincture of opium	2 drachms.
Tincture of cantharides	3½ drachms.

Mix. *Women's Hospital, London.*

Take of Spirit of camphor	1 part.
Vinegar of cantharides	1 part.
Acetic acid	1 part.

Mix. *Royal Chest Hospital.*

Mixtures of Camphor.

Take of Camphor water	3 ounces.
Compound tincture of lavender	1 ounce.
Tincture of opium	1 to 2 drachms.

Mix. Dose: ½ ounce every hour or two. In summer diarrhœa.

Bartholow.

Take of Spirit of camphor	2 drachms.
Tincture of capsicum	2 drachms.
Tincture of opium	2 drachms.
Tincture of ginger	2 drachms.

Mix. Dose: 20 to 40 minims every two or three hours, in diarrhœa.

Take of Camphor water .... 2 ounces.
Solution of acetate of ammonium............ 2 ounces.

Mix. Dose : ½ ounce every two hours. To quiet the restlessness, delirium, etc., of fevers. *Bartholow.*

Take of Spirit of ether .... ½ drachm.
Camphor water ....to 1 ounce.

Mix. One dose. *London Hospital.*

PILLS OF CAMPHOR.

Take of Camphor.................................... 3 grains.
Extract of henbane........................ 2 grains.
Alcohol .... 1 drop.

Make 1 pill. *London Hospital.*

Take of Camphor .... 2 grains.
Opium .... ½ grain.

Make 1 pill. *St. Bartholomew's Hospital.*

Take of Camphor .... 1 grain.
Opium .... 1 grain.
Extract of hops.......................... sufficient.

Make 1 pill. *Brompton Consumption Hospital.*

Take of Camphor .... 40 grains.
Lactucarium .... 40 grains.

Mix and divide into 20 pills.
Dose : 2 pills at bedtime. For chordee. *Ricord.*

Take of Camphor.................................... 30 grains.
Opium .... 10 grains.

Mix and divide into 10 pills.
Dose : 1 pill. For chordee. *Ricord.*

CAMPHORA MONOBROMATA—MONOBROMATED CAMPHOR.

Prepared by submitting camphor to the action of bromine, with gentle heat, dissolving the product in warm petroleum benzin, and crystallizing. It is in colorless, acicular crystals, of a camphoraceous odor and taste. Used in epilepsy, hysteria, delirium tremens, priapism, incontinence of urine, convulsions, etc. It has little, if any, advantage over the bromides of potassium, sodium, etc.

Dose : 2 to 6 grains, in pill or suspended in syrup or mucilage.

CANELLA (U. S., Br., Fr.)—CANELLA—WILD CINNAMON.

The bark of Canella alba, Murray (*Nat. ord.*, *Canellaceæ*), a tree indigenous to the West Indies.

Canella bark is an aromatic stimulant and tonic, but is seldom employed except as an adjunct to other more efficient drugs.

Dose: 5 to 10 grains. (See Powder of Aloes and Canella.)

AROMATIC POWDER OF CANELLA.

Take of Canella	1 part.
Ginger	1 part.
Long pepper	1 part.

Mix. Dose: 5 to 10 grains. *St. George's Hospital.*

CANNABIS AMERICANA (U. S.)—AMERICAN HEMP.
CANNABIS INDICA (U. S. et al. Ph.)—INDIAN HEMP.

The flowering tops of Cannabis sativa, L. (*Nat. ord.*, *Urticaceæ*), an annual herb indigenous to India, but cultivated and naturalized in this country. American hemp, though specifically the same plant as the Indian, differs from it in being less active, doubtless owing to climatic influences.

By evaporating alcoholic tinctures of hemp-tops, extracts are obtained which represent the active properties of the plant.

Cannabis is narcotic, anodyne, and antispasmodic. It has been used successfully in traumatic tetanus, in chorea, hysteria, neuralgia, and a variety of nervous diseases. Though less certain than opium as a hypnotic, it has the advantage of not producing as unpleasant after-effects.

Dose: Of the powdered drug, 2 to 20 grains; but it is seldom administered in this form.

PREPARATIONS.

Extractum Cannabis Americanæ (U. S.)—Extract of American Hemp.

Extractum Cannabis Indicæ (U. S. et al. Ph.)—Extract of Indian Hemp.

Take of American (or Indian) hemp	12 ounces.
Alcohol	sufficient.

Macerate the hemp in 12 ounces of alcohol for four days, then percolate until 2 pints are obtained, or the hemp is exhausted. Evaporate this on a water-bath to a proper consistence.

Dose: Of American hemp, ½ grain; of Indian hemp, ¼ to ½ grain.

Tinctura Cannabis (U. S., Br., Ger.)—Tincture of Hemp.

Take of Extract of Indian hemp .... 360 grains.
Alcohol .... 1 pint.

Dissolve and filter.

Dose: 5 to 15 minims.

MIXTURES OF HEMP.

Take of Tincture of hemp.......................... 10 minims.
Spirit of peppermint .... 1 minim.
Water.. 1 drachm.

Mix. One dose. *Hospital Formulary.*

Take of Extract of hemp .... 1 grain.
Ether .... 15 minims.
Camphor water .... 1 ounce.

Mix. One dose. *London Fever Hospital.*

PILLS OF HEMP.

Take of Extract of hemp .... ½ grain.
Liquorice..................................... sufficient.
Mucilage .... sufficient.

Make 1 pill. *Brompton Consumption Hospital.*

Take of Extract of hemp .... ¼ grain.
Extract of gentian........................ 2 grains.
Bread mass .... 1 grain.

Make 1 pill. *Royal Chest Hospital.*

CANTHARIS (U. S. et al. Ph.)—CANTHARIDES—SPANISH FLIES.

Cantharis vesicatoria, De Geer (*Ord. Coleoptera*), a beautiful insect, of a golden green color, which is found upon trees and shrubs in most parts of Europe. They are plunged into hot vinegar and water, or exposed to the vapor of hot vinegar, and then dried and preserved for use.

As found in the shops, they retain their natural form and color, and have an acrid, burning taste. Taken internally they stimulate the kidneys, and produce more or less irritation of the urinary passages. In large doses they produce strangury, priapism, hæmaturia, etc., and in excessive doses are an active poison.

Applied externally they vesicate, and may also produce their constitutional effects.

Internally, cantharides are employed in incontinence of urine, spermatorrhœa, gleet, etc.; externally, as a vesicant and rubefacient in a great variety of cases.

Dose: 1 to 2 grains.

PREPARATIONS.

Ceratum Cantharidis (U. S. et al. Ph.)—Cantharides Cerate.

Take of Cantharides .... 12 ounces.
Yellow wax.............................. 7 ounces.
Resin .... 7 ounces.
Lard .... 10 ounces.

To the wax, resin, and lard, previously melted together and strained, add the cantharides, and, by means of a water-bath, keep the mixture in a liquid state for half an hour, stirring occasionally. Then remove from the water-bath, and stir until cold.

Termed *Cantharides plaster* by the European Pharmacopœias.

It is the common blistering cerate, or *fly-blister.*

The following is a more elegant preparation :

Ceratum Extracti Cantharidis (U. S.)—Cerate of Extract of Cantharides.

Take of Cantharides .... 5 ounces.
Resin .... 3 ounces.
Yellow wax.............................. 6 ounces.
Lard.............................. 7 ounces.
Stronger alcohol.......... ....2½ pints, or sufficient.

Exhaust the cantharides with the alcohol, and evaporate the tincture to an extract. Mix this with the resin, wax, and lard, previously melted together, keep the mixture at a temperature of 212° for fifteen minutes, strain, and stir until cool.

Charta Cantharidis (U. S., Br., Fr.)—Cantharides Paper.

Take of White wax .... 4 ounces.
Spermaceti.............................. 1½ ounce.
Olive oil (by weight) .... 2 ounces.
Canada turpentine.............................. ½ ounce.
Cantharides .... ½ ounce.
Water.............................. 5 ounces.

Mix all together and boil gently for two hours, constantly stirring, and strain. Then coat strips of paper on one side, by passing them over the surface of the melted liquid.

Collodium cum Cantharide (U. S.)—Collodion with Cantharides—Cantharidal Collodion.

Take of Cantharides.............................. 8 ounces.
Pyroxylon..............................100 grains.
Canada turpentine ....320 grains.
Castor oil..............................160 grains.
Stronger ether.............................. 1½ pint.
Stronger alcohol.............................. sufficient.

Percolate the cantharides with the ether until 15 ounces have passed, then with sufficient alcohol to obtain ½ pint more. Allow this last to evaporate spontaneously until reduced to 1 ounce, then mix it with the first portion, add the other ingredients, and agitate until dissolved.

An admirable preparation, blistering with rapidity and certainty.

Linimentum Cantharidis (U. S.)—Liniment of Cantharides.

Take of Cantharides .... 1 ounce.
Oil of turpentine .... ½ pint.

Digest for three hours on a water-bath, and strain.

Tinctura Cantharidis (U. S. et al. Ph.)—Tincture of Cantharides.

Take of Cantharides .... 1 ounce.
Diluted alcohol .... sufficient.

Moisten, pack, and percolate to 2 pints.

Dose: 5 to 20 minims.

More than double the strength of the British preparation, and only about one-third of that of the French and German.

Unguentum Cantharidis (U. S., Br., Ger.)—Ointment of Cantharides.

Take of Cantharides cerate .... 120 grains.
Resin cerate .... 360 grains.

Mix.

Blistering Liquid (Br.).

Take of Cantharides .... 8 ounces.
Acetic acid .... 4 ounces.
Ether .... sufficient.

Mix the cantharides and acid, and, after twenty-four hours, percolate with the ether until 20 ounces are obtained.

Extract of Cantharides (Fr.).

Take of Cantharides .... 1 part.
Alcohol (60 per cent.) .... 8 parts.

Prepare a tincture by maceration, and evaporate it to the proper consistence.

Vinegar of Cantharides (Br.).

Take of Cantharides .... 2 ounces.
Glacial acetic acid .... 2 ounces.
Acetic acid .... 18 ounces, or sufficient.

Mix 13 ounces of acetic acid with the glacial acetic acid, and digest the cantharides in the mixture for two hours at a temperature of 200°; then transfer to a percolator, and when the liquid ceases to pass, pour 5 ounces of acetic acid over the residuum, express, filter, and add sufficient acetic acid to make 20 ounces.

MIXTURES OF CANTHARIDES.

Take of Tincture of cantharides	½ ounce.
Copaiba	½ ounce.
Tincture of chloride of iron	1 ounce.

Mix. Dose: 30 drops three times a day, in gleet. *Bumstead.*

Take of Tincture of cantharides	2 drachms.
Tincture of chloride of iron	6 drachms.

Mix. Dose: 10 drops three times a day, in gleet. *Bumstead.*

CAPSICUM (U. S. et al. Ph.)—CAYENNE PEPPER.

The fruit of Capsicum fastigiatum, Blume, C. annuum, L., and several other species of capsicum (*Nat. ord., Solanaceæ*), herbs or shrubs indigenous to tropical America, but widely cultivated.

Capsicum has a pungent odor, and a very hot, acrid taste. It is an active stimulant and irritant. Applied externally it quickly reddens the skin, and is often used as a rubefacient. Internally it is employed as an aid to digestion in dyspepsia, in delirium tremens when vomiting is a prominent symptom, in vomiting from other causes, sick headache, sea-sickness, etc. An infusion is often used as a stimulating gargle in relaxed conditions of the throat.

Dose: 1 to 5 grains.

PREPARATIONS.

Infusum Capsici (U. S.)—Infusion of Capsicum.

Take of Capsicum	½ ounce.
Boiling water	1 pint.

Macerate for two hours, and strain.

Dose: 1 to 4 drachms. Used chiefly as a gargle.

Tinctura Capsici (U. S., Br., Ger.)—Tincture of Capsicum.

Take of Capsicum	1 ounce.
Diluted alcohol	sufficient.

Moisten, pack, and percolate to 2 pints.

Dose: 10 to 60 minims.

The British tincture is slightly weaker than this, while the German is nearly three times stronger.

Oleoresina Capsici (U. S.)—Oleoresin of Capsicum.

Take of Capsicum	12 ounces.
Ether	sufficient.

Obtain 24 ounces of ethereal tincture by percolation, evaporate or distil off the ether, and strain.

Dose: 1 to 3 grains.

Gargles of Capsicum.

Take of Tincture of capsicum .... 1 drachm.
Water .... 8 ounces.

Mix. *St. George's Hospital.*

Take of Tincture of capsicum ....100 minims.
Diluted sulphuric acid .... 1 drachm.
Decoction of pale bark .... 10 ounces.

Mix. *St. Bartholomew's Hospital.*

Take of Tincture of capsicum .... 1 drachm.
Diluted acetic acid .... 2 drachms.
Water ....to 6 ounces.

Mix. *Royal Chest Hospital.*

Lotion of Capsicum and Cantharides.

Take of Tincture of capsicum .... 30 parts.
Tincture of cantharides .... 20 parts.
Cologne water .... 50 parts.

Mix. A stimulating lotion for alopecia. *G. H. Fox.*

CARBO ANIMALIS (U. S., Br., Ger.)—ANIMAL CHARCOAL—BONE-BLACK.

The residue of bones which have been exposed to a red heat without access of air.

PREPARATION.

Carbo Animalis Purificatus (U. S., Br.)—Purified Animal Charcoal.

Prepared by digesting bone black in diluted hydrochloric acid, washing the undissolved portion, and heating it to redness.

Animal charcoal is used only in making preparations.

CARBO LIGNI—WOOD CHARCOAL.

The residue left by the destructive distillation of wood.

Charcoal has the property of absorbing gases, the volume of gas absorbed exceeding many times that of the charcoal employed. This property renders it useful in dyspepsia attended with the formation of gas, and in flatulence generally. Externally it is employed as a dressing for foul and gangrenous ulcers, abscesses, etc.

Dose: 1 to 4 drachms.

PREPARATIONS.

CHARCOAL LOZENGES (Fr.).

Take of	Wood charcoal	10 parts.
	White sugar	30 parts.
	Mucilage of tragacanth	4 parts.

Make lozenges of 15 grains each.

CHARCOAL POULTICE (Br.).

Take of	Wood charcoal	½ ounce.
	Crumb of bread	2 ounces.
	Linseed meal	1½ ounce.
	Boiling water	10 ounces.

Macerate the bread in the water for ten minutes, and add the meal gradually, stirring the ingredients; then add half the charcoal, and sprinkle the remainder on the surface of the poultice.

CARDAMOMUM (U. S. et al. Ph.)—CARDAMOM.

The dried capsules of Elettaria Cardamomum, Maton (*Nat. ord., Zingiberaceæ*), a tall, flag-like perennial indigenous to India.

Cardamom seeds have an agreeable aromatic odor and taste, and possess aromatic and carminative properties. Used as an adjunct to other medicines.

Dose: 10 to 30 grains.

PREPARATIONS.

Tinctura Cardamomi (U. S.)—Tincture of Cardamom.

Take of	Cardamom	4 ounces.
	Diluted alcohol	sufficient.

Moisten, pack, and percolate to 2 pints.

Dose: 1 to 2 drachms.

Tinctura Cardamomi Composita (U. S., Br.)—Compound Tincture of Cardamom.

Take of	Cardamom	360 grains.
	Caraway	120 grains.
	Cinnamon	300 grains.
	Cochineal	60 grains.
	Clarified honey (by weight)	2 ounces.
	Diluted alcohol	sufficient.

Moisten the powders with the alcohol, then pack, and percolate to 2 pints and 6 ounces; add to this the honey, and filter.

Dose: 1 to 2 drachms.

The British Pharmacopœia employs raisins instead of the honey.

CARUM (U. S. et al. Ph.)—CARAWAY.

The fruit of Carum carui, L. (*Nat. ord., Umbelliferæ*), an annual herb indigenous to Asia, but cultivated everywhere.

Caraway has an agreeable aromatic odor and taste. Used as a stomachic and carminative in infantile cases, and as an adjunct to other medicines.

Dose: 30 to 60 grains.

PREPARATIONS.

Oleum Carui (U. S. et al. Ph.)—Oil of Caraway.

The oil distilled from caraway fruit.

Dose: 1 to 2 minims.

Caraway Water (Br.).

Take of Caraway fruit	1 pound.
Water	1 gallon.

Distil 1 gallon. Used as a vehicle.

Spirit of Caraway (Fr.).

Take of Caraway fruit	1 part.
Alcohol (80%)	8 parts.

Macerate two days, then distil off all the spirit employed.

Dose: 1 to 2 drachms.

CARYOPHYLLUS (U. S. et al. Ph.)—CLOVES.

The flower buds of Caryophyllus aromaticus, L. (*Nat. ord., Myrtaceæ*), a tree indigenous to the Moluccas, but widely cultivated in tropical countries.

Cloves have a strong, agreeable odor, a hot, acrid taste, and possess stimulant and aromatic properties. They are used chiefly as a flavoring ingredient, and as a condiment.

Dose: 5 to 10 grains.

PREPARATIONS.

Infusum Caryophylli (U. S., Br.)—Infusion of Cloves.

Take of Cloves	120 grains.
Boiling water	1 pint.

Macerate two hours in a covered vessel, and strain.

Dose: 1 to 2 ounces.

Oleum Caryophylli (U. S. et al. Ph.)—Oil of Cloves.

The oil distilled from cloves.

Dose: 1 to 5 minims. Often applied on cotton to the cavities of decayed teeth, for the relief of toothache.

TINCTURE OF CLOVES (Fr.).

Take of Cloves	1 part.
Alcohol (80%)	5 parts.

Macerate ten days, express, and filter.

Dose: 1 to 2 drachms.

SPIRIT OF CLOVES (Fr.).

Take of Cloves	1 part.
Alcohol (80%)	8 parts.

Macerate four days, then distil off all the spirit employed.

Dose: 1 to 2 drachms.

CASCARILLA (U. S. et al. Ph.)—CASCARILLA.

The bark of Croton Eluteria, Bennett (*Nat. ord., Euphorbiaceæ*), a small tree indigenous to the Bahamas.

Cascarilla has an aromatic odor, and a warm, spicy, bitter taste. It is used as a mild tonic in dyspepsia, etc.

Dose: 20 to 30 grains.

PREPARATIONS.

Infusum Cascarillæ (U. S., Br.)—Infusion of Cascarilla.

Take of Cascarilla	1 ounce.
Water	sufficient.

Moisten, pack, and percolate to 1 pint. Or, macerate the bark in 1 pint of boiling water for two hours, and strain.

Dose: 1 to 2 ounces.

EXTRACT OF CASCARILLA (Ger.).

Take of Cascarilla	1 part.
Boiling water	6 parts.

Macerate the bark in 4 parts of the water for twenty-four hours, and express. Digest the residue with 2 parts of water, express, mix the liquids, and evaporate to a thick extract.

TINCTURE OF CASCARILLA (Br., Fr., Ger.).

Take of Cascarilla	2½ ounces.
Proof spirit	20 ounces.

Macerate the bark for forty-eight hours in 15 ounces of the spirit, then percolate with enough more to make 20 ounces.

Dose: ½ to 2 drachms.

CASSIA FISTULA (U. S., Br., Fr.)—PURGING CASSIA.

The pulp of the fruit of Cassia fistula, L. (*Nat. ord., Leguminosæ*), a tree indigenous to the East Indies, but naturalized in the West Indies and South America. The fruit consists of long pods containing seeds imbedded in soft, black pulp, which has a sweet taste. It is a mild laxative. Used in the confection of senna.

Dose: 1 to 2 drachms, laxative; 1 to 2 ounces, purgative.

CASSIA MARILANDICA (U. S.)—AMERICAN SENNA.

The leaflets of Cassia Marilandica, L. (*Nat. ord., Leguminosæ*), an herbaceous perennial indigenous to the United States.

Used as a substitute for Alexandria senna, which it resembles in its action, though requiring to be administered in doses one-third larger.

Dose: 1 to 4 drachms.

PREPARATION.

INFUSION OF AMERICAN SENNA.

Take of American senna	1 ounce.
Coriander	1 drachm.
Boiling water	1 pint.

Macerate for an hour, and strain.

Dose: 2 to 6 ounces.

CASTANEA (U. S.)—CHESTNUT.

The leaves of Castanea vesca, L. (*Nat. ord., Cupuliferæ*), our common chestnut tree.

Chestnut leaves exert a sedative influence, which has proved of essential service in whooping-cough. When practicable they should be used while fresh. They may be gathered, for preservation, in July and August.

Dose: 10 to 60 grains.

PREPARATIONS.

FLUID EXTRACT OF CHESTNUT LEAVES.

Take of Chestnut leaves	16 ounces.
Glycerin	4 ounces.
Sugar	6 ounces.
Boiling water	sufficient.

Digest the leaves with water for twenty-four hours, express, and strain; repeat twice with sufficient water to cover the leaves, mix the infusions, add the glycerin and sugar, and evaporate to 1 pint.

Dose: ½ to 1 drachm. *Maisch.*

INFUSION OF CHESTNUT LEAVES.

Take of Chestnut leaves	1 ounce.
Boiling water	1 pint.

Infuse and strain.

Dose: 2 to 4 ounces.

CASTOREUM (U. S. et al. Ph.)—CASTOR.

A peculiar, concrete substance obtained from the preputial follicles of the beaver (*Castor fiber*, L.). It occurs in unctuous masses of a characteristic odor, and an acrid, bitter, nauseous taste.

Castor is stimulant and antispasmodic, and is used in hysteria and other nervous affections, typhoid conditions, etc.

Dose: 10 to 20 grains.

PREPARATIONS.

Tinctura Castorei (U. S. et al. Ph.)—Tincture of Castor.

Take of Castor	2 ounces.
Alcohol	2 pints.

Macerate for seven days, and filter.

Dose: ½ to 2 drachms.

The British preparation is made in the proportion of 1 to 20; the French and German, 1 to 10. The French Codex prepares also an ethereal tincture of the same strength.

CATECHU (U. S. et al. Ph.)—CATECHU.

An aqueous extract prepared from the wood of Acacia Catechu, Willd., and A. Suma, Kurz (*Nat. ord., Leguminosæ*), trees indigenous to India. The British Pharmacopœia employs *Pale Catechu*, derived from Uncaria Gambier, Roxb. (*Nat. ord., Rubiaceæ*).

Catechu consists mainly of tannic acid and extractive, and is powerfully astringent. Used in the later stages of diarrhœa and dysentery, after the active inflammatory symptoms have been subdued. Locally it is employed as an astringent in hemorrhages, relaxation of the throat, spongy gums, etc.

Dose: 5 to 30 grains.

Infusum Catechu Compositum (U. S., Br.)—Compound Infusion of Catechu.

Take of Catechu	½ ounce.
Cinnamon	60 grains.
Boiling water	1 pint.

Macerate for an hour, in a covered vessel, and strain.

Dose: 1 to 2 ounces. *Infusion of Catechu* (*Br.*).

Tinctura Catechu (U. S. et al. Ph.)—Tincture of Catechu.

Take of Catechu	3 ounces.
Cinnamon	2 ounces.
Diluted alcohol	sufficient.

Mix the powders, moisten, pack, and percolate to 2 pints.

Dose: ½ to 2 drachms.

The French and German preparations are made with catechu, 1 part, diluted alcohol, 5 parts.

Compound Powder of Catechu (Br.).

Take of Pale catechu	4 ounces.
Kino	2 ounces.
Rhatany	2 ounces.
Cinnamon	1 ounce.
Nutmeg	1 ounce.

Mix thoroughly, and reduce to a fine powder.

Dose: 20 to 40 grains.

Catechu Lozenges (Br., Fr.).

Take of Pale catechu	720 grains.
Refined sugar	25 ounces.
Gum arabic	1 ounce.
Mucilage of gum arabic	2 ounces.
Distilled water	sufficient.

Mix the powders, add the mucilage and sufficient water to form a proper mass, and divide into 720 lozenges.

Dose: 1 to 6 lozenges.

The French Codex employs tragacanth instead of gum arabic.

Mixtures of Catechu.

Take of Tincture of catechu .... 1 drachm.
Bicarbonate of potassium .... 10 grains.
Cinnamon water .... 1 ounce.

Mix. One dose. *British Skin Hospital.*

Take of Tincture of catechu .... 30 minims.
Chalk mixture ....to 1 ounce.

Mix. One dose. *University College Hospital.*

CERA—WAX.

CERA ALBA (U. S. et al. Ph.)—WHITE WAX.

Yellow wax bleached by exposure to light and moisture.

CERA FLAVA (U. S. et al. Ph.)—YELLOW WAX.

The prepared honey-comb of the honey-bee (*Apis mellifica*, L.). Wax is used in making cerates, ointments, and plasters.

CERATA—CERATES.

A cerate is a mixture of wax and some fatty or oily substance, of a consistence between that of an ointment and a plaster. It should be soft enough to be spread with ease, but sufficiently hard to adhere to the skin without melting.

Ceratum (U. S., Fr.)—Cerate—Simple Cerate.

Take of Lard .... 8 ounces.
White wax .... 4 ounces.

Melt together, and stir until cool.

The French Codex employs oil of sweet almond, 3 parts; white wax, 1 part.

CERII OXALAS (U. S., Br.)—OXALATE OF CERIUM.

Obtained from the mineral *cerite*, an ore of cerium, or by adding a solution of oxalate of ammonium to a solution of any salt of cerium.

It is a white powder, without odor or taste. Used to allay obstinate vomiting, especially that of pregnant women.

Dose: 1 to 4 grains.

Capsules of Oxalate of Cerium.

Take of Oxalate of cerium.......................... 80 grains.

Divide into 20 capsules.

Dose: 1 capsule every fourth hour. In the sickness of pregnancy.

"This remedy is usually administered in doses so small as to be quite inefficient; but, in the above manner, I have found it more frequently useful than any other drug." *Fordyce Barker.*

Pills of Oxalate of Cerium.

Take of Oxalate of cerium......................... 3 grains.
Extract of gentian......................... sufficient.

Make 1 pill. *London Chest Hospital.*

CETACEUM (U. S. et al. Ph.)—SPERMACETI.

A concrete substance obtained from the head of the spermaceti whale (*Physeter macrocephalus*, L.). When pure, it is in white, translucent, crystalline masses, unctuous to the touch, of an oily odor and insipid taste. It is used in the preparation of ointments and cerates, and internally as a demulcent.

Dose: 15 to 30 grains.

PREPARATIONS.

Ceratum Cetacei (U. S., Ger.)—Spermaceti Cerate.

Take of Spermaceti............................. 1 ounce.
White wax............................. 3 ounces.
Olive oil............................. 5 ounces.

Melt the spermaceti and wax together; then add the oil previously heated, and stir constantly until cool.

Spermatic Ointment (Br.).

Take of Spermaceti............................. 5 ounces.
White wax............................. 2 ounces.
Almond oil............................. 20 ounces.

Melt together with a gentle heat, and stir constantly until cool.

Saccharated Spermaceti (Ger.).

Take of Spermaceti............................. 1 part.
Sugar............................. 3 parts.

Mix, and rub into a very fine powder.

MIXTURE OF SPERMACETI.

Take of Spermaceti........................ 15 grains.
Chloric ether........................ 5 minims.
Compound tragacanth powder.............. 20 grains.
Pimento water........................to 1 ounce.

Mix. One dose. *Brompton Consumption Hospital.*

CETRARIA (U. S. et al. Ph.)—ICELAND MOSS.

The lichen, Cetraria Islandicus, Acharius (*Nat. ord., Lichenes*). Indigenous to the northern hemisphere, growing in high latitudes.

Iceland moss is demulcent and nutritious, owing to the large percentage of starch which it contains. It also contains a bitter, tonic principle, which improves the appetite and aids digestion. Chiefly used in chronic pulmonary complaints.

PREPARATIONS.

Decoctum Cetrariæ (U. S., Br.)—Decoction of Iceland Moss.

Take of Iceland moss........................ ½ ounce.
Water........................ sufficient.

Boil the moss in 1 pint of water for fifteen minutes, strain with compression, and add sufficient water through the strainer to make 1 pint.

Dose: 1 to 2 ounces.

The British Pharmacopœia directs the proportion of 1 to 20.

ICELAND MOSS WITHOUT BITTERNESS (Ger.).

Take of Iceland moss........................ 5 parts.
Tepid water........................ 30 parts.
Solution of carbonate of potassium........... 1 part.

Macerate for three hours, and then wash with cold water.

CHENOPODIUM (U. S., Fr., Ger.)—WORMSEED.

The fruit of Chenopodium Anthelminticum, L. (*Nat. ord., Chenopodiaceæ*), an annual herb indigenous to Tropical America, but naturalized in the United States, especially in the southern portions.

Though the fruit alone is officinal, all parts of the plant are efficient as an anthelmintic. The fresh juice, the seed, or the oil may be used for the destruction of lumbricoid worms.

Dose: 10 to 40 grains.

PREPARATIONS.

Oleum Chenopodii (U. S.)—Oil of Wormseed.

The oil distilled from wormseed.
Dose: 4 to 8 minims.

DECOCTION OF WORMSEED.

Take of	Wormseed leaves	1 ounce.
	Water (or milk)	1 pint.

Boil and strain.
Dose: 4 to 8 ounces.

CHIMAPHILA (U. S., Fr.)—PIPSISSEWA.

The leaves of Chimaphila umbellata, Nutt. (*Nat. ord., Ericaceæ*), a small shrub indigenous to the northern hemisphere.

Pipsissewa is diuretic, tonic, and astringent. Employed in urinary affections, as hæmaturia, cystitis, gleet, etc.

Dose: ½ to 1 drachm.

PREPARATIONS.

Decoctum Chimaphilæ (U. S.)—Decoction of Pipsissewa.

Take of	Pipsissewa	1 ounce.
	Water	sufficient.

Boil the pipsissewa in 1 pint of water for fifteen minutes, strain, and add sufficient water through the strainer to make 1 pint.

Dose: 2 to 4 ounces.

Extractum Chimaphilæ Fluidum (U. S.)—Fluid Extract of Pipsissewa.

Take of	Pipsissewa	16 ounces.
	Glycerin	4 ounces.
	Alcohol	sufficient.
	Water	sufficient.

Mix 8 ounces of alcohol, 3 of glycerin, and 5 of water, moisten the pipsissewa with 8 ounces of the mixture, and proceed according to the general formula, page 161. Finish the percolation with diluted alcohol, reserve 14 ounces, and add 1 ounce of glycerin to the remainder before evaporation.

Dose: ½ to 1 drachm.

CHIRETTA (U. S., Br.)—CHIRATA.

The entire plant Ophelia Chirata, Griseb. (*Nat. ord., Gentianaceæ*), an herb indigenous to India.

Its properties are very similar to those of gentian, and it is used for the same purposes.

Dose: ½ to 1 drachm.

PREPARATIONS.

INFUSION OF CHIRATA (Br.).

Take of Chirata	¼ ounce.
Distilled water at 120°	10 ounces.

Infuse for half an hour, and strain.

Dose: 1 to 2 ounces.

TINCTURE OF CHIRATA (Br.).

Take of Chirata	2½ ounces.
Proof spirit	sufficient.

By maceration and percolation, obtain 20 ounces of tincture.

Dose: ½ to 2 drachms.

CHLORAL (U. S., Br., Ger.)—HYDRATE OF CHLORAL.

By passing dry chlorine gas through absolute alcohol, and purifying the product by distillation, first over sulphuric acid and then over quicklime, anhydrous chloral is obtained. This is a thin, oily liquid, which, mixed with one-eighth its weight of water, forms a hydrate.

Hydrate of chloral occurs in crystalline masses, which attract moisture in a moist atmosphere, and slowly evaporate, without liquefying, in dry air. It is readily soluble in water, alcohol, chloroform, ether, and oils, has an aromatic odor, and a peculiar, pungent taste. It is a valuable hypnotic and antispasmodic. Used in insomnia, hysteria, convulsions, delirium tremens, mania, tetanus, etc. Applied externally, it exerts an antiseptic and deodorizing influence, and is used in skin diseases and as a surgical dressing.

Dose: 1 to 10 grains for children; 10 to 30 for adults. Generally administered in aqueous solution, flavored with an aromatic syrup.

PREPARATIONS.

Syrup of Chloral (Br.).

Take of	Hydrate of chloral	80 grains.
	Distilled water	4 drachms.
	Simple syrup	sufficient.

Dissolve the chloral in the water, and add syrup to make 1 ounce. Dose: ½ to 2 drachms.

Injection of Chloral.

Take of	Chloral	1 to 4 grains.
	Water	1 ounce.

Dissolve. An excellent injection for gonorrhœa. *Bartholow.*

Lotion of Chloral.

Take of	Chloral	20 parts.
	Glycerin	5 parts.
	Bay rum	50 parts.
	Water	to 200 parts.

Mix. Use for pityriasis capitis. *G. H. Fox.*

Mixtures of Chloral.

Take of	Chloral	15 grains.
	Fluid extract of conium seed	15 minims.
	Fluid extract of hyoscyamus	15 minims.
	Water	to 1 drachm.

Mix. One dose; to be taken thrice daily, after meals.

Hospital Formulary.

Take of	Chloral	1 drachm.
	Bromide of potassium	2 drachms.
	Syrup of wild cherry	1 ounce.
	Water	1 ounce.

Mix. Dose: 1 drachm three times a day, for an infant under one year. In the convulsive stage of whooping-cough. *H. S. Dessau.*

Take of	Chloral	3 drachms.
	Sulphate of morphia	4 grains.
	Cherry-laurel water	1 ounce.

Mix. Dose: 15 to 30 minims. For cholera, cholera morbus, etc.

Bartholow.

Take of	Hydrate of chloral	15 grains.
	Bromide of potassium	15 grains.
	Syrup of tolu	½ drachm.
	Water	to 1 ounce.

Mix. One dose. *Middlesex Hospital.*

Take of Chloral .... 20 grains.
Syrup of orange flowers .... 40 minims.
Syrup of tolu .... 40 minims.
Water .... to 1 ounce.

Mix. One dose. *London Throat Hospital.*

CHLOROFORMUM (U. S. et al. Ph.)—CHLOROFORM.

The United States Pharmacopœia recognizes *Chloroformum Cenale* (*Commercial Chloroform*), and from it prepares *Chloroformum Purificatum* (*Purified Chloroform*), while the European Pharmacopœias recognize only the pure article.

Chloroform is prepared by distilling a mixture of alcohol, chlorinated lime, and water. It is, when pure, a heavy, colorless, volatile liquid, of an ethereal odor, and a hot, saccharine taste. It is slightly soluble in water, freely soluble in alcohol and ether, and readily dissolves fats, oils, resins, iodine, bromine, etc.

The effects of chloroform are much like those of ether, but its action is more rapid and powerful. Taken into the stomach it acts as an anodyne and antispasmodic. It is, however, seldom used in this manner at the present day, having been superseded by chloral. Administered by inhalation, it quickly produces anæsthesia, and is largely used for this purpose in surgical and obstetrical operations, hepatic and renal colic, and other painful affections, though it is much more dangerous than ether. It has been used hypodermically in neuralgia, and is often employed externally as a topical anodyne.

Dose: 5 to 60 minims.

PREPARATIONS.

Linimentum Chloroformi (U. S., Br., Fr.)—Liniment of Chloroform.

Take of Purified chloroform (by weight) .... 3 ounces.
Olive oil (by weight) .... 4 ounces.

Mix.

The British Pharmacopœia directs: chloroform, liniment of camphor, each an equal measure; the French Codex: chloroform 1 part, oil of sweet almonds 9 parts.

Mistura Chloroformi (U. S.)—Chloroform Mixture.

Take of Purified chloroform (by weight) .... ½ ounce.
Camphor .... 60 grains.
The yolk of one egg.
Water .... 6 ounces.

Rub the yolk first by itself, then with the camphor dissolved in the chloroform, and lastly, with the water gradually added.

Dose: ½ to 1 ounce.

Spiritus Chloroformi (U. S., Br.)—Spirit of Chloroform.

Take of Purified chloroform (by weight)............. 1 ounce.
Alcohol.................................. 12 ounces.

Dissolve.

Dose : ½ to 1 drachm.

The British Pharmacopœia directs the proportion of 1 to 19, by measure.

Chloroform Water (Br.).

Take of Chloroform................................ 1 drachm.
Distilled water.......................... 25 ounces.

Put them together in a well-stopped bottle, and shake until the chloroform is entirely dissolved.

Dose : ½ to 2 ounces.

Chloroform Ointment (Fr.).

Take of Chloroform................................ 20 parts.
White wax........................... 10 parts.
Lard .... 90 parts.

Melt the wax and lard on a water-bath, in a large-necked bottle, and, when partially cool, add the chloroform, and shake until cold.

Take of Chloroform .... 2 drachms.
Glycerin................................. ½ ounce.
Cerate................................... 1½ ounce.

Mix. For pruritus ani. *J. W. Wright.*

Mixtures of Chloroform.

Take of Spirit of chloroform....................... 15 minims.
Compound tincture of camphor .... 20 minims.
Syrup of squill........................ ½ drachm.
Water................................to 1 ounce.

Mix. One dose. *Middlesex Hospital.*

Take of Chloroform............................ 5 minims.
Camphor.............................. 5 grains.
Mucilage............................ 60 minims.
Water................................to 1 ounce.

Mix. One dose. *Guy's Hospital.*

Take of Spirit of chloroform .... ½ ounce.
Compound tincture of cardamom............ 2 ounces.

Mix. Dose : ½ drachm in ½ ounce of water, every half-hour until relieved. In colic depending on flatulence. *A. A. Smith.*

Take of	Spirit of chloroform	1½ drachm.
	Diluted hydrocyanic acid	1½ drachm.
	Tincture of hyoscyamus	½ ounce.
	Camphor water	to 4 ounces.

Mix. Dose: 1 drachm in ½ ounce of water, every two hours. A cough mixture in phthisis, when it is undesirable to use opiates. *A. A. Smith.*

CHONDRUS (U. S., Fr., Ger.)—IRISH MOSS—CARRAGEEN.

The entire frond of Chondrus crispus, Greville (*Nat. ord., Algæ*), a seaweed growing on the rocks along the shore from New England northward, and also along the coast of Europe. Our present supply comes mainly from Massachusetts. It is demulcent, nutritive, and slightly alterative. Used chiefly in chronic catarrhal affections, whether of the pulmonary or urinary system.

Decoction of Irish Moss.

Take of	Irish moss	½ ounce.
	Water	1½ pint.

Boil to 1 pint, strain, and flavor with lemon juice and sugar.

Dose: 4 to 6 ounces.

CIMICIFUGA (U. S.)—CIMICIFUGA—BLACK SNAKEROOT.

The rhizome of Cimicifuga racemosa, Ell. (*Nat. ord., Ranunculaceæ*), an herbaceous perennial indigenous to North America.

It has a faint, unpleasant odor, a bitter, acrid taste, and acts as a stimulant, antispasmodic, and sedative. As a stimulant to the secretions it is used in chronic bronchitis, phthisis, amenorrhœa, dropsy, etc., while its sedative effect is beneficial in chorea. It is also used in rheumatism, lumbago, after-pains, etc.

Dose: 20 to 60 grains.

PREPARATIONS.

Extractum Cimicifugæ Fluidum (U. S.)—Fluid Extract of Cimicifuga.

Take of	Cimicifuga	16 ounces.
	Stronger alcohol	sufficient.

Moisten the cimicifuga with 4 ounces of the alcohol, and proceed according to the general formula, page 161.

Dose: 20 to 60 minims.

TINCTURE OF CIMICIFUGA.

Take of	Cimicifuga	1 part.
	Diluted alcohol	sufficient.

Moisten, pack, and percolate to 5 parts.

Dose: 1 to 2 drachms. *J. P. Remington, Report of Am. Ph. Ass'n.*

Resin of Cimicifuga—Cimicifugin—Macrotin.

Take of	Cimicifuga	16 parts.
	Alcohol (sp. gr. 0.835)	sufficient.

Moisten, pack, and percolate until 16 parts of tincture are obtained. Evaporate this to a thick, syrupy consistence, pour into warm water, stir well, and allow to settle. Decant the supernatant liquid, wash the precipitate with successive portions of warm water until the washings are nearly tasteless and colorless; then cool the resinous mass, break it into small pieces, dry by exposure to the air, and powder it.

It is of a light, yellowish brown color when powdered, and has a peculiar, smoky odor, and a sweetish taste.

Dose: ½ to 2 grains. *J. U. Lloyd.*

MIXTURE OF CIMICIFUGA.

Take of	Water	4 ounces.
	Syrup of orange peel	2½ ounces.
	Tincture of cinnamon	3 drachms.
	Tincture of nux vomica	1 drachm.
	Fluid extract of cimicifuga	1 ounce.
	Gallic acid	2 drachms.

Mix. Dose: ½ ounce every third hour, in a little water, commencing the day before the normal end of the menstrual period. For uterine and ovarian neuralgia, associated with prolonged menstruation that is continued four or five days beyond the normal period of the individual.

Fordyce Barker.

CINCHONA (U. S. et al. Ph.)—CINCHONA—PERUVIAN BARK.

The bark of all species of cinchona (*Nat. ord.*, *Rubiaceæ*), containing at least two per cent. of the proper cinchona alkaloids, which yield crystallizable salts.

CINCHONA FLAVA (U. S. et al. Ph.)—YELLOW CINCHONA.

The bark of Cinchona Calisaya, Weddell.

CINCHONA PALLIDA (U. S. et al. Ph.)—PALE CINCHONA.

The bark of Cinchona officinalis, Hooker.

CINCHONA RUBRA (U. S. et al. Ph.)—RED CINCHONA.

The bark of Cinchona succirubra, Pavon.

In addition to the above-named, several other species of cinchona contribute to supply the market with this valuable drug. All the species of the genus are trees or shrubs indigenous to South America, but some of them are now successfully cultivated in Java and Ceylon.

Cinchona is the most valuable of vegetable tonics and febrifuges. As a tonic, it is employed in nearly all cases of debility; as a febrifuge, at one time or another, in nearly all cases of fever, but especially in those of a miasmatic origin. Its power over intermittent fever is so remarkable as to entitle it to be considered a specific; and it is in this disease, particularly, that its effects are most strikingly displayed. But in other fevers, and in acute inflammatory affections, as pneumonia and rheumatism, in full doses, it lowers the pulse, reduces the temperature, and produces, at least, a temporary lull in the progress of the disease.

These effects are due to the presence of a number of alkaloids, which the bark contains, the most important being, in the order of their relative value, *quinia, quinidia, cinchonia, cinchonidia.* When the most decided effects of cinchona are desired, these alkaloids, or rather, some of their salts, are generally employed; while for the simple tonic effects, the bark, or some of its preparations, may be used.

Dose: 10 to 60 grains.

PREPARATIONS.

Decoctum Cinchonæ Flavæ (U. S., Br.)—Decoction of Yellow Cinchona.

Take of Yellow cinchona .... 1 ounce.
Water.................................... sufficient.

Boil fifteen minutes, strain, and add sufficient water through the strainer to make 1 pint.

Dose: 1 to 2 ounces.

Decoctum Cinchonæ Rubræ (U. S.)—Decoction of Red Cinchona.

Made in the same manner as the preceding preparation.

Extractum Cinchonæ (U. S., Fr., Ger.)—Extract of Cinchona.

Take of Yellow cinchona .... 12 ounces.
Alcohol.................................... 3 pints.
Water.................................... sufficient.

Macerate the cinchona in 20 ounces of the alcohol for four days, then percolate with the remainder, continuing the process with water until 3 pints are obtained. Reserve this, and continue the percolation with water until 6 pints more are obtained. Distil off the alcohol from the tincture, evaporate the residue and the infusion separately, to the consistence of thin honey, then mix, and evaporate to a proper consistence.

Dose: 10 to 30 grains.

Extractum Cinchonæ Fluidum (U. S., Br.)—Fluid Extract of Cinchona.

Take of		
	Yellow cinchona	16 ounces.
	Glycerin	4 ounces.
	Alcohol	sufficient.
	Water	sufficient.

Mix 8 ounces of alcohol, 3 of glycerin, and 5 of water, moisten the cinchona with 5 ounces of the mixture, and proceed according to the general formula, page 161. Continue the percolation with diluted alcohol until 2 pints are obtained, reserve 14 ounces, and add 1 ounce of glycerin to the remainder before evaporation.

Dose: 10 to 60 minims, or more, as an antiperiodic.

Liquid Extract of Yellow Cinchona (Br.).

Infusum Cinchonæ Flavæ (U. S., Br.)—Infusion of Yellow Cinchona.

Take of		
	Yellow cinchona	1 ounce.
	Aromatic sulphuric acid	1 drachm.
	Water	sufficient.

Mix the acid with 1 pint of water, moisten, pack, and percolate the cinchona with the mixture, continuing with the water until the filtered liquid measures 1 pint.

Dose: 1 to 2 ounces.

The British Pharmacopœia directs: cinchona ½ ounce; water, 10 ounces.

Infusum Cinchonæ Rubræ (U. S.)—Infusion of Red Cinchona.

Made in the same manner as the preceding preparation.

Tinctura Cinchonæ (U. S. et al. Ph.)—Tincture of Cinchona.

Take of		
	Yellow cinchona	6 ounces.
	Alcohol, 3 parts—Water, 1 part	sufficient.

Moisten, pack, and percolate to 2 pints.

Dose: ½ to 4 drachms.

Tinctura Cinchonæ Composita (U. S., Br., Ger.)—Compound Tincture of Cinchona.

Take of	Red cinchona	4 ounces.
	Bitter orange peel	3 ounces.
	Serpentaria	360 grains.
	Alcohol	sufficient.
	Water	sufficient.

Moisten, pack, and percolate to 2½ pints.

Dose: ½ to 4 drachms.

The British Pharmacopœia directs: pale cinchona, 2 ounces; bitter-orange peel, 1 ounce; serpentaria, ½ ounce; saffron, 60 grains; cochineal, 30 grains; proof spirit, 20 ounces; the German Pharmacopœia: pale cinchona, 6 parts; orange-peel, 2 parts; gentian, 2 parts; cinnamon, 1 part; diluted alcohol, 50 parts.

MIXTURE OF CINCHONA AND RHUBARB.

Take of	Compound tincture of cinchona	1½ ounce.
	Mixture of rhubarb and soda	1½ ounce.
	Aromatic syrup of rhubarb	½ ounce.

Mix. Dose: ½ ounce at night, or night and morning.

A tonic laxative, useful in gastric derangements and vertigo. The mixture of rhubarb and soda contains 15 grains of each in an ounce of peppermint water. *F. A. Burrall.*

CINCHONIÆ SULPHAS (U. S., Br., Fr.)—SULPHATE OF CINCHONIA.

Take of the mother-water remaining after the crystallization of the sulphate of quinia, a convenient quantity. Add solution of soda until the liquid is alkaline, collect, wash and dry the precipitate. Then wash it with successive small portions of alcohol to remove other alkaloïds, mix the residue with eight times its weight of water, heat, add diluted sulphuric acid until it becomes clear, boil with animal charcoal, filter while hot, and set aside to crystallize. Dry the crystals on bibulous paper.

It is in white, shining crystals, of a bitter taste, and soluble in 54 parts of water.

Sulphate of cinchonia is used in the same manner, and for the same purposes, as sulphate of quinia. It has the advantage of being more soluble and less bitter, but is less active than sulphate of quinia, and should be administered in doses one-third larger.

Dose: 1 to 20 grains.

Mixture of Sulphate of Cinchonia.

Take of Sulphate of cinchonia........................ 1 drachm.
Tincture of chloride of iron.................. 2 drachms.
Water ....to 4 ounces.

Mix. Dose: 1 drachm. *Hospital Formulary.*

Pills of Sulphate of Cinchonia.

Take of Sulphate of cinchonia........................ 1 drachm.
Extract of cinchona........................ sufficient.

Mix and divide into 20 pills.

Dose: 1 to 6 pills.

Capsules of Sulphate of Cinchonia.

Take of Sulphate of cinchonia........................ ½ drachm.
Carbonate of ammonium .... 15 grains.
Camphor .... 8 grains.
Sulphate of morphia........................ ½ grain.

Mix and divide into 4 capsules.

Dose: 1 capsule at bedtime. In acute bronchial catarrh.

Daniel Lewis.

CINCHONIDIÆ SULPHAS—SULPHATE OF CINCHONIDIA.

Cinchonidia exists only in certain varieties of cinchona, and the sulphate is obtained from the mother-water remaining from the crystallization of sulphate of quinia, when manufactured from these barks. It is in white, acicular crystals, resembling those of sulphate of quinia, of a bitter taste, and soluble in 85 parts of water. Used as a tonic and antiperiodic.

Dose: 1 to 20 grains.

CINNAMOMUM (U. S. et al. Ph.)—CINNAMON.

The inner bark of Cinnamomum Zeylanicum, Breyne (*Nat. ord., Lauraceæ*), a small evergreen tree indigenous to Ceylon, where it is largely cultivated, and attains its greatest perfection. It is also cultivated in India and other tropical countries, but the product is inferior.

Cinnamon is an aromatic stimulant, and, owing to its agreeable taste, is largely used as a flavoring agent.

Dose: 10 to 20 grains.

PREPARATIONS.

Aqua Cinnamomi (U. S. et al. Ph.)—Cinnamon Water.

Take of Oil of cinnamon	½ drachm.
Carbonate of magnesium	60 grains.
Distilled water	2 pints.

Rub the oil with the carbonate of magnesium, then with the water, added gradually, and filter. Or, mix 18 ounces of cinnamon with 16 pints of water, and distil 8 pints.

The European Pharmacopœias employ the latter process. Used as a vehicle.

Oleum Cinnamomi (U. S. et al. Ph.)—Oil of Cinnamon.

The oil distilled from cinnamon bark.

Dose: 1 to 2 minims.

Spiritus Cinnamomi (U. S., Fr.)—Spirit of Cinnamon.

Take of Oil of cinnamon	1 ounce.
Stronger alcohol	15 ounces.

Dissolve.

Dose: 10 to 30 minims.

The French Codex directs: cinnamon, 1 part; alcohol 8 parts; distil off the spirit.

Tinctura Cinnamomi (U. S. et al. Ph.)—Tincture of Cinnamon.

Take of Cinnamon	3 ounces.
Alcohol, 2 parts—Water, 1 part	sufficient.

Moisten, pack, and percolate to 2 pints.

Dose: 1 to 3 drachms.

Pulvis Aromaticus (U. S., Br., Ger.)—Aromatic Powder.

Take of Cinnamon	2 ounces.
Ginger	2 ounces.
Cardamom	1 ounce.
Nutmeg	1 ounce.

Rub together into a fine powder.

Dose: 10 to 30 grains.

The British Pharmacopœia directs: equal parts of cinnamon, ginger, and cardamom; the German Pharmacopœia: cinnamon 5 parts; cardamom 3 parts; ginger 2 parts.

Confectio Aromatica (U. S.)—Aromatic Confection.

Take of Aromatic powder	4 ounces.
Clarified honey (by weight)	4 ounces.

Rub together.

SYRUP OF CINNAMON (Fr., Ger.).

Take of Cinnamon water .... 50 parts.
Sugar .... 95 parts.

Dissolve by agitation, without heat, and filter.

COCCUS (U. S. et al. Ph.)—COCHINEAL.

Cochineal is a small insect (*Coccus cacti*, L.), indigenous to Mexico and Central America, which feeds upon a species of cactus.

Though formerly considered medicinal, it is now only used to color medicinal preparations.

PREPARATION.

TINCTURE OF COCHINEAL (Br., Fr.).

Take of Cochineal .... 2½ ounces.
Proof spirit .... 20 ounces.

Macerate for seven days, strain, press, filter, and add sufficient proof spirit to make 20 ounces.

CODEIA (Fr., Ger.)—CODEIA.

Codeia is an alkaloid which exists in opium, in combination with meconic acid, and is separated in the process for obtaining morphia.

It occurs in colorless crystals, of a bitter taste, and possessing narcotic properties. It is said to produce less unpleasant after-effects than opium or morphia.

Dose: ½ to 3 grains.

COLCHICUM—MEADOW SAFFRON.
COLCHICI RADIX (U. S., Br., Fr.)—COLCHICUM ROOT.
COLCHICI SEMEN (U. S. et al. Ph.)—COLCHICUM SEED.

The corm and seed of Colchicum autumnale, L. (*Nat. ord., Melanthaceæ*), a bulbous perennial indigenous to Europe. Its flowers resemble those of the crocus, and are produced in autumn, while the seed is not matured until the following summer.

Colchicum stimulates the secretions, and, in full doses, acts as an emeto-cathartic; in overdoses it is a dangerous poison. Its stimulant effect is followed by a sedative action, which is especially beneficial in rheumatism and gout. In this latter disease, particularly, large doses of colchicum produce the happiest effect, often relieving pain and reducing the inflammation very quickly.

Dose: Of the root, 2 to 8 grains; of the seed, 2 to 10 grains.

PREPARATIONS.

Extractum Colchici Aceticum (U. S., Br.)—Acetic Extract of Colchicum.

Take of	Colchicum root	12 ounces.
	Acetic acid	4 ounces.
	Water	sufficient.

Mix the acid with 1 pint of water, add the colchicum, pack, and percolate with water until the root is exhausted. Evaporate the liquid on a water-bath, to a proper consistence.

Dose: 1 to 2 grains.

The British preparation is made from the fresh corm. Another British preparation, *extract of colchicum*, is made by evaporating the juice.

Extractum Colchici Radicis Fluidum (U. S.)—Fluid Extract of Colchicum Root.

Take of	Colchicum root	16 ounces.
	Glycerin	4 ounces.
	Alcohol	sufficient.
	Water	sufficient.

Mix 12 ounces of alcohol, 3 of glycerin, and 1 of water, moisten the colchicum with 5 ounces of the mixture, and proceed according to the general formula, page 161. Finish the percolation with diluted alcohol, reserve 14 ounces, and add 1 ounce of glycerin to the remainder, before evaporation.

Dose: 2 to 8 minims.

Extractum Colchici Seminis Fluidum (U. S.)—Fluid Extract of Colchicum Seed.

Prepared in the same manner as the preceding preparation.

Dose: 2 to 10 minims.

Tinctura Colchici (U. S. et al. Ph.)—Tincture of Colchicum.

Take of	Colchicum seed	4 ounces.
	Diluted alcohol	sufficient.

Moisten, pack, and percolate to 2 pints.

Dose: 15 to 60 minims.

Vinum Colchici Radicis (U. S., Br., Fr.)—Wine of Colchicum Root.

Take of Colchicum root .... 12 ounces.
Sherry wine .... sufficient.

Moisten, pack, and percolate to 2 pints.

Dose: 10 to 20 minims.

The British Pharmacopœia directs the proportion of 1 to 5; the French Codex, 3 to 50.

Vinum Colchici Seminis (U. S., Fr., Ger.)—Wine of Colchicum Seed.

Take of Colchicum seed .... 4 ounces.
Sherry wine .... 2 pints.

Macerate for seven days, express and filter.

The French Codex directs the proportion of 3 to 50; the German Pharmacopœia, 1 to 10.

ALCOHOLIC EXTRACT OF COLCHICUM SEED (Fr.).

Take of Colchicum seed .... 1 part.
Alcohol (60%) .... 6 parts.

Digest the seed in half of the alcohol, express and filter. Treat the residue in like manner with the remainder of the alcohol, mix the liquids, distil off the spirit, concentrate on a water bath, dissolve in four times its weight of cold distilled water, filter, and evaporate to a thick extract.

Dose: ½ to 2 grains.

TINCTURE OF COLCHICUM ROOT (Fr.).

Take of Colchicum root .... 1 part.
Alcohol (60%) .... 5 parts.

Macerate for ten days, express and filter.

Dose: 10 to 50 minims.

VINEGAR OF COLCHICUM (Ger., Fr.).

Take of Colchicum seed .... 1 part.
Alcohol .... 1 part.
Pure vinegar .... 9 parts.

Digest for eight days, express and filter.

Dose: ½ to 1 drachm.

The French Codex directs: colchicum root, 1 part; white vinegar, 12 parts.

MIXTURES OF COLCHICUM.

Take of Wine of colchicum seed .... 3 drachms.
Aromatic spirit of ammonia .... 13 drachms.

Mix. Dose: 1 drachm every three hours, until some physiological effect is produced. In gout. *Bartholow.*

Take of Wine of colchicum seed .... ½ ounce.
Solution of acetate of ammonium............ 2½ ounces.
Infusion of parsley .... 5 ounces.

Mix. Dose : ½ ounce every four hours. In dropsy. *Bartholow.*

Take of Tincture of colchicum seed.................. 6 minims.
Bicarbonate of potassium................... 5 grains.
Pimento water............................. 1 ounce.

Mix. One dose. *British Skin Hospital.*

Take of Tincture of colchicum seed................. 15 minims.
Carbonate of magnesium................... 6 grains.
Sulphate of magnesium..................... 30 grains.
Peppermint water..........................to 1 ounce.

Mix. One dose. *University College Hospital.*

Pills of Colchicum.

Take of Acetic extract of colchicum................. 1 grain.
Extract of belladonna...................... ½ grain.
Extract of chamomile...................... 2 grains.

Make 1 pill. *St. Thomas's Hospital.*

Take of Acetic extract of colchicum.................. 2 grains.
Dover's powder............................ 3 grains.

Make 1 pill. *Middlesex Hospital.*

Take of Acetic extract of colchicum.:................ 1 grain.
Blue pill.................................. 2 grains.
Extract of gentian......................... 2 grains.

Make 1 pill. *Charing Cross Hospital.*

Take of Acetic extract of colchicum................. 1 grain.
Sulphate of quinia......................... 1 grain.

Make 1 pill. *London Ophthalmic Hospital.*

COLOCYNTHIS (U. S. et al. Ph.)—COLOCYNTH.

The pulp of the fruit of Citrullus Colocynthis, Royle (*Nat. ord., Cucurbitaceæ*), an herbaceous vine, with perennial root, indigenous to Asia and Africa. The fruit resembles an orange in size and general appearance, but has a hard rind.

Colocynth is a drastic hydragogue. On account of its extremely violent action, it is seldom used alone.

Dose : 5 to 10 grains.

PREPARATION.

Extractum Colocynthidis (U. S., Fr., Ger.)—Extract of Colocynth.

Take of	Colocynth	48 ounces.
	Diluted alcohol	sufficient.

Macerate the colocynth in 8 pints of the alcohol for four days, and express. Then percolate the residue with diluted alcohol until the tincture and expressed liquids, taken together, measure 16 pints. Mix them, distil off 10 pints of alcohol, and evaporate the residue to dryness on a water-bath.

Dose: 5 to 20 grains.

Extractum Colocynthidis Compositum (U. S., Br., Ger.)—Compound Extract of Colocynth.

Take of	Extract of colocynth	$3\frac{1}{2}$ ounces.
	Purified aloes	12 ounces.
	Resin of scammony	3 ounces.
	Cardamom	$1\frac{1}{2}$ ounce.
	Soap	3 ounces.

Mix. Dose: 5 to 30 grains.

The British preparation is made by macerating colocynth, together with the other ingredients, in proof spirit, and evaporating the tincture to a pilular consistence.

Pilulæ Catharticæ Compositæ (U. S.)—Compound Cathartic Pills.

Take of	Compound extract of colocynth	32 grains.
	Extract of jalap	24 grains.
	Mild chloride of mercury	24 grains.
	Gamboge	6 grains.

Mix the powders, form a mass with water, and divide into 24 pills.

Dose: 1 to 4 pills.

Compound Pill of Colocynth (Br., Fr.).

Take of	Colocynth	1 ounce.
	Barbadoes aloes	2 ounces.
	Scammony	2 ounces.
	Sulphate of potassium	$\frac{1}{4}$ ounce.
	Oil of cloves	2 drachms.
	Distilled water	sufficient.

Mix the powders, add the oil of cloves, and beat into a mass with water.

Dose: 5 to 10 grains.

Pill of Colocynth and Hyoscyamus (Br.).

Take of	Compound pill of colocynth	2 ounces.
	Extract of hyoscyamus	1 ounce.

Beat together.

Dose: 5 to 10 grains.

Tincture of Colocynth (Ger.).

Take of Colocynth.................................. 1 part.
        Alcohol.................................. 10 parts.

Prepare a tincture by maceration.

Dose: 15 to 30 minims.

Pills of Colocynth.

Take of Compound colocynth pill.................. 2 grains.
        Blue pill.................................. 1½ grain.
        Ipecacuanha.............................. ⅛ grain.
        Extract of henbane....................... 1 grain.

Make 1 pill. *University College Hospital.*

Take of Compound colocynth pill.................. 4 grains.
        Croton oil................................ ½ minim.

Make 1 pill. *King's College Hospital.*

Take of Compound extract of colocynth .... 30 grains.
        Resin of podophyllum .... 2 grains.
        Extract of nux vomica .... 3 grains.

Mix, and divide into 10 pills.

Dose: 1 pill at bedtime. In habitual constipation.

COLLODIUM (U. S. et al. Ph.)—COLLODION.

Take of Pyroxylon..................................200 grains.
        Stronger ether..............................12½ ounces.
        Alcohol................................ 3½ ounces.

Mix the ether and alcohol in a suitable bottle, add the pyroxylon, and dissolve.

Used as a dressing for abrasions, wounds, etc., and as a vehicle for vesicants and irritants.

Collodium Flexile (U. S., Br., Ger.)—Flexible Collodion.

Take of Collodion .... 1 pint.
        Canada turpentine...........................320 grains.
        Castor oil...................................160 grains.

Mix.

See also *Carbolized Collodion.*

CONIUM—POISON HEMLOCK.
CONII FRUCTUS (U. S., Br., Fr.)—CONIUM SEED.
CONII FOLIA (U. S. et al. Ph.)—CONIUM LEAVES.

The unripe fruit and leaves of Conium maculatum, L. (*Nat. ord., Umbelliferæ*) a tall, biennial herb indigenous to the Old World, but naturalized here.

Though the seed and leaves are recognized by the United States Pharmacopœia, the latter are generally considered unreliable except when fresh.

Conium is a powerful sedative of the centres of motion, and, in sufficient doses, paralyzes them. Its action is, in fact, directly the reverse of that of strychnia. It is used in tetanus, chorea, epilepsy, spasmodic croup, acute mania, and other nervous and mental diseases.

Too great stress cannot be laid upon the necessity of using an efficient preparation of this plant. Its active principle, *conia*, is volatile, and is lost from the leaves and seed by long keeping.

Dose: of the fruit, ½ to 4 grains; of the leaves, 3 to 8 grains.

PREPARATIONS.

Extractum Conii (U. S. et al. Ph.)—Extract of Conium.

Take of Conium leaves, fresh........................ 12 ounces.

Bruise the leaves, sprinkling on a little water, and express the juice; heat this to the boiling point, filter, and evaporate to a proper consistence, either in a vacuum with the aid of heat, or in shallow vessels, at the ordinary temperature, by means of a current of air directed over the surface of the liquid.

Dose: 1 to 3 grains.

An unreliable preparation, often wholly inert.

Extractum Conii Alcoholicum (U. S., Fr.)—Alcoholic Extract of Conium.

Take of Conium leaves, recently dried................ 12 ounces.
Alcohol.................................... 1 pint.
Diluted alcohol.............................. sufficient.

Moisten the conium with the alcohol, then percolate with diluted alcohol until 1 pint of tincture is obtained; allow this to evaporate spontaneously to 3 ounces, continue the percolation until 2 pints more are obtained or the conium is exhausted; evaporate this, at or below 160°, to the consistence of syrup, add the 3 ounces of reserved liquid, and continue the evaporation, at or below 120°, until reduced to a proper consistence.

Dose: ½ to 2 grains.

Scarcely more reliable than the preceding.

The French Codex prepares also an alcoholic extract of conium seed.

Extractum Conii Fructus Fluidum (U. S.)—Fluid Extract of Conium Seed.

Take of Conium seed .... 16 ounces.
Glycerin .... 4 ounces.
Hydrochloric acid ....180 grains.
Alcohol .... sufficient.
Water .... sufficient.

Mix 8 ounces of alcohol, 3 of glycerin, and 5 of water, moisten the conium with 4 ounces of the mixture, and proceed according to the general formula, page 161. Finish the percolation with diluted alcohol, reserve 14 ounces, and add the acid and one ounce of glycerin to the remainder before evaporation.

Dose: 3 to 5 minims.

Succus Conii (U. S., Br., Fr.)—Juice of Conium.

Take of Fresh conium leaves .... a convenient quantity.
Alcohol .... sufficient.

Bruise the leaves, express, and to every 5 measures of juice add 1 measure of alcohol. After seven days, filter.

Dose: 30 to 60 minims.

The British Pharmacopœia directs the proportion of 3 to 1. The French Codex makes a similar preparation by macerating the fresh leaves in an equal weight of alcohol for ten days.

Tinctura Conii (U. S., Br., Fr.)—Tincture of Conium.

Take of Conium leaves, recently dried.............. 4 ounces.
Diluted alcohol .... sufficient.

Moisten, pack, and percolate to 2 pints.

Dose: ½ to 2 drachms.

The British Pharmacopœia directs: conium seed, 2½ ounces; proof spirit, 20 ounces; the French Codex: conium leaves, 1 part; alcohol (60%), 5 parts. The French Codex prepares also an ethereal tincture.

Conium Ointment (Ger.).

Take of Extract of conium.......................... 1 part.
Wax ointment .... 9 parts.

Mix.

Conium Plaster (Fr., Ger.).

Take of Alcoholic extract of conium.................. 90 parts.
Elemi.............................. 20 parts.
White wax.............................. 10 parts.

Melt the elemi and wax together, then add the extract of conium.

Hemlock Poultice (Br.).

Take of Hemlock leaves........................... 1 ounce.
Linseed meal........................... 3 ounces.
Boiling water........................... 10 ounces.

Mix the hemlock and meal, and add them to the water gradually, with constant stirring.

Mixtures of Conium.

Take of Extract of conium........................... 5 grains.
Carbonate of sodium........................... 7½ grains.
Spirit of pimento........................... 30 minims.
Decoction of liquorice...........................to 1 ounce.

Mix. For a dose, three or four times daily. *Guy's Hospital.*

Take of Juice of conium........................... 30 minims.
Extract of henbane........................... 3 grains.
Mucilage........................... 2 drachms.
Water...........................to 1 ounce.

Mix. One dose. *Royal Chest Hospital.*

Pills of Conium.

Take of Extract of conium........................... 4 grains.
Powder of ipecacuanha........................... 1 grain.

Make 1 pill. *Guy's Hospital.*

Take of Extract of conium........................... 4 grains.
Ipecacuanha........................... ½ grain.
Hydrochlorate of morphia........................... ⅛ grain.

Make 1 pill. *Brompton Consumption Hospital.*

Take of Extract of conium........................... 3 grains.
Sulphate of zinc........................... 2 grains.

Make 1 pill. *London Chest Hospital.*

COPAIBA (U. S. et al. Ph.)—COPAIBA—BALSAM OF COPAIVA.

The oleoresin obtained from Copaifera officinalis, L., and other species of copaiferà (*Nat. ord., Leguminosæ*), large trees indigenous to the warmer regions of South America.

Copaiba is a clear, transparent liquid of a yellowish color, a characteristic odor, and a disagreeable, nauseous taste. It is a stimulant whose effect is chiefly upon the mucous membranes, and particularly those of the genito-urinary organs. Used in diseases of the mucous membranes, as chronic bronchitis, gonorrhœa, gleet, vesical irritation, etc.

Dose: 10 to 40 minims or grains.

PREPARATIONS.

Oleum Copaibæ (U. S., Br.)—Oil of Copaiba.

The oil obtained from copaiba by distillation.

Dose: 10 to 15 minims.

Pilulæ Copaibæ (U. S., Fr.)—Pills of Copaibæ.

Take of Copaiba	2 ounces.
Magnesia, recently prepared	60 grains.

Mix, allow the mass to concrete, and divide into 200 pills.

Dose: 2 to 6 pills.

Electuary of Copaiba, Cubebs, and Catechu (Fr.).

Take of Copaiba	1 part.
Cubeb	1 part.
Catechu	1 part.

Mix the copaiba and catechu, then add the cubebs.

Dose: ½ to 1 drachm.

Mixtures of Copaiba.

Take of Copaiba	1 ounce.
Solution of potassa	2 drachms.
Extract of liquorice	½ ounce.
Spirit of nitrous ether	1 ounce.
Syrup of gum arabic	6 ounces.
Oil of wintergreen	16 drops.

Mix the copaiba and solution of potassa, and the liquorice and spirit of nitrous ether separately, then unite, and add the other ingredients.

Dose: ½ ounce after meals. In gonorrhœa. *Bumstead.*

Take of Oil of copaiba	1 drachm.
Oil of cubebs	1 drachm.
Alum	2 drachms.
White sugar	4 drachms.
Mucilage	3 drachms.
Water	2 ounces.

Mix. Dose: 1 drachm three times a day. In gonorrhœa.

Bumstead.

Take of Copaiba	10 drachms.
Tincture of cantharides	2 drachms.
Tincture of chloride of iron	2 drachms.

Mix. Dose: ½ to 1 drachm. In gonorrhœa. *Bumstead.*

Take of Copaiba................................ 15 minims.
Mucilage................................ 2 drachms.
Water................................to 1 ounce.

Mix. One dose. *St. Thomas's Hospital.*

Take of Copaiba................................ 20 minims.
Solution of potassa........................ 20 minims.
Tincture of opium........................ 5 minims.
Peppermint water........................ 1 ounce.

Mix. One dose. *Great Northern Hospital.*

Take of Copaiba................................ 20 minims.
Mucilage................................ 40 minims.
Diluted sulphuric acid.................... 10 minims.
Acid infusion of roses....................to 1 ounce.

Mix. One dose. *St. George's Hospital.*

Take of Copaiba................................ 25 parts.
Syrup of tolu................................ 25 parts.
Syrup of gum arabic........................ 25 parts.
Cinnamon water.............................. 25 parts.

Mix. Dose: 1 or 2 drachms three times a day, between meals.

G. H. Fox.

CUPRUM—COPPER.

Metallic copper is not used medicinally.

CUPRI SUBACETAS (U. S., Fr., Ger.)—SUBACETATE OF COPPER—VERDIGRIS.

Prepared by exposing copper to the action of acetic vapors. It occurs in bluish green masses, of an acetic odor, and a disagreeable, coppery taste. In small doses, verdigris is astringent and tonic; in overdoses a violent corrosive poison; applied topically, it is a mild escharotic, and is occasionally used in venereal ulcers and warts.

Dose: $\frac{1}{8}$ to $\frac{1}{4}$ grain.

PREPARATION.

Verdigris Plaster (Fr.).

Take of Subacetate of copper........................ 1 part.
Turpentine................................ 1 part.
Burgundy pitch................................ 2 parts.
Yellow wax................................ 4 parts.

Rub the copper and turpentine together, and add to the pitch and wax previously melted.

Said to be a good application for corns and warts.

CUPRI SULPHAS (U. S. et al. Ph.)—SULPHATE OF COPPER—BLUE VITRIOL.

Obtained on a large scale from copper pyrites, and may be prepared by dissolving copper in hot sulphuric acid. It occurs in beautiful blue crystals, readily soluble in water, and having an acid, styptic taste.

In small doses it is tonic and astringent; in large doses, a prompt and efficient emetic. It is employed topically as an astringent and mild escharotic.

Dose: $\frac{1}{8}$ to $\frac{1}{4}$ grain, tonic; 3 to 5 grains, emetic.

Collyrium of Sulphate of Copper.

Take of Sulphate of copper........................ 2 grains.
Water................................ 1 ounce.

Dissolve. Stimulant and astringent. Used in a number of the hospitals of this country and Europe.

Injections of Sulphate of Copper.

Take of Sulphate of copper........................ 12 grains.
Water................................ 4 to 6 ounces.

Dissolve. Used in gonorrhœa. *Bumstead.*

Take of Sulphate of copper........................ 4 grains.
Sulphate of morphia........................ 8 grains.
Solution of subacetate of lead............... 1 drachm.
Rose water........................ 4 ounces.

Mix. Used in gonorrhœa after the acute stage. *Bartholow.*

Mixtures of Sulphate of Copper.

Take of Sulphate of copper........................ $\frac{1}{2}$ grain.
Sulphate of magnesium..................... 1 ounce.
Diluted sulphuric acid..................... 1 drachm.
Water................................ 4 ounces.

Mix. Dose: $\frac{1}{2}$ ounce every four hours. In dysentery. *Bartholow.*

Pills of Sulphate of Copper.

Take of Sulphate of copper........................ 1 grain.
Sulphate of morphia........................ 1 grain.
Sulphate of quinia........................ 24 grains.

Mix, and make 24 pills.

Dose: 1 pill three times a day. In chronic diarrhœa and chronic dysentery. *Bartholow.*

Take of Sulphate of copper........................ ¼ grain.
Opium........................... ½ grain.
Extract of gentian .... 2 grains.

Make 1 pill. Dose: 1 or 2 pills. *Brompton Consumption Hospital.*

Take of Sulphate of copper .... ½ grain.
Extract of opium.......................... ½ grain.

Make 1 pill. *Royal Free Hospital.*

CUPRUM AMMONIATUM (U. S.)—AMMONIATED COPPER.

Take of Sulphate of copper .... ½ ounce.
Carbonate of ammonium ....360 grains.

Rub together, wrap in bibulous paper, and dry with a gentle heat.

This salt has a deep blue color, an ammoniacal odor, and is freely soluble in water. It has been used with asserted benefit in epilepsy, chorea, and other nervous diseases.

Dose: ¼ to ½ grain.

CORIANDRUM (U. S. et al. Ph.)—CORIANDER.

The fruit of Coriandrum sativum, L. (*Nat. ord., Umbelliferæ*), an annual herb indigenous to the Mediterranean and Caucasian regions, but naturalized throughout Europe.

Coriander is an aromatic stimulant and carminative. Used as an adjunct to other medicines.

Dose: 20 to 60 grains.

PREPARATIONS.

Oil of Coriander (Br.).

The oil obtained from coriander by distillation.

Dose: 2 to 10 minims. Chiefly used for flavoring.

Spirit of Coriander (Fr.).

Take of Coriander................................ 1 part.
Alcohol (80%)............................. 8 parts.

Macerate two days, and distil off the spirit.

CORNUS FLORIDA (U. S.)—DOGWOOD.

The bark of Cornus florida, L. (*Nat. ord.*, *Cornaceæ*), the flowering dogwood of North America. The bark of Cornus sericea, L., and Cornus circinata, L'Her., though not officinal, possesses similar properties.

Dogwood possesses bitter tonic properties similar to those of cinchona. It was formerly much used as an antiperiodic in intermittent fever, and often with success. It may be employed with benefit as a tonic in convalescence from acute diseases.

Dose : ½ to 1 drachm.

PREPARATIONS.

Decoctum Cornus Floridæ (U. S.)—Decoction of Dogwood.

Take of Dogwood	1 ounce.
Water	sufficient.

Boil the dogwood in 1 pint of water for fifteen minutes, strain, and add sufficient water through the strainer to make 1 pint.

Dose : 1 to 2 ounces.

Extractum Cornus Floridæ Fluidum (U. S.)—Fluid Extract of Dogwood.

Take of Dogwood	16 ounces.
Glycerin	4 ounces.
Alcohol	sufficient.
Water	sufficient.

Mix 8 ounces of alcohol, 3 of glycerin, and 5 of water, moisten the dogwood with 5 ounces of the mixture, and proceed according to the general formula, page 161. Finish the percolation with diluted alcohol, reserve 14 ounces, and add 1 ounce of glycerin to the remainder before evaporation.

Dose : ½ to 1 drachm.

CREASOTUM (U. S. et al. Ph.)—CREASOTE.

Creasote is one of the products of the destructive distillation of wood. It is a colorless, oily, neutral liquid, of a strong, smoky odor, and a very caustic taste. Its medicinal effects and uses are similar to those of carbolic acid, which has superseded it to a very great extent. Often applied topically for the relief of toothache, and occasionally administered internally in obstinate vomiting.

Dose : 1 to 3 minims, largely diluted.

PREPARATIONS.

Aqua Creasoti (U. S., Ger.)—Creasote Water.

Take of Creasote	1 drachm.
Distilled water	1 pint.

Mix, agitate till the creasote is dissolved, and filter.

Dose: 1 to 4 drachms.

Inhalation of Creasote (Br.).

Take of Creasote	12 minims.
Boiling water	8 ounces.

Mix, and inhale the vapor through a suitable apparatus.

Creasote Mixture (Br.).

Take of Creasote	16 minims.
Glacial acetic acid	16 minims.
Spirit of juniper	½ drachm.
Syrup	1 ounce.
Distilled water	15 ounces.

Mix the creasote and acid, then add the water, and lastly the other ingredients.

Dose: 1 to 2 ounces.

Ointment of Creasote (Br.).

Take of Creasote	1 drachm.
Simple ointment	1 ounce.

Mix.

Bath of Creasote.

Take of Creasote	¼ ounce.
Glycerin	2 ounces.
Water	30 gallons.

Mix. *British Skin Hospital.*

Mixture of Creasote.

Take of Creasote	1 minim.
Tincture of opium	2 minims.
Spirit of chloroform	15 minims.
Glycerin	1 drachm.
Water	to one ounce.

Mix. One dose. *Royal Chest Hospital.*

Pill of Creasote.

Take of Creasote	1 minim.
Compound asafetida pill	2 grains.
Compound rhubarb pill	2 grains.

Make 1 pill. *St. Bartholomew's Hospital.*

CROCUS (U. S. et al. Ph.)—SAFFRON.

The stigmas of Crocus sativus, L. (*Nat. ord., Iridaceæ*), a bulbous plant indigenous to Greece and Asia Minor, but so long cultivated as to have become naturalized in many other countries.

Saffron is stimulant and antispasmodic. Formerly very much used, it has latterly fallen into disrepute with physicians, though it is still largely employed as a popular remedy. Hot decoctions are often administered to hasten the eruption of exanthematous diseases, to allay the pains of dysmenorrhœa, and to favor the return of the menses in amenorrhœa. It is also used as a coloring agent in pharmacopœial preparations.

Dose: 10 to 20 grains.

PREPARATION.

TINCTURE OF SAFFRON (Br., Fr., Ger.).

Take of Saffron	1 ounce.
Proof spirit	20 ounces.

Macerate, percolate, and add proof spirit to make 20 ounces.

Dose: 1 to 3 drachms. Used chiefly for coloring.

CUBEBA (U. S. et al. Ph.)—CUBEB.

The unripe fruit of Piper cubeba, L. (*Cubeba officinalis, Miq.; Nat. ord., Piperaceæ*), a woody climber indigenous to Java, Borneo, and Sumatra.

Cubebs resemble black pepper in general appearance, but are lighter colored. They have a strongly aromatic taste, with some bitterness and acridity.

Cubebs have a stimulant action upon the mucous membranes, especially those of the genito-urinary system. They are largely used in the treatment of gonorrhœa, cystitis, and other genito-urinary diseases, and in chronic bronchitis, etc.

Dose: 10 grains to 3 drachms.

PREPARATIONS.

Extractum Cubebæ Fluidum (U. S.)—Fluid Extract of Cubeb.

Take of Cubeb	16 ounces.
Stronger alcohol	sufficient.

Moisten the cubeb with 5 ounces of the alcohol, and proceed according to the general formula, page 161.

Dose: ½ to 1 drachm.

Oleoresina Cubebæ (U. S.)—Oleoresin of Cubeb.

Take of Cubeb	12 ounces.
Ether	sufficient.

Moisten, pack, and percolate to 24 ounces; distil off most of the ether, and allow the remainder to evaporate spontaneously.

Dose: 5 to 30 minims.

Oleum Cubebæ (U. S., Br.)—Oil of Cubeb.

The oil obtained from cubeb by distillation.

Dose: 5 to 20 minims.

Tinctura Cubebæ (U. S., Br.)—Tincture of Cubeb.

Take of Cubeb	4 ounces.
Diluted alcohol	sufficient.

Moisten, pack, and percolate to 2 pints.

Dose: ½ to 2 drachms.

Trochisci Cubebæ (U. S.)—Troches of Cubeb.

Take of Oleoresin of cubeb	½ ounce.
Oil of sassafras	1 drachm.
Liquorice	4 ounces.
Gum arabic	2 ounces.
Sugar	3 ounces.
Syrup of tolu	sufficient.

Mix the powders, add the oleoresin and oil, then the syrup, and divide into 480 troches.

CONFECTION OF CUBEB.

Take of Cubeb	4 drachms.
Copaiba	2 drachms.
Treacle	2 drachms.

Mix. Dose: 2 drachms. *St. Mary's Hospital.*

PILLS OF CUBEB AND COPAIBA.

Take of Cubeb	2 ounces.
Subnitrate of bismuth	2 ounces.
Copaiba	2 ounces.
Magnesia	1 drachm.
Oil of peppermint	20 drops.

Mix, and divide into pills of 5 grains each. *Bumstead.*

POWDER OF CUBEB AND IRON.

Take of Cubeb	2 drachms.
Carbonate of iron	½ drachm.

Mix. To be taken three times a day, in gonorrhœa. *Bumstead.*

See also *Copaiba.*

CYPRIPEDIUM (U. S.)—CYPRIPEDIUM—LADIES' SLIPPER.

The rhizome and rootlets of Cypripedium pubescens, Willd., and of Cypripedium parviflorum, Salisb. (*Nat. ord., Orchidaceæ*), herbaceous perennials indigenous to North America, growing in bogs and low woods.

Cypripedium is a nervous stimulant, producing effects similar to those of valerian.

Dose: 15 to 30 grains.

DELPHINIUM STAPHISAGRIA (Fr.)—STAVESACRE.

The seeds of Delphinium Staphisagria, L. (*Nat. ord., Ranunculaceæ*), a biennial herb, native of the south of Europe and Asia Minor, but extensively naturalized in other countries.

It owes its medicinal activity to an alkaloid termed *delphinia*, which may be obtained by exhausting an alcoholic extract with boiling water acidulated with sulphuric acid, and then precipitating the alkaloid with ammonia.

The powdered seeds are sometimes used to destroy pediculi, while the tincture has been employed both externally and internally in rheumatism, neuralgia, and skin diseases. Delphinia is employed for the same purposes.

Dose: of the powdered seeds, 1 to 3 grains; of delphinia, $\frac{1}{16}$ to $\frac{1}{8}$ grain.

PREPARATIONS.

Ointment of Delphinia.

Take of Delphinia .... 30 grains.
Olive oil .... 1 drachm.
Lard .... 1 ounce.

Rub the delphinia first with the oil, then add the lard. *Royle.*

Solution of Delphinia.

Take of Delphinia .... 40 grains.
Rectified spirit .... 2 ounces.

Dissolve. For external use. *Royle.*

Tincture of Stavesacre.

Take of Stavesacre .... 1 part.
Alcohol .... 5 parts.

Macerate two weeks, and filter.

Dose: 10 to 15 minims.

DIGITALIS (U. S. et al. Ph.)—DIGITALIS—FOXGLOVE.

The leaves of Digitalis purpurea, L. (*Nat. ord., Scrophulariaceæ*), a perennial herb indigenous to Europe, but cultivated in this country, mainly for ornamental purposes. For medicinal use the leaves should be collected from plants growing without cultivation.

Digitalis, in large doses, is a cardiac poison. In doses which fall just short of fatal effect, it causes faintness, nausea, vomiting, and diarrhœa, together with irregularity of the heart's action, and a notable fall of bodily temperature. In smaller doses it reduces the frequency of the heart's pulsations, while it adds to their force. It also has a powerful diuretic action.

Digitalis is administered in certain cases of organic disease of the heart, especially where there are feebleness and irregularity of its action, with or without dropsical effusion. It is also used in dropsy dependent upon renal congestion, in certain hemorrhages, in delirium tremens, and sometimes in acute inflammatory diseases.

The effects of digitalis depend upon a principle termed *digitalin*. As at present obtained, this is a somewhat unreliable preparation.

Dose: Of the powdered leaves, ½ to 3 grains.

Extractum Digitalis (U. S., Fr., Ger.)—Extract of Digitalis.

Take of Digitalis, recently dried	12 ounces.
Alcohol	1 pint.
Diluted alcohol	sufficient.

Moisten the digitalis with the alcohol, then percolate with diluted alcohol until 1 pint is obtained. Allow this to evaporate spontaneously to 3 ounces. Continue the percolation until 2 pints more are obtained, or the digitalis is exhausted. Evaporate this, at or below 160°, to a syrupy consistence, mix with the 3 ounces first obtained, and evaporate, at or below 120°, to the proper consistence.

Dose: ¼ to ½ grain.

The French Codex prepares an alcoholic and an aqueous extract from the dried leaves; the German Pharmacopœia, an alcoholic extract from the fresh leaves.

Extractum Digitalis Fluidum (U. S.)—Fluid Extract of Digitalis.

Take of Digitalis	16 ounces.
Glycerin	4 ounces.
Alcohol	sufficient.
Water	sufficient.

Mix 12 ounces of alcohol, 3 of glycerin, and 1 of water, moisten the digitalis with ½ pint of the mixture, and proceed according to the general formula, page 161. Finish the percolation with diluted alcohol, reserve 14 ounces, and add 1 ounce of glycerin to the remainder, before evaporation.

Dose: 1 to 3 minims.

Infusum Digitalis (U. S., Br.)—Infusion of Digitalis.

Take of Digitalis, recently dried .... 60 grains.
Tincture of cinnamon .... 1 ounce.
Boiling water .... ½ pint.

Macerate the digitalis in the water for two hours, in a covered vessel, strain, and add the tincture of cinnamon.

Dose: 1 to 4 drachms.

The British Pharmacopœia directs: digitalis, 30 grains; boiling distilled water, 10 ounces.

Tinctura Digitalis (U. S. et al. Ph.)—Tincture of Digitalis.

Take of Digitalis, recently dried .... 4 ounces.
Diluted alcohol .... sufficient.

Moisten, pack, and percolate to 2 pints.

Dose: 10 to 20 minims.

The British Pharmacopœia directs the proportion of 2½ to 20; the French Codex, 1 to 4; while the German Pharmacopœia employs fresh leaves and tops, 5 parts; alcohol, 6 parts. The French Codex also prepares a tincture similar to the German, from the fresh plant.

MIXTURES OF DIGITALIS.

Take of Infusion of digitalis .... 2 ounces.
Tincture of rhatany .... 1 ounce.
Fluid extract of ergot .... 1 ounce.

Mix. Dose: ½ ounce as required. In hemorrhages. *Bartholow.*

Take of Tincture of digitalis .... 16 drops.
Chloride of ammonium .... 16 grains.
Syrup of tolu .... 1 ounce.
Water .... 1 ounce.

Mix. Dose: 1 drachm every two hours, for a child of one year. In bronchitis. *J. Lewis Smith.*

Take of Infusion of digitalis .... 8 ounces.
Bromide of potassium .... 1 ounce.

Mix. Dose: ½ ounce morning and night, and after a week, at night only. In the spermatorrhœa of plethora. *Bartholow.*

Take of Tincture of digitalis .... 10 minims.
Spirit of nitrous ether .... 30 minims.
Bitartrate of potassium .... 40 grains.
Water .... to 1 ounce.

Mix. One dose. *London Fever Hospital.*

Take of Tincture of digitalis........................ 5 minims.
Vinegar of squill.......................... 20 minims.
Acetate of potassium........................ 20 grains.
Water ....to 1 ounce.

Mix. One dose. *British Skin Hospital.*

Take of Tincture of digitalis......................... 15 minims.
Diluted phosphoric acid..................... 15 minims.
Sulphate of quinia.......................... 5 grains.
Water .... 1 ounce.

Mix. One dose. Antipyretic. *Fothergill.*

Take of Tincture of digitalis .... 10 minims.
Spirit of nitrous ether .... ½ drachm.
Infusion of buchu......................... 1 ounce.

Mix. Take three times a day, in cardiac debility with scanty urine. *Fothergill.*

PILLS OF DIGITALIS.

Take of Digitalis .... ½ grain.
Sulphate of iron .... ½ grain.
Capsicum .... ¼ grain
Compound rhubarb pill .... 1½ grain.

Make 1 pill. *Samaritan Hospital.*

Take of Digitalis .... ½ grain.
Squill .... 1 grain
Blue pill.................................. 1 grain.
Liquorice .... 1 grain.
Treacle ....to 5 grains.

Make 1 pill. *Middlesex Hospital.*

DIGITALINUM (U. S., Br., Fr.)—DIGITALIN.

A concentrated tincture of digitalis is mixed with acetic acid and animal charcoal, filtered, nearly neutralized with ammonia, and precipitated with a solution of tannin. The precipitate is mixed with oxide of lead, dried, mixed with animal charcoal, digested with alcohol, filtered, and evaporated to dryness. Lastly, the powdered product is washed with ether to remove impurities.

Digitalin is a white, or yellowish-white powder, without odor, and having a very bitter taste.

Dose: $\frac{1}{60}$ to $\frac{1}{30}$ grain.

DULCAMARA (U. S. et al. Ph.)—BITTERSWEET.

The young branches of Solanum dulcamara, L. (*Nat. ord.*, *Solanaceæ*), a plant with a perennial root, and an herbaceous, climbing stem, indigenous to Europe, but naturalized and very common in this country.

Bittersweet is narcotic, diuretic, diaphoretic, and alterative. Used in chronic skin diseases, chronic rheumatism, bronchitis, etc.

Dose: 30 to 60 grains.

PREPARATIONS.

Decoctum Dulcamaræ (U. S.)—Decoction of Bittersweet.

Take of Bittersweet............................ 1 ounce.
Water.................................... sufficient.

Boil in 1 pint of water for fifteen minutes, strain, and add sufficient water through the strainer to make 1 pint.

Dose: 1 to 2 ounces.

Similar to the *Infusion of Bittersweet, Br.*

Extractum Dulcamaræ (U. S., Fr., Ger.)—Extract of Bittersweet.

Take of Bittersweet............................ 12 ounces.
Diluted alcohol........................... sufficient.

Exhaust the bittersweet by percolation, distil off the alcohol, strain, and evaporate to a proper consistence.

Dose: 5 to 10 grains.

Extractum Dulcamaræ Fluidum (U. S.)—Fluid Extract of Bittersweet.

Take of Bittersweet............................ 16 ounces.
Glycerin.................................. 4 ounces.
Alcohol................................... sufficient.
Water, each............................... sufficient.

Mix 8 ounces of alcohol, 3 of glycerin, and 5 of water, moisten the bittersweet with 6 ounces of the mixture, and proceed according to the general formula, page 161. Finish the percolation with diluted alcohol, and add 1 ounce of glycerin to the remainder, before evaporation.

Dose: 30 to 60 minims.

ELATERIUM (U. S., Br.)—ELATERIUM.

A peculiar, pulverulent substance deposited by the juice of Ecballium Elaterium, Richard (*Nat. ord.*, *Cucurbitaceæ*), a prostrate annual plant, indigenous to the south of Europe, growing in waste places, but cultivated in Germany, France, and England.

The juice is expressed from the nearly ripe fruit, set aside to deposit, and the deposit subsequently dried on linen filters.

Elaterium occurs in thin, friable flakes, of a pale greenish color, and an acrid, bitterish taste. It is a very active hydragogue cathartic, and is much used in dropsy, especially when dependent upon renal disease.

Its active principle, *elaterin*, is much more reliable than the crude drug, and is generally preferred.

Dose : $\frac{1}{16}$ to $\frac{1}{2}$ grain.

PREPARATIONS.

COMPOUND POWDER OF ELATERIUM (Br.).

Take of Elaterium .... 10 grains.
Sugar of milk .... 90 grains.

Rub to a fine powder.

Dose : $\frac{1}{2}$ to 5 grains.

PILLS OF ELATERIUM.

Take of Elaterium .... $\frac{1}{12}$ grain.
Extract of henbane .... 4 grains.

Make 1 pill. *St. Mary's Hospital.*

Take of Elaterium .... $\frac{1}{6}$ grain.
Extract of belladonna .... $\frac{1}{4}$ grain.
Capsicum .... 1 grain.
Extract of jalap .... $3\frac{1}{2}$ grains.

Make 1 pill. *Charing Cross Hospital.*

ELATERINUM—ELATERIN.

The best method of obtaining it is to exhaust elaterium with chloroform, then add ether, which will cause the elaterin to deposit as a white, crystalline powder. *Flückiger and Hanbury.*

Dose : $\frac{1}{40}$ to $\frac{1}{8}$ grain.

SOLUTION OF ELATERIN.

Take of Elaterin .... 1 grain.
Nitric acid .... 4 drops.
Alcohol .... 1 ounce.

Mix. Dose : 30 to 40 minims. *U. S. Dispensatory.*

ELIXIR SIMPLEX—SIMPLE ELIXIR.

Take of	Oil of orange	5 parts.
	Oil of Ceylon cinnamon	2 parts.
	Sugar in coarse powder	1,000 parts.
	Precipitated phosphate of calcium	30 parts.
	Alcohol (stronger alc.)	sufficient.
	Distilled water	sufficient.

Dissolve the oils in sufficient alcohol to make the solution weigh 300 parts. Dissolve the sugar in 1,700 parts of distilled water by agitation, without heat. Add the latter solution gradually, and in small portions at a time, to the alcoholic solution of the oils, constantly stirring, until a permanent milkiness makes its appearance. Then reverse the proceeding, by gradually pouring the milky mixture into the remainder of the syrup, under constant stirring. Rub the precipitated phosphate of calcium with a small quantity of the syrup to a smooth, thin paste, mix this thoroughly with the rest of the syrup, and filter through a well-wetted white filter. Return the first portions, until the filtrate runs off clear. When all the liquid has passed, wash the filter with a mixture of alcohol, 1 part, and distilled water, 6 parts, until the whole product weighs 3,000 parts.

Charles Rice, Report of Am. Ph. Ass'n.

This elixir furnishes an admirable vehicle for the administration of many remedies. If physicians would employ such a vehicle, adding to it extemporaneously such remedies as are desired, instead of prescribing the compound elixirs so much in vogue, they would often save themselves much disappointment.

ELEMI (Br., Ger., Fr.)—ELEMI.

A concrete, resinous exudation, the botanical source of which is undetermined, but is probably Canarium commune, L. (*Nat. ord., Amyridaceæ*), chiefly imported from Manilla.

It occurs as a soft, unctuous, adhesive mass, becoming harder and more resinous by age, of a yellowish-white color, and a fragrant, fennel-like odor. Little used in this country, but extensively employed in Europe. Its properties are of a terebinthinate character.

PREPARATION.

Ointment of Elemi (Br., Ger.).

Take of	Elemi	¼ ounce.
	Simple ointment	1 ounce.

Melt, strain, and stir until cool.

EMETIA—EMETINE.

An alkaloid occurring in ipecacuanha. It may be obtained by drying the powdered bark of the root with a little milk of lime, and exhausting the mixture with boiling chloroform, petroleum-benzin, or ether. It is a white powder, turning brown on exposure to light.

Flückiger and Hanbury.

Dose: $\frac{1}{8}$ to $\frac{1}{2}$ grain, emetic; $\frac{1}{80}$ to $\frac{1}{20}$ grain, diaphoretic and expecto rant.

ERGOTA (U. S. et al. Ph.)—ERGOT.

The compact mycelium or spawn of Claviceps purpurea, Tulasne (*Nat. ord., Fungi*), a fungus developed in the heads of numerous plants of the order Gramineæ. For medicinal use it is obtained almost exclusively from rye, Secale cornutum, L. It has a peculiar, nauseous odor, and a bitterish, acrid taste.

Ergot of rye acts specifically upon the uterus, both in the impregnated and unimpregnated state. It is used to increase the uterine contractions in childbirth, when these are too feeble to expel the fœtus, provided there be no mechanical obstacle, and, after delivery, to restrain or prevent hemorrhage; also in menorrhagia, metrorrhagia, and other uterine affections. It is also used in other hemorrhages, as in those from the nose, lungs, stomach, intestines, bladder, etc.

Dose: 20 to 60 grains.

PREPARATION.

Extractum Ergotæ Fluidum (U. S., Br.)—Fluid Extract of Ergot.

Take of Ergot	16 ounces.
Glycerin	4 ounces.
Acetic acid	$\frac{1}{2}$ ounce.
Alcohol	sufficient.
Water	sufficient.

Mix 8 ounces of alcohol, 3 of glycerin, and 5 of water, moisten the ergot with 4 ounces of the mixture, and proceed according to the general formula, page 161. Finish the percolation with diluted alcohol, reserve 14 ounces, and add the acid and 1 ounce of glycerin to the remainder, before evaporation.

Dose: 20 to 60 minims.

This is the best preparation of ergot for general use.

The British preparation is termed *Liquid Extract of Ergot;* it contains neither acid nor glycerin.

Vinum Ergotæ (U. S.)—Wine of Ergot.

Take of Fluid extract of ergot	4 ounces.
Sherry wine	28 ounces.

Mix, and filter.

Dose: 1 to 3 drachms.

This preparation has no advantage over extemporaneous mixtures of wine and fluid extract of ergot. When the stomach is irritable, the administration of wine or other stimulant with ergot is often useful.

Extract of Ergot (Ger.)—Ergotin.

Take of Ergot	1 part.
Diluted alcohol	1 part.
Distilled water	4 parts.

Macerate the ergot for six hours in 2 parts of the water, strain, and express; repeat the operation with the remainder of the water, mix the liquids, evaporate to the consistence of thin syrup, and add the alcohol. After twenty-four hours, filter, and evaporate to a thick extract.

Dose: 5 to 10 grains.

Infusion of Ergot (Br.).

Take of Ergot	¼ ounce.
Boiling distilled water	10 ounces.

Infuse in a covered vessel, for half an hour, and strain.

Dose: 1 to 2 ounces.

Tincture of Ergot (Br.).

Take of Ergot	5 ounces.
Proof spirit	20 ounces.

Macerate the ergot in 15 ounces of the spirit, for forty-eight hours, then percolate with the remainder of the spirit, express, and add enough more to make 20 ounces.

Dose: 10 minims to 1 drachm.

Hypodermic Injection of Ergotin.

Take of Ergotin	30 grains.
Water	13 drachms.
Glycerin	13 drachms.

Mix. Dose: 1 drachm. *National Dispensatory.*

Mixtures of Ergot.

Take of Fluid extract of ergot	20 minims.
Carbonate of ammonium	4 grains.
Spirit of chloroform	15 minims.
Camphor water	to 1 ounce.

Mix. One dose. *University College Hospital.*

Take of Fluid extract of ergot........................ 20 minims.
Aromatic sulphuric acid.................... 10 minims.
Cinnamon water............................to 1 ounce.

Mix. One dose. *Samaritan Hospital.*

Take of Fluid extract of ergot........................ 15 minims.
Sulphate of magnesium .... 1 drachm.
Gallic acid .... 10 grains.
Diluted sulphuric acid...................... 5 minims.
Water ....to 1 ounce.

Mix. One dose. *Royal Chest Hospital.*

Take of Fluid extract of ergot...................... 15 minims.
Tincture of perchloride of iron.............. 15 minims.
Spirit of chloroform .... 15 minims.
Infusion of quassia........................ 1 ounce.

Mix. One dose. *St. Mary's Hospital.*

Take of Fluid extract of ergot...................... ½ ounce.
Tincture of chloride of iron................ 2 drachms.
Tincture of nux vomica.................... 2 drachms.

Mix. Dose: 30 minims three times daily.

In subinvolution of the uterus, metrorrhagia, etc.

Take of Fluid extract of ergot...................... 3½ ounces.
Tincture of digitalis........................ ½ ounce.

Mix. Dose: 1 drachm three times a day. In dilatation of the heart without valvular lesions. *Bartholow.*

PILLS OF ERGOT.

Take of Extract of ergot.......................... 1 grain.
Tannic acid.............................. 2 grains.
Extract of opium.......................... ¼ grain.

Make 1 pill. *London Chest Hospital.*

SUPPOSITORIES OF ERGOT.

Take of Aqueous extract of ergot (Squibb's).......... 40 grains.
Oil of theobroma.......................... 50 grains.

Mix and divide into 12 rectal suppositories.

One morning and evening, for two days before menstruation comes on, then 1 three times a day, until metrorrhagia ceases.

In metrorrhagia at the climacteric, when not the result of organic disease requiring surgical measures, though I have found this treatment successful in metrorrhagia due to submucous fibroids.

The patient should keep the recumbent posture during the first two or three days of the flow, after which it is generally unnecessary, except the hour after each suppository is used: *Fordyce Barker.*

See also *Mixtures of Bromide of Potassium.*

ERIGERON (U. S.)—FLEABANE.

The leaves and tops of Erigeron Canadense, L., Canada fleabane, and of E. Philadelphicum, L., Philadelphia fleabane (*Nat. ord., Compositæ*), both indigenous herbs, and found everywhere as common weeds.

Erigeron acts as a diuretic, and is used in dropsy, and genito-urinary diseases.

Dose : ½ to 1 drachm.

PREPARATIONS.

Extractum Erigerontis Canadensis Fluidum (U. S.)—Fluid Extract of Canada Erigeron.

Take of Canada erigeron........................... 16 ounces.
Alcohol................................... sufficient.

Moisten the erigeron with ½ pint of alcohol, and proceed according to the general formula, page 161.

Dose : ½ to 1 drachm.

Oleum Erigerontis Canadensis (U. S.)—Oil of Canada Erigeron.

The oil obtained from Canada erigeron by distillation.

Dose : 5 to 10 minims.

ERYTHROXYLON—COCA.

The leaves of Erythroxylon coca, Lam. (*Nat. ord., Erythroxylaceæ*), a shrub indigenous to the mountains of Peru and Bolivia. The leaves are chewed by the natives, apparently for the same reasons that other nations use tobacco. It is said to satisfy hunger, strengthen the weak, and to solace men under misfortune. Those who use it are said to undergo fatigue and privation with it alone, better than others without it, but abundantly supplied with food. Many of the statements concerning its virtues lack confirmation, and its true place in the materia medica is not at present decided.

Dose : ½ to 2 drachms.

PREPARATION.

TINCTURE OF COCA.

Take of Coca .... 1 part.
Diluted alcohol .... sufficient.

Moisten, pack, and percolate to 5 parts.

Dose: 1 to 2 drachms. *J. P. Remington, Report Am. Ph. Ass'n.*

EUCALYPTUS—BLUE GUM TREE.

The leaves of Eucalyptus globulus, Labill. (*Nat. ord., Myrtaceæ*), a tree indigenous to Tasmania and Victoria, and recently introduced into Europe, and also into California and the Southern States.

Eucalyptus leaves, and the oil obtained from them, have been used with asserted benefit in intermittent fevers, but their antiperiodic powers are doubted by many. The drug seems to possess tonic and stimulant properties which may ultimately give it an established position, but at present little is positively known of it.

Dose: 15 to 60 grains.

OIL OF EUCALYPTUS.

The oil obtained from eucalyptus by distillation.

Dose: 10 to 20 minims.

TINCTURE OF EUCALYPTUS.

Take of Eucalyptus .... 1 part.
Alcohol .... sufficient.

Moisten, pack, and percolate to 5 parts.

Dose: 1 to 3 drachms. *J. P. Remington, Report Am. Ph. Ass'n.*

EUONYMUS (U. S.)—WAHOO.

The bark of Euonymus atropurpureus, Jacq. (*Nat. ord., Sapindaceæ*), a shrub indigenous to North America. Wahoo, in moderate doses, is a mild cathartic, somewhat resembling rhubarb in its action, and may be usefully employed in habitual constipation. In large doses it may act as a hydragogue, and also stimulate the kidneys to increased activity.

Dose: 1 to 2 drachms.

PREPARATIONS.

INFUSION OF EUONYMUS.

Take of Euonymus .... 1 ounce.
Boiling water .... 1 pint.

Infuse for half an hour, and strain.

Dose: 2 to 4 ounces.

TINCTURE OF EUONYMUS.

Take of Euonymus	1 part.
Diluted alcohol	2 parts.

Macerate two weeks, and filter.

Dose: 1 to 2 drachms.

EUPATORIUM (U. S.)—THOROUGHWORT—BONESET.

The leaves and flowering tops of Eupatorium perfoliatum, L. (*Nat. ord., Compositæ*), an herbaceous perennial indigenous to North America, growing in damp and swampy places.

Boneset has a strongly bitter taste, and possesses diaphoretic, expectorant, and tonic properties. As a diaphoretic it is used in acute bronchial affections, rheumatism, intermittent and remittent fevers, etc., the warm, infusion being taken freely. Very large doses of the warm infusion act as an emetic. The infusion taken quite cold is used as a tonic.

Dose: 1 to 2 drachms.

PREPARATIONS.

Infusum Eupatorii (U. S.)—Infusion of Thoroughwort.

Take of Thoroughwort	1 ounce.
Boiling water	1 pint.

Macerate two hours, and strain.

Dose: 1 to 2 ounces.

TINCTURE OF THOROUGHWORT.

Take of Fresh thoroughwort	1 part.
Alcohol	2 parts.

Macerate two weeks, and filter.

Dose: 1 to 2 drachms.

EXTRACTA FLUIDA (U. S.)—FLUID EXTRACTS.

Unless otherwise directed, the fluid extracts should be prepared according to the following process: 16 ounces of the powdered drug are moistened with a specified quantity of the menstruum, and packed in a percolator. The surface of the powder is then covered with a disk of paper, and the remaining portion of 16 ounces of menstruum is poured upon it. When the liquid begins to drop from the percolator, the lower orifice is closed with a cork, the upper one covered closely, and it is set aside in a moderately warm place for four days.

The cork is then removed, more menstruum is gradually poured on, and the percolation continued until 24 ounces are obtained. Of these the first 14 ounces are reserved, and the remainder, having been carefully evaporated to 2 ounces, is mixed with the reserved portion, and filtered if necessary.

FEL BOVINUM (Br., Fr., Ger.)—OX BILE.

PREPARATIONS.

Purified Ox Bile (Br., Ger.).

Take of Ox bile	1 pint.
Rectified spirit	2 pints.

Mix by agitation in a bottle, set aside for twelve hours, decant the clear liquid, and evaporate it on a water-bath to a suitable consistence for making pills.

Dose: 5 to 10 grains.

Used in dyspepsia with deficiency of biliary secretion, in habitual constipation, and in solution, per enema, to soften hardened fæces.

Pills of Ox Bile and Aloes.

Take of Purified ox bile	20 grains.
Extract of aloes	30 grains.
Resin of podophyllum	2½ grains.

Mix, and divide into 10 pills.

Dose: 1 pill at night in chronic constipation; 1 pill night and morning in acute constipation. *Hospital Formulary.*

FERRUM (U. S. et al. Ph.)—IRON.

In medicine, as in the arts, iron is the most useful of metals.

Being a normal constituent of the blood, it may be considered a food as well as an important curative agent. That it is more than a food is evinced by the fact that, in many instances, it is requisite to administer it in quantities many times in excess of any theoretical estimate of the needs of the system, in order to obtain its important curative effects. In fact, the therapeutics of iron, based upon the theory of simply supplying the system with the quantity of the drug which can be assimilated, are delusive, as clinical experience has frequently demonstrated.

Iron acts as a tonic, not only by increasing the relative proportion of red globules of the blood when these are deficient, but also, probably, by a direct stimulant action upon the mucous membranes with which it is brought in contact, whereby digestion and assimilation are improved.

Most of the preparations of iron are powerfully astringent, as well as tonic, and often this fact should be considered in choosing which one to use in a given case.

The organic salts of iron are less astringent than the inorganic, and of the latter, the proto-salts less than the per-salts. Several of the astringent preparations, as the acetate, sulphate, perchloride, and pernitrate, are used to check hemorrhages from the lungs, stomach, and bowels; and some of them, especially the perchloride and subsulphate, are used as topical astringents and styptics.

Though the restorative effects of iron are most marked in anæmia, it is used with no less benefit in many other conditions of depression. It is very generally used in convalescence from acute diseases, during the course of wasting diseases, as phthisis, scrofula, necrosis of the bones, etc.; and in diphtheria, erysipelas, neuralgia, chorea, and in many other diseases in which, though its action may not be understood, its effects are none the less satisfactory.

FERRUM REDACTUM (U. S. et al. Ph.)—REDUCED IRON.

Take of Subcarbonate of iron....................... 30 ounces.

Wash and dry it thoroughly, and enclose it in an iron reduction tube, placed in a charcoal furnace. Connect the tube with a hydrogen generator so arranged as to pass a stream of dry hydrogen through it. Then heat the tube to dull redness, and maintain the temperature until the subcarbonate is reduced.

The process will require from five to eight hours.

It is a tasteless powder, of an iron-gray color, and without metallic lustre. It is a favorite preparation of iron with many physicians, especially in cases of anæmia where other chalybeates disagree.

Dose: 3 to 6 grains.

PREPARATIONS.

Lozenges of Reduced Iron (Br.).

Take of Reduced iron..............................	720 grains.
Refined sugar	25 ounces.
Gum arabic	1 ounce.
Mucilage of gum arabic....................	2 ounces.
Distilled water	sufficient.

Mix the powders, add the mucilage and sufficient water to form a mass, and divide into 720 lozenges. Dry in a hot air-chamber.

Dose: 1 to 6 lozenges.

Pill of Reduced Iron.

Take of Reduced iron.............................. 3 grains.
 Balsam of Peru .... ½ minim.
 Starch.................................. sufficient.

Make 1 pill. *St. Bartholomew's Hospital.*

Take of Reduced iron.............................. 20 grains.
 Sulphate of quinia........................ 20 grains.
 Digitalis................................ 20 grains.
 Squill .... 10 grains.

Mix and divide into 20 pills.

Dose : a pill three or four times a day. In mitral regurgitation.

Bartholow.

FERRI ACETAS—ACETATE OF IRON.

PREPARATIONS.

Solution of Acetate of Iron (Ger.).

Take of Solution of persulphate of iron.............. 10 parts.
 Water of ammonia......................... 8 parts.
 Diluted acetic acid .... 6 parts.
 Distilled water.......................... sufficient.

Dilute the iron solution with 30 parts of distilled water, and add the water of ammonia, previously diluted with 160 parts of distilled water. Wash the precipitate with distilled water, express until the weight shall amount to 5 parts, and dissolve this in the acetic acid. After several days, filter, and add sufficient distilled water to make the whole weigh 10 parts.

Dose : 15 to 30 minims.

Tincture of Acetate of Iron (Br.).

Take of Solution of persulphate of iron.............. 2½ ounces.
 Acetate of potassium...................... 2 ounces.
 Rectified spirit.......................... sufficient.

Dissolve the acetate in 10 ounces, and add the solution of persulphate to 8 ounces of the spirit, mix the solutions, filter, and add sufficient spirit through the filter to make 20 ounces.

Dose : 5 to 30 minims.

Ethereal Tincture of Acetate of Iron (Ger.).

Take of Solution of acetate of iron.................. 9 parts.
 Alcohol.................................. 2 parts.
 Acetic ether............................. 1 part.

Mix.

Dose : 20 to 30 minims.

FERRI ARSENIAS (Br.)—ARSENIATE OF IRON.

Take of		
	Sulphate of iron	9 ounces.
	Arseniate of sodium, dried at 300°	4 ounces.
	Acetate of sodium	3 ounces.
	Boiling distilled water	sufficient.

Dissolve the arseniate and acetate of sodium in 2 pints, and the sulphate of iron in 3 pints of the water, mix the solutions, collect and thoroughly wash the precipitate; express strongly, and dry in a warm air-chamber, at or below 100°.

It is a tasteless, amorphous powder, of a green color, and insoluble in water. It combines, to some extent, the tonic properties of both iron and arsenic.

Dose: $\frac{1}{16}$ to $\frac{1}{2}$ grain.

Pills of Arseniate of Iron.

Take of		
	Arseniate of iron	$\frac{1}{16}$ grain.
	Extract of gentian	sufficient.

Make 1 pill. *Brompton Consumption Hospital.*

Take of		
	Arseniate of iron	2 grains.
	Extract of cinchona	12 grains.

Mix, and divide into 12 pills.

Dose: 1 pill three times a day after meals. In chlorosis. *Bartholow.*

FERRI CARBONAS SACCHARATA (Br., Ger.)—SACCHARATED CARBONATE OF IRON.

Take of		
	Sulphate of iron	2 ounces.
	Carbonate of ammonium	1¼ ounce.
	Boiling distilled water	2 gallons.
	Refined sugar	1 ounce.

Dissolve the sulphate and the carbonate, each in $\frac{1}{2}$ gallon of water, mix the solutions, and set aside for twenty-four hours. Then decant, wash the precipitate with the remainder of the water, express, and rub it with the sugar. Finally, dry at a temperature not exceeding 212°.

It is in small, coherent lumps, of a gray color, with a sweet, very feeble chalybeate taste.

Dose: 5 to 20 grains.

PREPARATIONS.

Pilula Ferri Carbonatis (U. S. et al. Ph.)—Pill of Carbonate of Iron.

Take of Sulphate of iron	8 ounces.
Carbonate of sodium	9 ounces.
Clarified honey, by weight	3 ounces.
Sugar	2 ounces.
Boiling water	2 pints.
Syrup	sufficient.

Dissolve the salts separately, each in 1 pint of water, and having added 2 ounces of syrup to the iron solution, filter both solutions. Mix the solutions, when cold, in a bottle just large enough to hold them, and set aside to settle. Then decant, wash the precipitate with a mixture of water and syrup, in the proportion of 1 pint to 1 ounce, drain, and express. Lastly, mix the precipitate immediately with the clarified honey and sugar, and evaporate on a water-bath until it weighs 8 ounces.

Dose: 2 to 10 grains.

The French and German preparations are made in a similar manner, while the British Pharmacopœia directs: saccharated carbonate of iron, 1 ounce; confection of roses, ¼ ounce.

Under the name of *Vallet's Mass*, this has attained a well-merited popularity.

Mistura Ferri Composita (U. S., Br.)—Compound Mixture of Iron.

Take of Myrrh	60 grains.
Sugar	60 grains.
Carbonate of potassium	25 grains.
Sulphate of iron	20 grains.
Spirit of lavender	½ ounce.
Rose water	7½ ounces.

Rub the myrrh, sugar, and carbonate with the rose water, then with the spirit of lavender, and lastly, with the sulphate of iron. Keep in a well-stopped bottle.

Dose: 1 to 2 ounces.

This valuable preparation, known as *Griffith's Mixture*, is highly esteemed as a restorative in anæmia and chlorosis.

The British preparation is very similar.

Pills of Carbonate of Iron, Arsenic, and Quinia.

Take of Pill of carbonate of iron	60 grains.
Arsenious acid	1 grain.
Sulphate of quinia	20 grains.

Mix, and divide into 40 pills.

Dose: 2 pills three times a day. In the anæmia of chronic malarial poisoning. *Bartholow.*

FERRI SUBCARBONAS (U. S., Fr.)—SUBCARBONATE OF IRON.

Take of Sulphate of iron .... 8 ounces.
Carbonate of sodium...................... 9 ounces.
Water .... 8 pints.

Dissolve the salts separately, each in 4 pints of water, mix the solutions, and, after the precipitate has subsided, decant, wash it well with water, and dry it on bibulous paper.

It is a reddish-brown, odorless, tasteless, and insoluble powder. Formerly much used in anæmia, neuralgia, etc., but has been superseded by more eligible preparations. Mixed with water, it may be used as a substitute for the hydrated oxide in arsenical poisoning, when the latter is not at hand.

Dose: 5 to 30 grains.

PREPARATIONS.

Trochisci Ferri Subcarbonatis (U. S.)—Troches of Subcarbonate of Iron.

Take of Subcarbonate of iron........................ 5 ounces.
Vanilla .... 30 grains.
Sugar .... 15 ounces.
Mucilage of tragacanth.................... sufficient.

Rub the vanilla first with a part of the sugar, then with the iron and the remainder, until thoroughly mixed, then with the mucilage form a mass, to be divided into 480 troches.

Dose: 1 to 6 troches.

Emplastrum Ferri (U. S., Br.)—Iron Plaster.

Take of Subcarbonate of iron...................... 3 ounces.
Lead plaster............................ 24 ounces.
Burgundy pitch.......................... 6 ounces.

To the pitch and plaster, previously melted together, add the subcarbonate, and stir until cold.

The British preparation is made with hydrated oxide, instead of subcarbonate of iron.

FERRI CHLORIDUM (U. S., Ger.)—CHLORIDE OF IRON.

Take of Iron, in the form of wire. .... 2 ounces.
Hydrochloric acid (by weight) .... 12 ounces.
Nitric acid (by weight) ....1 ounce or sufficient.

Add the iron to 8 ounces of the hydrochloric acid in a flask, and heat gently until the acid is saturated. Filter, add the remainder of the hydrochloric acid, heat nearly to the boiling-point, and add the nitric acid in successive portions until red fumes are no longer evolved, and a drop of the liquid ceases to yield a blue precipitate with ferridcyanide of potassium. Then evaporate on a sand-bath until reduced to 8 ounces and 360 grains, and set aside, covered with glass, until it solidifies. Lastly, break in pieces, and preserve in a well-stopped bottle.

It is in orange-yellow, crystalline pieces, very deliquescent, and soluble in water, alcohol, and ether.

Chloride of iron is used topically as a styptic, and, properly diluted, as an astringent. It is seldom employed internally.

Dose: 1 to 5 grains.

PREPARATIONS.

Liquor Ferri Chloridi (U. S. et al. Ph.)—Solution of Chloride of Iron.

Take of Iron in the form of wire	3 ounces.
Hydrochloric acid (by weight)	17½ ounces.
Nitric acid	sufficient.
Distilled water	sufficient.

Proceed as in the preceding preparation, except that the iron is to be dissolved in 11 ounces of hydrochloric acid, and at the end of the process, instead of evaporating, the solution is to be diluted with sufficient distilled water to make 1 pint.

It is a reddish-brown liquid, of an acid and strongly styptic taste. Its chief use is in the preparation of tincture of chloride of iron, for which it is occasionally substituted. It is also employed topically as an astringent and styptic.

Dose: 2 to 10 minims.

The British *Solution of Perchloride of Iron* has the strength of its tincture, while its *Strong Solution of Perchloride of Iron* is similar to the above.

Tinctura Ferri Chloride (U. S., Br.)—Tincture of Chloride of Iron.

Take of Solution of chloride of iron	½ pint.
Alcohol	1½ pint.

Mix, and preserve in a well-stopped bottle.

Dose: 10 to 30 minims.

British Pharmacopœia directs: strong solution of perchloride of iron, 5 ounces; rectified spirit, 15 ounces.

Tincture of chloride of iron is universally conceded to be one of the best preparations of iron for general use. In nearly all conditions of debility, however induced, its administration is beneficial; and in many cases of grave disease, as diphtheria, erysipelas, and septicæmia, it is absolutely indispensable.

Ethereal Tincture of Sesquichloride of Iron (Ger.).

Take of	Solution of sesquichloride of iron	1 part.
	Spirit of ether	14 parts.

Mix, and expose to the sunlight until the brown-yellow color has disappeared, then set in a shady place until it has assumed a yellowish or brownish-yellow color. Used like the preceding.

Syrup of Chloride of Iron (Fr.).

Take of	Solution of chloride of iron	15 parts.
	Syrup	985 parts.

Mix. It is a changeable preparation, and should not be prepared except at the time of dispensing.

Mixtures of Tincture of Iron and Chlorate of Potassium.

Take of	Tincture of chloride of iron	2 drachms.
	Chlorate of potassium	2 drachms.
	Syrup	4 ounces.

Mix. Dose: 1 drachm hourly. In diphtheria. *J. Lewis Smith.*

Mixtures of this character are in general use in the treatment of diphtheria.

Take of	Tincture of chloride of iron	½ ounce.
	Chlorate of potassium	1½ drachm.
	Glycerin	2 ounces.
	Water	2 ounces.

Mix. Dose: 1 drachm every three hours. In aphthous stomatitis, assuming an ulcerative form. Will cure the most obstinate cases in from five to seven days. *V. P. Gibney.*

Take of	Tincture of chloride of iron	15 minims.
	Chloric ether	5 minims.
	Chlorate of potassium	5 grains.
	Water	to 1 ounce.

Mix. One dose. *Brompton Consumption Hospital.*

Take of	Tincture of chloride of iron	3 to 4 drachms.
	Chlorate of potassium	1 to 1½ drachm.
	Solution of acetate of ammonium	1½ to 2 ounces.
	Water	to 8 ounces.

Mix. Dose: ½ ounce every hour. In diphtheria. *T. M. Lownds.*

Mixture of Tincture of Iron and Cinchonia.

Take of	Tincture of chloride of iron	2 drachms.
	Sulphate of cinchonia	1 drachm.
	Water	to 4 ounces.

Mix. Dose: 1 drachm. *Hospital Formulary.*

Mixture of Tincture of Iron and Quinia.

Take of Tincture of chloride of iron................ 2 drachms.
Solution of sulphate of quinia (15 grains to 1 ounce)................................ 2 ounces.

Mix. Dose: 1 drachm. *Hospital Formulary.*

Mixture of Tincture of Iron and Nux Vomica.

Take of Tincture of chloride of iron................ 4 drachms.
Tincture of nux vomica.................... 2 drachms.
Tincture of cantharides .... 2 drachms.

Mix. Dose: 20 minims three times a day. In gleet.

Mixtures of Tincture of Iron and Glycerin.

Take of Tincture of chloride of iron................2 to 3 drachms.
Glycerin .... 1½ ounce.
Water .... 1½ ounce.

Mix. Dose: 1 drachm every two hours for children. In recurring naso-facial erysipelas of strumous origin. *V. P. Gibney.*

Take of Tincture of chloride of iron................1 to 1½ drachm.
Glycerin .... 1 ounce.
Water .... 1 ounce.

Mix. Dose: 1 drachm half-hourly, alternating with a solution of chlorate of potassium. In diphtheria. *C. E. Billington.*

FERRI CITRAS (U. S., Ger.)—CITRATE OF IRON.

Take of solution of citrate of iron, a convenient quantity.

Evaporate, at or below 140°, to the consistence of syrup, and spread upon glass plates to dry.

It is in thin, transparent, garnet-colored scales, of a mildly acid, chalybeate taste. Well suited to persons of delicate stomachs.

Dose: 2 to 5 grains.

The German Pharmacopœia dissolves freshly prepared hydrated oxide of iron in a solution of citric acid.

PREPARATIONS.

Liquor Ferri Citratis (U. S.)—Solution of Citrate of Iron.

Take of Citric acid...................... 5 ounces and 6 drachms.
Solution of tersulphate of iron..... 1 pint.
Water of ammonia............... 20 ounces.
Distilled water.................. sufficient.

Dilute the water of ammonia with 2 pints, and the iron solution with 3 pints of distilled water, mix the solutions, collect and wash the precipitate.

Place half of it in a capsule heated to 140°, add the citric acid, and stir until the precipitate is nearly dissolved; then add enough of the reserved portion to saturate the acid. Lastly, filter the solution, and evaporate it, at or below 140°, to 1 pint.

Dose: 5 to 10 minims. Used like the preceding.

FERRI ET AMMONII CITRAS (U. S. et al. Ph.)—CITRATE OF IRON AND AMMONIUM.

Take of Solution of citrate of iron .... 1 pint.
Water of ammonia........................ 6 ounces.

Mix, and evaporate, at or below 140°, to the consistence of syrup, and spread upon glass plates to dry.

It is in garnet-red, translucent scales, of a slightly ferruginous taste, and wholly soluble in water. It is a mild chalybeate, and generally agrees well with the stomach.

Dose: 2 to 5 grains.

PREPARATIONS.

Wine of Citrate of Iron (Br.).

Take of Citrate of iron and ammonium..............160 grains.
Orange wine.......................... 20 ounces.

Dissolve, and let the solution remain three days in a closed vessel, shaking it occasionally, then filter.

Dose: 1 to 4 drachms.

Mixtures of Citrate of Iron and Ammonium.

Take of Citrate of iron and ammonium.............. 32 grains.
Carbonate of ammonium.................. 32 grains.
Syrup.............................. 2 ounces.
Anise water.......................... 2 ounces.

Mix. Dose: 1 drachm for children. *J. Lewis Smith.*

Take of Citrate of iron and ammonium.............. 5 grains.
Camphor water.......................... 1 ounce.

Dissolve. One dose. *London Throat Hospital.*

FERRI ET AMMONII SULPHAS (U. S., Ger.)—SULPHATE OF IRON AND AMMONIUM.

Take of Solution of tersulphate of iron .... 2 pints.
Sulphate of ammonium .... 4½ ounces.

Heat the solution to the boiling point, add the sulphate of ammonium, stirring until dissolved, then set aside to crystallize. Wash the crystals quickly in very cold water, wrap them in bibulous paper, and dry them in the open air.

It is in colorless or violet tinted crystals, of an astringent taste, and wholly soluble in water. Used in chronic catarrhal affections.

Dose 3 to 15 grains.

INHALATION OF SULPHATE OF IRON AND AMMONIUM.

Take of Sulphate of iron and ammonium	20 grains.
Water	8 ounces.

Mix. Use by means of a steam atomizer. Astringent.

G. M. Lefferts.

FERRI ET AMMONII TARTRAS (U. S.)—TARTRATE OF IRON AND AMMONIUM.

Take of Tartaric acid	12 ounces.
Solution of tersulphate of iron	2½ pints.
Carbonate of ammonium	sufficient.
Water	sufficient.

Dissolve 6 ounces of the acid in 2 pints of distilled water, and carefully neutralize it with carbonate of ammonium; then add the remainder of the acid, dissolved in ½ pint of distilled water. From the iron solution, prepare hydrated oxide of iron according to the formula for that substance, and add it to the solution of bitartrate of ammonium, kept at 140°, until it is no longer dissolved. Then filter the solution, evaporate, at or below 140°, to the consistence of syrup, and spread upon glass plates to dry.

It is in thin, transparent, garnet-red scales, of a sweetish, chalybeate taste, and soluble in water.

It is a mild tonic, and generally acceptable to the stomach.

Dose: 5 to 10 grains.

FERRI ET POTASSII TARTRAS (U. S., Br., Fr.)—TARTRATE OF IRON AND POTASSIUM.

Take of Solution of tersulphate of iron	1 pint.
Bitartrate of potassium	7 ounces.
Distilled water	4 pints.

From the iron solution, prepare hydrated oxide of iron according to the formula for that substance, and add it, until it is no longer dissolved, to the bitartrate of potassium, mixed with the water and heated to 140°. Then filter the solution, and evaporate, at or below 140°, to the consistence of syrup, and spread upon glass plates to dry.

It is in thin, transparent scales, of a ruby-red color, of a pleasant, ferruginous taste, and readily soluble in water. It is less liable to constipate, or to disorder the stomach, than any other ferruginous preparation. From its agreeable taste, it is generally well suited to children and persons of delicate stomachs.

Dose: 10 to 30 grains. *Tartrated Iron, Br.*

PREPARATIONS.

Lozenges of Tartrate of Iron and Potassium (Fr.).

Take of Tartrate of iron and potassium	50 parts.
White sugar	1,000 parts.
Vanilla sugar	30 parts.
Mucilage of tragacanth	100 parts.

Make lozenges weighing 15 grains each.

Syrup of Tartrate of Iron and Potassium (Fr.).

Take of Tartrate of iron and potassium	25 parts.
Cinnamon water	25 parts.
Syrup	950 parts.

Dissolve the tartrate in the cinnamon water, filter, and mix with the syrup.

Mixtures of Tartrate of Iron and Potassium.

Take of Tartrate of iron and potassium	1 drachm.
Compound tincture of cinchona	4 ounces.

Mix. Dose: 1 drachm. *Hospital Formulary.*

Take of Tartrate of iron and potassium	20 grains.
Iodide of potassium	10 grains.
Water	to 1 ounce.

Mix. One dose. *King's College Hospital.*

FERRI ET QUINIÆ CITRAS (U. S., Br.)—CITRATE OF IRON AND QUINIA.

Take of Solution of citrate of iron	10 ounces.
Sulphate of quinia	1 ounce.
Diluted sulphuric acid	sufficient.
Water of ammonia	sufficient
Distilled water	sufficient.

Triturate the sulphate of quinia with 6 ounces of water, add sufficient diluted sulphuric acid to dissolve it, then carefully add water of ammonia in slight excess. Wash the precipitated quinia, add it to the solution of citrate of iron, heated to 120°, and stir until it is dissolved. Lastly, evaporate, at or below 140°, to the consistence of syrup, and spread upon glass plates to dry.

It occurs in yellowish-green scales, of a bitter, chalybeate taste, and slowly soluble in water. On account of its tardy solubility, it is an eligible form in which to administer iron and quinia in powder without offending the taste.

Dose: 5 to 10 grains.

Syrup of Citrate of Iron and Quinia.

Take of Citrate of iron and quinia	2 drachms.
Syrup of lemons	3 ounces.

Mix. Dose: 1 drachm. An agreeable form of administration, especially for children.

Wine of Citrate of Iron and Quinia.

Take of Citrate of iron and quinia	1 drachm.
Sherry wine..............................	2 ounces.

Mix. Dose: 1 drachm. *Hospital Formulary.*

FERRI ET STRYCHNIÆ CITRAS (U. S.)—CITRATE OF IRON AND STRYCHNIA.

Take of Citrate of iron and ammonium..............	500 grains.
Strychnia	5 grains.
Citric acid...............................	5 grains.
Distilled water	9 drachms.

Dissolve the citrate in 1 ounce, and the strychnia, together with the citric acid, in 1 drachm of the distilled water. Mix the solutions, evaporate, at or below 140°, to the consistence of syrup, and spread upon glass plates to dry.

In general appearance, this salt resembles citrate of iron and ammonium, but differs in its very bitter taste. It possesses the tonic properties of iron and strychnia, but is in no way superior to extemporaneous mixtures of the two.

Dose: 3 to 5 grains.

FERRI HYPOPHOSPHIS—HYPOPHOSPHITE OF IRON.

Prepared by adding a solution of hypophosphite of sodium to one of chloride or sulphate of iron.

It is a white, amorphous powder, insoluble in cold water, and nearly tasteless.

It is believed, on theoretical rather than clinical grounds, to possess the combined virtues of iron and phosphorus. Used in diseases of the nervous system and in phthisis.

Dose: 5 to 10 grains.

FERRI IODIDUM (Br., Fr.)—IODIDE OF IRON.

Take of Fine iron wire	1½ ounce.
Iodine	3 ounces.
Distilled water	15 ounces.

Put the iodine, iron, and 12 ounces of water into a flask, and having heated the mixture gently for ten minutes, raise the heat and boil until the froth becomes white. Strain quickly into a dish of polished iron, washing the filter with the remainder of the water, and boil until a drop solidifies on cooling; then pour upon a porcelain dish, and, when it has solidified, break into pieces, and preserve in a well-stopped bottle.

It is crystalline, green with a tinge of brown, inodorous, deliquescent, and almost entirely soluble in water.

Used as a tonic and alterative in scrofula, diseases of the skin, etc.

It is an unstable preparation, and on this account is not included in the U. S. Pharmacopœia.

Dose: 1 to 5 grains.

PREPARATIONS.

Syrupus Ferri Iodidi (U. S. et al. Ph.)—Syrup of Iodide of Iron.

Take of Iodine	2 ounces.
Iron, in the form of wire	300 grains.
Distilled water	3 ounces.
Syrup	sufficient.

Mix the iodine, iron, and water in a flask, and shake occasionally until the solution has acquired a green color and lost the smell of iodine. Then filter it into a bottle containing 1 pint of syrup heated to 212°, shake thoroughly, and, when cool, add sufficient syrup to make the product measure 20 ounces. Preserve in two-ounce vials, well-stopped.

Dose: 20 to 40 minims.

This is, without doubt, the best form in which to administer iodine and iron in combination. It is largely employed in strumous affections, as caries and necrosis of the bones, tuberculosis, etc.

Pilulæ Ferri Iodidi (U. S., Br., Fr.)—Pills of Iodide of Iron.

Take of		
	Iodine	300 grains.
	Iron, in the form of wire	120 grains.
	Sugar	192 grains.
	Liquorice root	192 grains.
	Liquorice	48 grains.
	Gum arabic	48 grains.
	Reduced iron	96 grains.
	Water	1½ ounce.

Mix the iodine with 10 drachms of the water in a flask, add the iron gradually, agitating until the solution is of a light green color ; then filter into a capsule containing the reduced iron, washing the filter with the remainder of the water. Evaporate until a pellicle forms, add the remaining powders, previously mixed together, and continue the evaporation until it is reduced to a pilular consistence. Divide into 384 pills, and coat them with balsam of tolu, dissolved in ether.

Dose : 1 to 5 pills.

The British and French processes yield a similar product. They are known as *Blancard's Pills.*

Saccharated Iodide of Iron (Ger.).

Take of		
	Powdered iron	3 parts.
	Distilled water	10 parts.
	Iodine	8 parts.
	Sugar of milk	40 parts.

Mix the iron, water, and iodine, and, when reaction has ceased, filter into a capsule containing the sugar, mix well, and evaporate to dryness.

Dose : 10 to 30 grains.

Solution of Iodide of Iron (Ger.).

Take of		
	Powdered iron	3 parts.
	Distilled water	18 parts.
	Iodine	8 parts.

Mix in a glass flask, and heat gently until reaction has ceased, then filter.

This solution is prepared extemporaneously, for addition to mixtures, etc. Eight parts of the iodine employed correspond to ten parts of iodide of iron.

FERRI LACTAS (U. S., Fr., Ger.)—LACTATE OF IRON.

Take of		
	Lactic acid	1 ounce.
	Iron, in the form of filings	½ ounce.
	Distilled water	sufficient.

Mix the acid with 1 pint of distilled water in an iron vessel, add the iron, digest until reaction has ceased, adding distilled water to preserve the measure ; filter while hot, and set aside to crystallize.

It is a yellowish, or greenish crystalline powder, of a sweetish, ferruginous taste, and sparingly soluble in water. It is mild and unirritating in action.

Dose: 1 to 2 grains.

EFFERVESCING POWDER OF LACTATE OF IRON.

Take of Lactate of iron............................ 45 grains.
Tartaric acid............................ 2 drachms.
Bicarbonate of sodium.................... 3 drachms.

Mix, and preserve in a well-stopped bottle.

Dose: 10 to 20 grains. In gastric ulcer. *Lebert.*

CAPSULES OF LACTATE OF IRON, NUX VOMICA, AND CANNABIS.

Take of Lactate of iron........................... 50 grains.
Extract of nux vomica.................... 5 grains.
Extract of cannabis Indica.................10 to 15 grains.
Extract of belladonna..................... 3 grains.
Resin of podophyllum.................... 2 to 1 grain.

Mix well, and divide into 20 capsules.

Dose: 1 capsule after each meal. For anæmia with sluggish state of the bowels, nervous depression, and wandering neuralgic pains. The cannabis and the podophyllum in the above formula may be increased or diminished according to the effects produced. I begin with 10 grains of the former, and 2 grains of the latter, but after one week I find the patient tolerates 15 grains of the former, and requires only 1 grain of the latter. *Fordyce Barker.*

FERRI OXALAS (U. S.)—OXALATE OF IRON.

Take of Sulphate of iron.......................... 2 ounces.
Oxalic acid.............................436 grains.
Distilled water......................... sufficient.

Dissolve the sulphate in 30 ounces, and the acid in 15 ounces of distilled water, filter and mix the solutions. Decant, and when the precipitate has subsided, wash, and dry it with a gentle heat.

It is a lemon-yellow, crystalline powder, insoluble in water, and, therapeutically, of little importance.

Dose: 2 to 3 grains.

FERRI NITRAS—NITRATE OF IRON.

PREPARATION.

Liquor Ferri Nitratis (U. S., Br.)—Solution of Nitrate of Iron.

Take of Iron, in the form of wire.................. 2½ ounces.
Nitric acid (by weight).................... 5 ounces.
Distilled water......................... sufficient.

Mix the iron with 12 ounces of distilled water, and add, in small portions at a time, 3 ounces of nitric acid previously mixed with 6 ounces of distilled water, moderating the reaction by setting the vessel in cold water, to prevent the occurrence of red fumes. When effervescence has nearly ceased, agitate until a portion of the liquid, being filtered, is of a pale green color. Then filter, pour into a capacious capsule, heat to 130°, and add the remainder of the acid. When effervescence has ceased, continue the heat until no more gas escapes, then add sufficient distilled water to make the liquid measure 36 ounces.

It is a transparent liquid, of a pale amber color, and sp. gr. 1.060 to 1.070. It is tonic and astringent, and is used in hemorrhages, chronic diarrhœa, etc.

Dose: 8 to 20 minims. *Solution of Pernitrate of Iron, Br.*

FERRI OXIDUM HYDRATUM (U. S. et al. Ph.)—HYDRATED OXIDE OF IRON.

Take of Solution of tersulphate of iron .... 1 pint.
Water of ammonia .... 20 ounces.
Water .... sufficient.

To the water of ammonia, mixed with 2 pints of water, add, stirring constantly, the iron solution previously mixed with 2 pints of water. Wash the precipitate until the washings are nearly tasteless, then mix with sufficient water to make the product measure 1½ pint.

Used as an antidote to arsenic.

Dose: ½ to 2 ounces.

Termed, by the British Pharmacopœia, *Moist Peroxide of Iron;* and, when dried, at or below 212°, *Hydrated Peroxide of Iron.*

PREPARATIONS.

Saccharated Oxide of Iron (Ger.).

Take of Solution of sesquichloride of iron .... 20 parts.
Syrup .... 20 parts.
Solution of caustic soda .... 40 parts.
White sugar .... sufficient.
Distilled water .... sufficient.

Mix the iron solution and the syrup, add the soda solution, stirring constantly, and set aside for twenty-four hours. Then pour the clear liquid into 300 parts of boiling distilled water, and allow it to settle. Collect the precipitate, wash it, free it of most of the water, mix with 90 parts of sugar, and evaporate to dryness. Then mix with sufficient sugar to make the product 100 parts.

It forms a reddish powder, of a sweet, ferruginous taste, and wholly soluble in 5 parts of water.

Dose: 10 to 20 grains.

FERRI PHOSPHAS (U. S. et al. Ph.)—PHOSPHATE OF IRON.

Take of Sulphate of iron .... 5 ounces.
Phosphate of sodium .... 6 ounces.
Water .... 8 pints.

Dissolve the salts separately, each in 4 pints of water, mix the solutions, and set aside until the precipitate has subsided. Then decant, wash the precipitate with hot water, and dry it with a gentle heat.

It is a bluish powder, odorless, tasteless, and insoluble. It is theoretically assumed to produce the combined effects of iron and phosphorus—an assumption which is more than doubtful.

Dose: 5 to 10 grains.

PREPARATIONS.

SYRUP OF PHOSPHATE OF IRON (Br.).

Take of Granulated sulphate of iron .... 224 grains.
Phosphate of sodium .... 200 grains.
Acetate of sodium .... 74 grains.
Diluted phosphoric acid .... 5½ ounces.
Refined sugar .... 8 ounces.
Distilled water .... 8 ounces.

Dissolve the sulphate of iron in 4 ounces of the water, and the phosphate and acetate of sodium in the remainder. Mix the solutions, collect and wash the precipitate. Then press it strongly between folds of bibulous paper, dissolve in the phosphoric acid, filter, add the sugar, and dissolve without heat. The product should measure 12 ounces.

Dose: 1 drachm.

PILL OF PHOSPHATE OF IRON.

Take of Phosphate of iron .... 1 grain.
Extract of colocynth .... 1 grain.
Extract of hyoscyamus .... 1 grain.
Bread crumbs .... 2 grains.

Make 1 pill. *Charing Cross Hospital.*

FERRI PYROPHOSPHAS (U. S., Fr., Ger.)—PYROPHOSPHATE OF IRON.

Take of Phosphate of sodium .... 7½ ounces.
Solution of tersulphate of iron .... 7 ounces, or sufficient.
Citric acid .... 2 ounces.
Water of ammonia .... 5½ ounces, or sufficient.
Water .... sufficient.

Heat the phosphate in a porcelain capsule until dry, then, in an iron capsule, to incipient redness, without fusion. Then dissolve, with the aid of heat, in 3 pints of water, filter, cool to 50°, and add solution of tersulphate of iron until it ceases to produce a precipitate. Collect, wash, and transfer the precipitate to a weighed capsule. Add water of ammonia to the citric acid until it is saturated and dissolved, mix the solution with the precipitate in the weighed capsule, and evaporate until the liquid is reduced to 16 ounces (by weight). Spread this on glass plates to dry. Lastly, preserve in a well-stopped bottle, protected from light.

It is in apple-green scales, having an acidulous, slightly saline taste, and wholly soluble in water. It is an excellent chalybeate.

Dose: 2 to 5 grains.

PREPARATIONS.

Syrup of Pyrophosphate of Iron (Fr.).

Take of	Pyrophosphate of iron	1 part.
	Distilled water	2 parts.
	Syrup	97 parts.

Dissolve the pyrophosphate in the water, filter, and mix with the syrup.

Dose: 1 to 2 drachms.

Mixture of Pyrophosphate of Iron.

Take of	Pyrophosphate of iron	1 drachm.
	Sulphate of quinia	1 drachm.
	Strychnia	1 grain.
	Diluted phosphoric acid	2 drachms.
	Syrup of ginger	2 ounces.
	Water	to 4 ounces.

Mix. Dose: 1 drachm. *Hammond.*

FERRI SULPHAS (U. S. et al. Ph.)—SULPHATE OF IRON.

Take of	Iron, in the form of wire	12 ounces.
	Sulphuric acid (by weight)	18 ounces.
	Water	8 pints.

Mix the acid and water, add the iron, and heat until effervescence ceases. Pour off the solution, add 30 grains of sulphuric acid, and filter through paper, allowing the lower end of the funnel to touch the bottom of the receiving vessel. Then evaporate by heat until sufficiently concentrated, and set aside in a covered vessel to crystallize. Drain the crystals, dry upon bibulous paper, and keep in a well-stopped bottle.

It is in transparent, bluish-green crystals, which are efflorescent, and wholly soluble in water. It is tonic and powerfully astringent. Used for the general tonic effects of iron, and to restrain undue secretion, especially from mucous surfaces.

Dose: 1 to 3 grains.

PREPARATIONS.

Ferri Sulphas Exsiccata (U. S., Br., Ger.)—Dried Sulphate of Iron.

Take of Sulphate of iron .... sufficient.

Heat to 300° as long as it loses weight.

Dose : ½ to 2 grains, in pill.

Liquor Ferri Subsulphatis (U. S.)—Solution of Subsulphate of Iron.

Take of Sulphate of iron	12 ounces.
Sulphuric acid	510 grains.
Nitric acid	780 grains.
Distilled water	sufficient.

Mix the acids with ½ pint of distilled water in a capacious porcelain capsule, heat to the boiling point, and add the sulphate, one-fourth at a time, stirring after each addition until effervescence ceases. Keep it in ebullition until nitrous vapors are no longer perceptible, and the color assumes a deep, ruby-red tint. When nearly cold, add enough distilled water to make it measure 12 ounces.

It is an inodorous, syrupy liquid, of a ruby-red color, and of an extremely astringent taste, without causticity.

Used topically as an astringent and styptic.

Known as *Monsel's Styptic.*

Liquor Ferri Tersulphatis (U. S., Br., Ger.)—Solution of Tersulphate of Iron.

Take of Sulphate of iron	12 ounces.
Sulphuric acid	1020 grains.
Nitric acid	840 grains.
Water	sufficient.

Proceed as in preparing solution of subsulphate of iron, adding, at the close of the operation, sufficient distilled water to make the product measure 1½ pint.

It is a dark, reddish-brown liquid, almost odorless, and of an acid, and extremely styptic taste. Used in preparations.

Pilulæ Ferri Compositæ (U. S.)—Compound Pills of Iron.

Take of Myrrh	36 grains.
Sulphate of iron	18 grains.
Carbonate of sodium	18 grains.
Syrup	sufficient.

Rub the myrrh, first with the carbonate, then with the sulphate, make a mass with syrup, and divide into 24 pills.

Dose : 2 to 6 pills.

Blaud's Ferruginous Pills (Fr.).

Take of Dried sulphate of iron	30 parts.
Dried carbonate of potassium	30 parts.
Gum arabic	5 parts.
Water	30 parts.
Simple syrup	15 parts.

Dissolve the gum in the water on a water-bath, add the syrup and iron, then the carbonate of potassium, and evaporate, stirring constantly, to a proper consistence. Divide into pills weighing 6 grains each.

Dose: 1 to 3 pills thrice daily. Believed by many to be one of the best means of supplying the system with iron in anæmia, chlorosis, etc.

Hooper's Pills.

Take of Barbadoes aloes	8 ounces.
Sulphate of iron	4 ounces.
Extract of black hellebore	2 ounces.
Myrrh	2 ounces.
Soap	2 ounces.
Canella	1 ounce.
Ginger	1 ounce.

Beat into a mass with water, and divide into pills of 2½ grains each.
Dose: 2 to 3 pills. Laxative and emmenagogue.

Pills of Sulphate of Iron and Quinia.

Take of Sulphate of iron	1 grain.
Sulphate of quinia	1 grain.
Extract of chamomile	1 grain.

Make 1 pill. *Royal Chest Hospital.*

Take of Dried sulphate of iron	1½ drachm.
Sulphate of quinia	1 drachm.

Mix, and make 30 pills.

Dose: 1 pill three times a day, or 4 or 5 during the day. In enlarged spleen of malarial origin. *Bartholow.*

Pills of Iron and Valerian.

Take of Sulphate of iron	15 grains.
Valerianate of quinia	15 grains.
Strychnia	1 grain.
Extract of rhubarb	40 grains.

Mix, and divide into 30 pills.
Dose: 1 pill. *Samaritan Hospital.*

Pill of Iron and Zinc.

Take of	Sulphate of iron	1 grain.
	Sulphate of zinc	1 grain.
	Ipecacuanha	1 grain.
	Extract of conium	2 grains.

Make 1 pill. *London Chest Hospital.*

Pills of Iron and Croton Oil.

Take of	Sulphate of iron	12 grains.
	Croton oil	1 minim.
	Rhubarb pill	20 grains.

Mix, and divide into 12 pills.

Dose: 1 or 2 pills. *Samaritan Hospital.*

Mixtures of Sulphate of Iron.

Take of	Sulphate of iron	4 grains.
	Aromatic sulphuric acid	20 minims.
	Distilled water	1 ounce.

Mix. Dose: 1 drachm. *Ellis.*

Take of	Sulphate of iron	3 grains.
	Diluted sulphuric acid	5 minims.
	Infusion of quassia	1 ounce.

Mix. One dose. *Charing Cross Hospital.*

Take of	Sulphate of zinc	2 grains.
	Sulphate of magnesium	60 grains.
	Diluted sulphuric acid	5 minims.
	Peppermint water	1 ounce.

Mix. One dose. *London Throat Hospital.*

Inhalation of Sulphate of Iron.

Take of	Sulphate of iron	20 to 40 grains.
	Water	8 ounces.

Mix. Use by means of a steam atomizer. Astringent.

G. M. Lefferts.

FERRI SULPHURETUM (U. S., Br., Fr.)—SULPHURET OF IRON.

Prepared by heating iron filings with sulphur in a crucible. It is used only for the preparation of sulphuretted hydrogen, which it evolves when mixed with diluted sulphuric or hydrochloric acid.

FERRI VALERIANAS—VALERIANATE OF IRON.

Prepared by adding a solution of valerianate of sodium to one of sulphate of iron as long as a precipitate is produced, washing and drying the precipitate.

It is a dark brownish-red powder, having the odor, and some of the taste of valerian.

Used in anæmia attended with hysterical manifestations.

Dose: ½ to 2 grains.

FICUS (U. S. et al. Ph.)—FIG.

The fleshy receptacle of the fruit of Ficus Carica, L. (*Nat. ord., Artocarpaceæ*), a tree indigenous to Asia, but long cultivated in subtropical regions of the Old World, and now, to some extent, in the Southern United States.

Figs are nutritive and slightly laxative. They enter into the composition of confection of senna.

FILIX MAS (U. S. et al. Ph.)--FERN—MALE FERN.

The rhizome of Aspidium Filix-mas, Swartz (*Nat. ord., Filices*). This fern is indigenous to the Old World, where it is very common, and also to this country, being found from Lake Superior westward.

The root has a disagreeable odor, and a sweetish, afterward bitter, acrid taste. It yields an oleoresin, to which its medicinal effects are chiefly due.

Male fern is used solely as a remedy for tape-worm, and is one of the best.

The rhizome of *Aspidium marginale*, Willd., our common shield fern, has also been used as a tænicide, and with good results.

Dose: 1 to 3 drachms.

PREPARATIONS.

Oleoresina Filicis (U. S. et al. Ph.)—Oleoresin of Fern.

Take of Male fern	12 ounces.
Ether	sufficient.

Moisten, pack, and percolate to 24 ounces. Recover the greater part of the ether by distillation, and expose the residue in a capsule until the remaining ether has evaporated. Keep in a well-stopped bottle.

Dose: 20 to 40 minims.

Termed *extract*, *liquid extract*, and *ethereal extract*, by the German, British, and French pharmacopœias, respectively.

Confection of Male Fern.

Take of Oleoresin of fern	3 drachms.
Ether	1½ drachm.
Powdered valerian	2 drachms.
Purified honey	1½ ounce.

Mix. Dose: 3 or 4 drachms every half-hour. For tape-worm.

R. Tauszky.

Mixture of Male Fern.

Take of Liquid extract of male fern	1 drachm.
Glycerin	1 drachm.
Mucilage	1 drachm.
Water	to 1 ounce.

Mix. One dose. *Middlesex Hospital.*

FŒNICULUM (U. S. et al. Ph.)—FENNEL.

The fruit of Fœniculum vulgare, Gaertn. (*Nat. ord., Umbelliferæ*), an herbaceous perennial indigenous to Southern Europe, and extensively cultivated in France and Germany.

Fennel has an aromatic odor, and a sweetish taste. It is aromatic and carminative, and is used chiefly as an adjunct to other medicines.

Dose: 20 to 30 grains.

PREPARATIONS.

Aqua Fœniculi (U. S. et al. Ph.)—Fennel Water.

Take of Oil of Fennel	½ drachm.
Carbonate of magnesium	60 grains.
Distilled water	2 pints.

Rub the oil with the carbonate, then with the water, gradually added, and filter.

Used as a vehicle. The European pharmacopœias prepare it by distilling fennel with water. The United States Pharmacopœia also permits it to be prepared in this manner.

Oleum Fœniculi (U. S., Fr., Ger.)—Oil of Fennel.

The oil obtained from fennel by distillation.

Dose: 5 to 15 minims.

Spirit, or Essence of Fennel (Fr.).

Take of Fennel	1 part.
Alcohol (80%)	8 parts.

Distill off the alcohol.

Dose: 20 to 30 minims.

FRASERA (U. S.)—AMERICAN COLUMBO.

The root of Frasera Carolinensis, Walt. (*Nat. ord., Gentianaceæ*), a biennial or triennial, indigenous to the United States. It has a very bitter taste, and possesses tonic properties analogous to those of gentian and columbo, and is occasionally substituted for them.

Dose: 30 to 60 grains.

Infusion of American Columbo.

Take of American Columbo	1 ounce.
Boiling water	1 pint.

Infuse one hour, and strain.

Dose: 1 to 2 ounces. *Wood.*

Tincture of American Columbo.

Take of American Columbo	1 part.
Diluted alcohol	5 parts.

Macerate two weeks, and filter.

Dose: 1 to 2 drachms.

GALBANUM (U. S. et al. Ph.)—GALBANUM.

A gum-resin obtained from Ferula galbaniflua, Boiss. et Buhse, and other species of Ferula (*Nat. ord., Umbelliferæ*), tall, herbaceous plants, indigenous to Persia. The drug is met with in drops or tears cohering in a mass, though sometimes of a semifluid consistence. It has a peculiar, aromatic odor, and a bitter, acrid taste.

It is employed internally as a stimulating expectorant, and externally, in the form of a plaster, as an application to indolent swellings, etc.

Dose: 5 to 20 grains.

PREPARATIONS.

Pilulæ Galbani Compositæ (U. S.)—Compound Pills of Galbanum.

Take of Galbanum	36 grains.
Myrrh	36 grains.
Asafetida	12 grains.
Syrup	sufficient.

Beat together and divide into 24 pills.

Dose: 1 to 5 pills. *Compound Pill of Asafetida, Br.*

Emplastrum Galbani Compositum (U. S.)—Compound Galbanum Plaster.

Take of	Galbanum	8 ounces.
	Turpentine	1 ounce.
	Burgundy pitch	3 ounces.
	Lead plaster	36 ounces.

Mix, melt, and strain the galbanum and turpentine, add the pitch, and then the plaster, previously melted.

GALBANUM PLASTER (Br.).

Take of	Galbanum	1 ounce.
	Ammoniacum	1 ounce.
	Yellow wax	1 ounce.
	Lead plaster	8 ounces.

Mix, melt, and strain the galbanum and ammoniacum, then add the plaster and wax, previously melted together.

GALBANUM PLASTER WITH SAFFRON (Ger.).

Take of	Purified galbanum	24 parts.
	Turpentine	6 parts.
	Lead plaster	24 parts.
	Yellow wax	8 parts.
	Saffron, powdered	1 part.

Dissolve the galbanum in the turpentine by means of a steam-bath, add the plaster and wax, previously melted together, then the saffron, previously rubbed to a pulp with a little alcohol.

GALLA (U. S. et al. Ph.)—NUTGALL.

Excrescences, caused by the sting of an insect, upon the young branches of Quercus infectoria, Olivier (*Nat. ord., Cupuliferæ*), a small oak of Greece, Asia Minor, Cyprus, and Syria.

They are spherical, two-fifths to four-fifths of an inch in diameter, hard and brittle, and having a bitter, astringent taste, due to the tannic, and gallo-tannic acids which they contain.

Galls are important as being the source from which tannic and gallic acids are derived. They are used both externally and internally as an astringent.

Dose: 10 to 20 grains.

PREPARATIONS.

Tinctura Gallæ (U. S. et al. Ph.)—Tincture of Nutgall.

Take of Nutgall	4 ounces.
Diluted alcohol	sufficient.

Moisten, pack, and percolate to 2 pints.

Dose: ½ to 1 drachm.

Unguentum Gallæ (U. S., Br.)—Ointment of Nutgall.

Take of Nutgall	60 grains.
Lard	420 grains.

Mix thoroughly.

Applied to hemorrhoids, prolapsus ani, etc.

Ointment of Galls and Opium (Br.).

Take of Ointment of galls (80 grains to 1 ounce)	1 ounce.
Opium, in powder	32 grains.

Mix thoroughly.

Used like the preceding. Will often give great relief in inflamed and irritable hemorrhoids.

Decoction of Galls.

Take of Galls	2½ ounces.
Water	40 ounces.

Boil to 20 ounces. *St. George's Hospital.*

GAMBOGIA (U. S. et al. Ph.)—GAMBOGE.

A gum-resin obtained from Garcinia Morella, Desv. (*Nat. ord., Guttiferæ*), a middle-sized tree, indigenous to Siam, Cambogia, and Cochin China.

Gambogia is of an orange-yellow color, and has a disagreeable, acrid taste. It is a very active hydragogue cathartic, but, on account of the violence of its action, it is seldom used, except in combination with other remedies.

Dose: 1 to 3 grains.

PREPARATIONS.

Compound Pill of Gamboge (Br.).

Take of Gamboge	1 ounce.
Barbadoes aloes	1 ounce.
Compound powder of cinnamon	1 ounce.
Hard soap	2 ounces.
Syrup	sufficient.

Mix the powders, add the syrup, and beat into a mass.

Dose: 5 to 10 grains.

COMPOUND PILLS OF GAMBOGE (Fr.).

Take of Barbadoes aloes	20 parts.
Gamboge	20 parts.
White honey	10 parts.
Oil of anise	1 part.

Make a mass, and divide into pills of 3 grains each.

Dose: 1 to 2 pills.

COMPOUND PILLS OF GAMBOGE—MILLER'S PILLS.

Take of Scammony	20 grains.
Aloes	20 grains.
Gamboge	20 grains.
Calomel	20 grains.
Cream of tartar	20 grains.
Extract of dandelion	sufficient.

Mix, and divide into 20 pills.

Dose: 1 to 2 pills. *Hospital Formulary.*

GAULTHERIA (U. S., Fr.)—WINTERGREEN.

The leaves of Gaultheria procumbens, L. (*Nat. ord.*, *Ericaceæ*), a very small shrub with a creeping stem, indigenous to North America.

It has an agreeable odor, an aromatic, and slightly astringent taste, and is used as a flavoring agent. The volatile oil is the preparation generally employed.

Dose: ½ to 2 drachms.

PREPARATIONS.

Oleum Gaultheriæ (U. S.)—Oil of Gaultheria.

The oil obtained from wintergreen by distillation.

Dose: 2 to 10 minims.

GAULTHERIA WATER.

Take of Oil of gaultheria	16 minims.
Carbonate of magnesium	1 drachm.
Water	1 pint.

Rub the oil with the carbonate, then with the water added gradually, and filter.

Used as a vehicle. *Griffith.*

SPIRIT, OR ESSENCE OF GAULTHERIA.

Take of Oil of Gaultheria	1 ounce.
Alcohol	15 ounces.

Dissolve.

Dose: ½ to 1 drachm. A drachm, taken in a wineglass of sweetened water three times a day, is useful in common colds. *F. A. Burrall.*

GELSEMIUM (U. S.)—YELLOW JESSAMINE.

The root of Gelsemium sempervirens, Ait. (*Nat. ord., Loganiaceæ*), a shrubby, climbing vine, indigenous to the Southern States.

Gelsemium has a peculiar odor, and an agreeable, bitter taste. It is an arterial sedative, lowering the pulse, and producing a corresponding depression of the nervous system. In overdoses it is a very dangerous poison. It is used in the early stages of acute inflammatory affections, as pleurisy, pneumonia, and rheumatism, and in some spasmodic diseases, as tetanus, whooping-cough, spasmodic asthma, etc.

Dose: 2 to 5 grains.

PREPARATIONS.

Extractum Gelsemii Fluidum (U. S.)—Fluid Extract of Gelsemium.

Take of Gelsemium	16 ounces.
Alcohol	sufficient.

Moisten the gelsemium with 4 ounces of alcohol, and proceed according to the general formula, page 161.

Dose: 2 to 5 minims.

Tincture of Gelsemium.

Take of Gelsemium	1 part.
Diluted alcohol	sufficient.

Moisten, pack, and percolate to 5 parts.

Dose: 10 to 20 minims. *J. P. Remington, Report Am. Ph. Ass'n.*

GENTIANA (U. S. et al. Ph.)—GENTIAN.

The root of Gentiana lutea, L. (*Nat. ord., Gentianaceæ*), an herbaceous perennial, indigenous to Southern and Central Europe.

It has a feeble odor, an intensely bitter taste, and is a simple bitter without any astringency. As a tonic it is well suited to cases of gastric derangement, and to stimulate the appetite in convalescence from acute diseases.

Some of our indigenous species of gentian, as G. Catesbæi and G. Andrewsii, have also been used medicinally, and with good effect.

Dose: $\frac{1}{2}$ to 1 drachm.

PREPARATIONS.

Extractum Gentianæ (U. S. et al. Ph.)—Extract of Gentian.

Take of Gentian	12 ounces.
Water	sufficient.

Exhaust the gentian by percolation with water, boil the infusion to three-fourths of its bulk, strain, and evaporate to a proper consistence.

Dose: 10 to 30 grains.

Extractum Gentianæ Fluidum (U. S.)—Fluid Extract of Gentian.

Take of Gentian	16 ounces.
Glycerin	4 ounces.
Alcohol	sufficient.
Water	sufficient.

Mix 8 ounces of alcohol, 3 of glycerin, and 5 of water, moisten the gentian with 4 ounces of the mixture, and proceed according to the general formula, page 161.

Dose: ½ to 1 drachm.

Infusum Gentianæ Compositum (U. S., Br.)—Compound Infusion of Gentian.

Take of Gentian	½ ounce.
Bitter-orange peel	60 grains.
Coriander	60 grains.
Alcohol	2 ounces.
Water	sufficient.

Mix the alcohol with 14 ounces of water, moisten the mixed powders with 3 drachms of the menstruum, then pack, and percolate with the remainder and sufficient water to make 1 pint.

Dose: 1 to 2 ounces.

A British preparation, almost identical with this, is termed *Gentian Mixture*, while the preparation having this title is made with gentian, 60 grains; bitter orange peel, 60 grains; fresh lemon peel, ¼ ounce; boiling distilled water, 10 ounces.

Tinctura Gentianæ Composita (U. S., Br., Fr.)—Compound Tincture of Gentian.

Take of Gentian	2 ounces.
Bitter orange peel	1 ounce.
Cardamom	½ ounce.
Diluted alcohol	sufficient.

Moisten, pack, and percolate to 2 pints.

Dose: ½ to 2 drachms.

The French Codex directs: gentian, 10 parts; carbonate of sodium, 3 parts; alcohol (60%), 300 parts.

Tincture of Gentian (Fr., Ger.).

Take of Gentian .... 1 part.
Alcohol (60%) .... 5 parts.

Macerate for ten days, express, and filter.

Dose: ½ to 2 drachms.

Syrup of Gentian (Fr.).

Take of Gentian .... 10 parts.
Boiling water .... 100 parts.
White sugar .... sufficient.

Macerate the gentian in the water for six hours, express, filter, and add sugar in the proportion of 19 parts to 10 parts of filtered liquid.

Dose: 2 to 4 drachms.

Wine of Gentian (Fr.).

Take of Gentian .... 3 parts.
Alcohol (60%) .... 6 parts.
Red wine .... 100 parts.

Macerate the gentian in the alcohol for twenty-four hours, add the wine, macerate for ten days, express and filter.

Dose: 1 to 2 ounces.

Mixtures of Gentian.

Take of Diluted hydrochloric acid .... 10 minims.
Compound infusion of gentian .... to 1 ounce.

Mix. One dose. *Brompton Consumption Hospital.*

Take of Diluted nitro-hydrochloric acid .... 12 minims.
Spirit of chloroform .... 10 minims.
Compound infusion of gentian .... to 1 ounce.

Mix. One dose. *University College Hospital.*

Take of Diluted hydrocyanic acid .... 3 minims.
Bicarbonate of sodium .... 15 grains.
Compound infusion of gentian .... to 1 ounce.

Mix. One dose. *Brompton Consumption Hospital.*

Take of Infusion of gentian .... 6 drachms.
Infusion of senna .... 3 drachms.
Compound tincture of cardamom .... 1 drachm.

Mix. One dose. *St. George's Hospital.*

Pill of Gentian and Iron.

Take of Extract of gentian .... 3 grains.
Sulphate of iron .... 1 grain.

Make 1 pill. *Guy's Hospital.*

PILL OF GENTIAN AND ZINC.

Take of Extract of gentian	3 grains.
Sulphate of zinc	1 grain.
Columbo	sufficient.

Make 1 pill. *Guy's Hospital.*

GERANIUM (U. S.)—GERANIUM—CRANESBILL.

The rhizome of Geranium maculatum, L. (*Nat. ord., Geraniaceæ*), an herbaceous perennial, indigenous to North America, being a prominent and beautiful feature of moist woodlands during its season of bloom, from April till July.

The rhizome is from one to three inches long, and from a quarter to a half inch in diameter. It has an astringent taste, and contains tannic and gallic acids.

It is an excellent astringent, and may be employed in dysentery, diarrhœa, etc., with as good effects as catechu or kino.

Dose: ½ to 1 drachm.

Extractum Geranii Fluidum (U. S.)—Fluid Extract of Geranium.

Take of Geranium	16 ounces.
Glycerin	4 ounces.
Alcohol	sufficient.
Water	sufficient.

Mix 8 ounces of alcohol, 3 of glycerin, and 5 of water, moisten the geranium with 4 ounces of the mixture, and proceed according to the general formula, page 161.

Dose: ½ to 1 drachm.

EXTRACT OF GERANIUM.

Take of Geranium, bruised	1 pound.
Water	1 gallon.

Boil to one-half, strain, and evaporate to a proper consistence. Very similar to rhatany, and may be given in the same cases and in the same doses.

Dose: 5 to 20 grains. *Griffith.*

DECOCTION OF GERANIUM.

Take of Geranium	1 ounce.
Water	1½ pint.

Boil to 1 pint.

Dose: 1 to 2 ounces. *Wood.*

TINCTURE OF GERANIUM.

Take of Geranium, fresh .... 1 part.
Alcohol .... 2 parts.

Macerate two weeks, and filter.

Dose: 1 to 2 drachms. Particularly suited to the treatment of such discharges as continue after the removal of their exciting cause.

Bigelow.

GLYCERINUM (U. S. et al. Ph.)—GLYCERIN.

A sweet principle obtained from fats and fixed oils, by decomposing them into their proximate principles.

It is a thick, syrupy, colorless liquid, without odor, and having a very sweet taste. It is largely used as a solvent, and to sweeten mixtures when from any reason it is desirable to avoid the administration of sugar.

A class of officinal preparations, termed glycerites (*glycerita*), are solutions of medicinal substances in glycerin, made by trituration.

Dose: ½ to 2 drachms.

LOTIONS OF GLYCERIN.

Take of Glycerin .... 1 drachm.
Water .... to 1 ounce.

Mix. *British Skin Hospital.*

Take of Glycerin .... ½ ounce.
Saccharated solution of lime .... ½ ounce.

Mix. *Samaritan Hospital.*

GLYCYRRHIZA (U. S. et al. Ph.)—LIQUORICE ROOT.

The root and subterraneous stem of Glycyrrhiza glabra, L. (*Nat. ord., Leguminosæ*), an herbaceous perennial indigenous to Southern Europe, but extensively cultivated in other regions.

Liquorice root contains sugar and albuminous matter, and, in addition, a sweet principle termed glycyrrhizin (*glycyrrhizinum*).

It is demulcent and slightly laxative, and is used in acute catarrhal affections of the pulmonary and urinary organs. The extract is much used to cover the taste of nauseous medicines.

Dose: 1 to 3 drachms.

PREPARATIONS.

Extractum Glycyrrhizæ (U. S. et al. Ph.)—Liquorice.

An aqueous extract, made on a large scale by boiling the root with water, straining, and evaporating until it solidifies on cooling.

Dose: 1 to 2 drachms.

Extractum Glycyrrhizæ Fluidum (U. S., Br.)—Fluid Extract of Liquorice Root.

Take of Liquorice root........................... 16 ounces.
Glycerin .... 4 ounces.
Alcohol........................... sufficient.
Water .... sufficient.

Mix 8 ounces of alcohol, 3 of glycerin, and 5 of water, moisten the root with 4 ounces of the mixture, and proceed according to the general formula, page 161.

Dose: ½ to 1 drachm.

Mistura Glycyrrhizæ Composita (U. S.)—Compound Mixture of Liquorice—Brown Mixture.

Take of Liquorice .... ½ ounce.
Sugar .... ½ ounce.
Gum arabic .... ½ ounce.
Camphorated tincture of opium............ 2 ounces.
Wine of antimony........................... 1 ounce.
Spirit of nitrous ether .... ½ ounce.
Water........................... 12 ounces.

Rub the powders with the water gradually added, then add the other ingredients, and mix.

Dose: 1 to 5 drachms.

An excellent and popular cough mixture.

Trochisci Glycyrrhizæ et Opii (U. S.)—Troches of Liquorice and Opium.

Take of Extract of opium........................... 24 grains.
Liquorice .... 2 ounces.
Gum arabic........................... 1 ounce.
Sugar .... 3 ounces.
Oil of anise........................... 15 minims.

Rub the powders together, then add the oil, and with water form a mass, to be divided into 480 troches.

Dose: 1 to 5 troches. In cough, and irritation of the throat.

LIQUORICE PASTE (Fr., Ger.).

Take of Liquorice .... 1 part.
White sugar........................... 1 part.
Gum arabic .... 2 parts.
Water .... 6 parts.

Dissolve the liquorice in the water, and strain; add the gum, previously washed, and when dissolved, strain, and add the sugar. Evaporate until nearly solid, roll into sheets, cut them into strips, and dry. The German process is different, but the product is quite similar. Another similar French preparation contains a minute quantity of extract of opium.

Compound Liquorice Powder (Ger.).

Take of Senna	2 parts.
Liquorice root	2 parts.
Fennel seed	1 part.
Washed sulphur	1 part.
White sugar	6 parts.

Mix. Dose: ½ to 1 drachm. A mild and excellent laxative.

Syrup of Liquorice (Ger.).

Take of Peeled liquorice root	4 parts.
Water	18 parts.
White sugar	12 parts.
Clarified honey	12 parts.

Macerate the root in the water for one night, express, filter, and evaporate to 7 parts, then add the sugar and honey.

Used as a vehicle.

Take of Fluid extract of liquorice root	2 ounces.
Syrup	14 ounces.

Mix. Dose: 1 to 2 drachms. *Charles Rice.*

Pectoral Elixir (Ger.).

Take of Purified liquorice	2 parts.
Anisated spirit of ammonia	2 parts.
Fennel water	6 parts.

Dissolve the liquorice in the fennel water, then add the anisated spirit.

Decoction of Liquorice.

Take of Liquorice root	1½ ounce.
Anise	½ ounce.
Water	20 ounces.

Boil fifteen minutes, and strain.

Dose: 1 to 3 ounces. *Brompton Consumption Hospital.*

GOSSYPIUM (U. S. et al. Ph.)—COTTON.

The hairy filaments attached to the seeds of Gossypium herbaceum, L., and other species of Gossypium (*Nat. ord., Malvaceæ*).

The cotton plant is indigenous to the tropical and subtropical regions of Asia and Africa, but is extensively cultivated in this country.

Cotton is used in the preparation of pyroxylon (*soluble gun cotton*), and as a surgical dressing. Freed from impurities, it is an admirable application to recent wounds, burns, etc., effectually preserving them from contact with the air while it absorbs the discharges.

PREPARATIONS.

Pyroxylon (U. S. et al. Ph.)—Soluble Gun Cotton.

Take of Cotton, freed from impurities	½ ounce.
Nitric acid (by weight)	3½ ounces.
Sulphuric acid (by weight)	4 ounces.

Mix the acids gradually, and when the temperature has fallen to 90°, add the cotton. Allow it to macerate fifteen hours, then wash first with cold, afterward with boiling water, and dry on a water-bath.

For its uses, see *Collodion.*

Medicated Cotton.

Take of Boracic acid	60 grains.
Glycerin	20 minims.
Water	6 drachms.
Cotton wool, in a thin sheet	60 grains.

Mix the acid, glycerin, and water, and dissolve with the aid of heat. Saturate the cotton with the solution, and dry with a moderate heat.

Antiseptic and disinfectant. Used as a nasal plug, or respirator, in affections of the nose and naso-pharyngeal region. It may also be used as a surgical dressing. *London Throat Hospital.*

Take of Tannic acid	30 grains.
Glycerin	10 minims.
Water	6 drachms.
Cotton wool, in a thin sheet	60 grains.

Dissolve the acid in the glycerin and water, saturate the cotton with the solution, and dry with a moderate heat. Astringent. Used like the preceding. *London Throat Hospital.*

Take of Solution of perchloride of iron	½ ounce.
Glycerin	10 minims.
Cotton wool, in a thin sheet	60 grains.

Mix the glycerin with the iron solution, saturate the cotton with the mixture, and dry by exposure to the air. Astringent and styptic.

London Throat Hospital.

Take of Iodoform	70 grains.
Pure ether	10 drachms.
Absolute alcohol	2 drachms.
Glycerin	10 minims.
Cotton wool, in a thin sheet	60 grains.

Dissolve the iodoform in the ether, add the alcohol and glycerin, previously mixed, saturate the cotton with the solution, and dry by exposure to the air. Prepare in a room without fire or artificial light. Stimulant and antiseptic. *London Throat Hospital.*

GOSSYPII RADICIS CORTEX (U. S.)—BARK OF COTTON ROOT.

The bark of the root of Gossypium herbaceum, L., and of other species of Gossypium. See *Gossypium.*

Cotton root bark has an effect upon the uterus like that of ergot. It has been long used by the negresses of the Southern States to produce abortion. It is used medicinally in amenorrhœa and dysmenorrhœa.

Dose: 10 to 30 grains.

PREPARATIONS.

Extractum Gossypii Radicis Fluidum (U. S.)—Fluid Extract of Cotton Root.

Take of Bark of cotton root	16 ounces.
Glycerin	4 ounces.
Alcohol	sufficient.
Water	sufficient.

Mix 4 ounces of alcohol, 3 of glycerin, and 5 of water, moisten the bark with 4 ounces of the mixture, and proceed according to the general formula, page 161.

Dose: 10 to 30 minims.

Decoction of Cotton Root.

Take of Bark of cotton root	4 ounces.
Water	2 pints.

Boil to 1 pint.

Dose: 2 to 4 ounces every twenty or thirty minutes as an oxytocic.

National Dispensatory.

GRANATI FRUCTUS CORTEX (U. S. et al. Ph.)—POMEGRANATE RIND.

GRANATI RADICIS CORTEX (U. S. et al. Ph.)—BARK OF POMEGRANATE ROOT.

The rind of the fruit and bark of the root of Punica Granatum, L. (*Nat. ord., Granateæ*), a shrub or small tree indigenous to Asia, but cultivated in many subtropical countries.

The bark of pomegranate root is powerfully astringent, owing to the large percentage of tannin which it contains, and which is its chief constituent. Its most important use, however, is not as an astringent, but as a remedy for tape-worm. The rind of the fruit is used as an astringent in diarrhœa, etc.

Dose: of the rind, 20 to 30 grains; of the root bark, 10 to 30 grains.

PREPARATIONS.

DECOCTION OF POMEGRANATE ROOT (Br., Fr.).

Take of Pomegranate root bark .... 2 ounces.
Distilled water .... 40 ounces.

Boil down to 20 ounces, and strain.

Dose: 1 to 2 ounces.

ALCOHOLIC EXTRACT OF POMEGRANATE ROOT (Fr.).

Take of Pomegranate root bark .... 1 part.
Alcohol (60%) .... 6 parts.

Exhaust the bark by percolation with the alcohol, and evaporate to a soft extract.

Dose: 10 to 20 grains.

MIXTURE OF POMEGRANATE ROOT AND MALE FERN.

Take of Pomegranate root .... 4 ounces.

Macerate twenty-four hours in 8 ounces of water, then boil to 3 ounces, and add of

Oleoresin of fern .... 3 drachms.
Ether .... 1 drachm.
Fluid extract of valerian .... 2 drachms.
Croton oil .... 1 minim.
Honey .... 1½ ounce.

Mix. Dose: After a fast of twenty-four hours, eating only herring and onions, or garlic, take one-third of the mixture, and repeat in fifteen minutes. An hour later, take a dose of castor oil, and if the worm is not expelled within three hours, take the remainder of the mixture. For tapeworm. *R. Tauszky.*

GRINDELIA.—GRINDELIA.

The leaves and tops of Grindelia robusta, Nutt. and G. squarrosa, Dunal (*Nat. ord., Compositæ*), herbaceous perennials, indigenous to the Pacific coast of North America. The medicinal properties of the two are probably nearly identical. At present grindelia is little more than a subject of experiment. It has been found very beneficial in some cases of asthma, and is reported to have been serviceable in bronchitis and whooping-cough.

The fluid extract, applied externally, was asserted to cure Rhus poisoning, but like many other remedies for this affection, it has not substantiated the claims made for it. It is no better than any other application which protects the inflamed surface from the air.

Dose: 15 to 60 grains.

Fluid Extract of Grindelia.

Take of Grindelia .... 16 parts.
Alcohol, 3 parts—water, 2 parts .... sufficient.

Moisten the grindelia with 6 ounces of the menstruum, and proceed according to the general formula, page 161.

Dose: 15 to 60 minims.

Tincture of Grindelia.

Take of Grindelia .... 1 part.
Alcohol .... sufficient.

Moisten, pack, and percolate to 5 parts.

Dose: ½ to 2 drachms. *J. P. Remington, Rept. Am. Ph. Ass'n.*

Mixture of Grindelia.

Take of Fluid extract of grindelia .... 4 ounces.
Fluid extract of rhubarb .... 1 ounce.
Fluid extract of senna .... 1 ounce.

Mix. Dose: 2 drachms every half hour during the spasm of hay fever, and afterward at intervals of three hours. *Napheys.*

GUAIACI LIGNUM (U. S. et al. Ph.)—GUAIACUM WOOD.

The wood of Guaiacum officinale, L. (*Nat. ord., Zygophylleæ*), a tree indigenous to the West Indies and Central America.

It is very heavy, and contains about twenty-five per cent. of resin (*Guaiaci resina*), its most important and valuable constituent. Both the wood and resin are used medicinally.

Guaiacum is stimulant and alterative, and is used mainly in diseases of a chronic and obstinate character, such as syphilis, chronic skin diseases, chronic rheumatism, etc. It has also been used considerably in amenorrhœa and dysmenorrhœa.

The resin, being much more eligible than the wood, is generally employed. The latter enters into the composition of *Compound Decoction of Sarsaparilla*, which see.

PREPARATIONS.

Decoction of Guaiacum Wood (Fr.).

Take of Guaiacum wood .... 5 parts.
Water .... sufficient.

Boil the wood for an hour in sufficient water to obtain 100 parts.

Dose: 1 to 2 ounces.

TINCTURE OF GUAIACUM WOOD (Fr.).

Take of Guaiacum wood	1 part.
Alcohol (60%)	5 parts.

Macerate ten days, express and filter.
Dose: 2 to 4 drachms.

GUAIACI RESINA (U. S. et al. Ph.)—GUAIAC.

The resin obtained from the wood of Guaiacum officinale, L., by exudation, by incision, by heat, or by decoction. Medicinal properties and uses the same as of Guaiacum wood.

Dose: 10 to 30 grains.

PREPARATIONS.

Tinctura Guaiaci (U. S., Br., Ger.)—Tincture of Guaiac.

Take of Guaiac	6 ounces.
Alcohol	sufficient.

Moisten, pack, and percolate to 2 pints.
Dose: ½ to 1 drachm

Tinctura Guaiaci Ammoniata (U. S., Br., Ger.)—Ammoniated Tincture of Guaiac.

Take of Guaiac	6 ounces.
Aromatic spirit of ammonia	2 pints.

Macerate seven days, and filter.
Dose: ½ to 1 drachm.

GUAIACUM MIXTURE (Br.).

Take of Guaiacum resin	½ ounce.
Refined sugar	½ ounce.
Gum arabic	¼ ounce.
Cinnamon water	20 ounces.

Triturate the guaiac with the sugar and gum, adding the cinnamon water gradually.
Dose: ½ to 2 ounces.

Take of Ammoniated tincture of guaiacum	30 minims.
Mucilage	1 drachm.
Water	to 1 ounce.

Mix. One dose. *London Ophthalmic Hospital.*

Gargle of Guaiac.

Take of	Ammoniated tincture of guaiac	3 drachms.
	Solution of potassa	3 drachms.
	Tincture of opium	2 drachms.
	Cinnamon water	to 8 ounces.

Mix. Use as a gargle every hour, in clergyman's sore throat.

Garner.

Compound Powder of Guaiac.

Take of	Guaiac	15 grains.
	Carbonate of magnesium	15 grains.
	Precipitated sulphur	15 grains.
	Gum arabic	15 grains.
	Bicarbonate of potassium	22 grains.

Mix. Dose: ½ to 1 drachm. *St. George's Hospital.*

GUARANA—PAULLINIA.

The powdered seeds of Paullinia sorbilis, Martius (*Nat. ord., Sapindaceæ*), a climbing vine indigenous to the region of the Amazon.

It has a bitter and astringent taste, and contains caffein as its most important constituent. In Brazil it is used in much the same manner and for the same purposes as we use coffee. When first introduced to the medical profession, it was claimed to possess very important properties, but it is now seldom employed except in nervous and sick headaches.

Dose: 10 to 30 grains.

Fluid Extract of Guarana.

Take of	Guarana	16 parts.
	Diluted alcohol	sufficient.

Moisten the guarana with 6 parts of diluted alcohol, and proceed according to the general formula, page 161.

Dose: 10 to 30 minims.

Tincture of Guarana.

Take of	Guarana	1 part.
	Alcohol	sufficient.

Moisten, pack, and percolate to 5 parts.

Dose: 1 to 2 drachms. *J. P. Remington, Report of Am. Ph. Ass'n.*

HÆMATOXYLON (U. S. et al. Ph.)—LOGWOOD.

The heart-wood of Hæmatoxylon Campechianum, L. (*Nat. ord., Leguminosæ*), a medium-sized tree, indigenous to the region about the bay of Campeachy in Yucatan, and other portions of Central America, from whence it has been introduced into many of the West India Islands.

Logwood has a feeble, rather unpleasant odor, and a sweetish, astringent taste. It is a mild astringent, well suited to the later stages of infantile diarrhœa.

Dose: ½ to 2 drachms, in decoction.

PREPARATIONS.

Decoctum Hæmatoxyli (U. S., Br.)—Decoction of Logwood.

Take of Logwood.................................. 1 ounce.
Water.................................. 2 pints.

Boil to 1 pint, and strain.

Dose: 1 to 2 ounces.

Extractum Hæmatoxyli (U. S., Br., Ger.)—Extract of Logwood.

Take of Logwood .... 12 ounces.
Water .... 8 pints.

Boil to 4 pints, strain, and evaporate to dryness.

Dose: 10 to 30 grains.

Mixtures of Logwood.

Take of Decoction of logwood.................................. 5 drachms.
Lime water.................................. 3 drachms.

Mix. One dose. *Brompton Consumption Hospital.*

Take of Bicarbonate of potassium.................................. 20 grains.
Tincture of opium.................................. 5 minims.
Tincture of catechu.................................. 15 minims.
Decoction of logwood..................................to 1 ounce.

Mix. One dose. *Royal Free Hospital.*

Take of Extract of logwood .... 10 grains.
Wine of opium.................................. 5 minims.
Wine of ipecacuanha .... 10 minims.
Chalk mixture..................................to 1 ounce.

Mix. One dose. *Guy's Hospital.*

HAMAMELIS (U. S.)—WITCH-HAZEL.

The bark of the small branches of Hamamelis Virginica, L. (*Nat. ord., Hamamelaceæ*), a shrub from six to twelve or more feet high, common throughout the United States and Canada, growing in moist woods, and blooming late in the fall.

Hamamelis possesses astringent properties of a valuable character, and which can scarcely be attributed solely to the comparatively small percentage of tannin which it contains. Clinical experience has demonstrated its value in hemorrhage from the lungs, stomach, bowels, uterus, etc., and, used internally and topically, in hemorrhoids.

Dose: ½ to 2 drachms.

Decoction of Witch-Hazel.

Take of Witch-hazel .... 1 ounce.
Water .... 1 pint.

Boil and strain. Dose: 2 to 4 ounces. *N. S. Davis.*

Fluid Extract of Witch-Hazel.

Take of Witch-hazel .... 16 parts.
Alcohol, 1 part—Glycerin, 1 part—Water, 4 parts .... sufficient.

Moisten the witch-hazel with six ounces of the menstruum, and proceed according to the general formula, page 161.

Report of Am. Ph. Ass'n.

Dose: 1 to 2 drachms.

Most of the commercial extracts of witch-hazel are unreliable. This formula is believed to yield a product entitled to confidence.

Injection of Witch-Hazel.

Take of Tincture of witch-hazel .... 1 drachm.
Water .... 1 pint.

Mix. Use after stool, in hemorrhoids.

Tincture of Witch-Hazel.

Take of Witch-hazel, fresh .... 1 part.
Diluted alcohol .... 2 parts.

Macerate two weeks, express and filter.

Dose: 5 to 30 minims.

A tincture made in this manner by the author, has invariably yielded good results.

HUMULUS (U. S. et al. Ph.)—HOPS.

The strobiles of Humulus Lupulus, L. (*Nat. ord.*, *Urticaceæ*), an herbaceous climbing vine, with a perennial root, indigenous to both the Old World and the New.

Hops have a strong, aromatic odor, and an agreeable, bitter taste. The seeds are covered with yellowish glands, which are the active portion, and are termed lupulin (*Lupulina*).

Hops are tonic and mildly narcotic. Internally they are employed to allay pain, to relieve restlessness, and to calm morbid excitement of the sexual organs. Topically, they are employed as a fomentation in painful swellings.

For internal administration, lupulin is generally used.

Dose: 20 to 60 grains.

PREPARATIONS.

Infusum Humuli (U. S., Br.)—Infusion of Hops.

Take of Hops.................................... ½ ounce.
Boiling water.............................. 1 pint.

Macerate for two hours, and strain.

Dose: 1 to 2 ounces.

Tinctura Humuli (U. S., Br.)—Tincture of Hops.

Take of Hops.................................... 5 ounces.
Diluted alcohol.......................... sufficient.

Moisten, pack, and percolate to 2 pints.

Dose: 1 to 3 drachms.

Extract of Hops (Br., Fr.).

Take of Hops.................................... 1 pound.
Rectified spirit............................ 30 ounces.
Distilled water............................ sufficient.

Macerate the hops in the spirit for seven days, express, filter, and evaporate to a soft extract. Boil the residual hops with the water for an hour, express, strain, and evaporate to a soft extract. Mix the two extracts, and evaporate below 140° to a proper consistence for making pills.

Dose: 5 to 15 grains.

Inhalations of Hops.

Take of Dried carbonate of sodium................ 20 grains.
Water at 140°.............................. 1 pint.
Dissolve, and add of
Extract of hops............................ 1 drachm.

The vapor to be inhaled. Sedative. *G. M. Lefferts.*

Take of Hops .... 1½ ounce.
Hot water.......................... .... 20 ounces.

Mix. The vapor to be inhaled.

Brompton Consumption Hospital.

HYDRARGYRUM (U. S. et al. Ph.)—MERCURY.

A heavy metal which is fluid at ordinary temperatures, but crystallizes and becomes solid at 39°. It is slowly vaporized at the temperature of the air, but rapidly at its boiling-point, 662°.

Mercury in the fluid, metallic state is nearly, or quite inert, but all of its compounds are active. Some of them are violent corrosive poisons, and all are more or less irritating. In small and repeated doses, all of them stimulate the secretory and excretory organs, and thus exert an alterative influence. By this increased activity of the glandular system, accumulations of liquids, swellings and indurations, often rapidly disappear. But carried too far, this influence is disorganizing and destructive, producing excessive salivation, swelling of the tongue and gums, loosening of the teeth, fetid breath, and a profound cachexia.

The immediate effects of large doses of the various preparations of mercury differ greatly, some acting as purgatives, sedatives, etc., while others act as corrosive poisons.

In its various forms, mercury is extensively employed as an alterative, especially in the many manifestations of syphilis. It is also used as a cholagogue, sialagogue, cathartic, etc., and for sedative effect.

Few drugs have been so badly misused, while fewer yet are more valuable if used with wise discrimination.

PREPARATIONS.

Hydrargyrum cum Creta (U. S., Br.)—Mercury with Chalk.

Take of Mercury .... 3 ounces.
Prepared chalk .... 5 ounces.

Rub together until globules cease to be visible.

Dose: 2 to 6 grains.

This is an excellent preparation, and is especially valuable in diseases of children requiring the employment of mercury.

According to Dr. Piffard, sugar of milk is a far better triturant for mercury than chalk; and a preparation made with it instead of chalk is correspondingly more active. He examined many specimens of each microscopically, and invariably found the mercury in a much more finely divided state when sugar of milk had been employed, than when chalk was used.

Emplastrum Hydrargyri (U. S. et al. Ph.)—Mercurial Plaster.

Take of Mercury	6 ounces.
Olive oil, (by weight)	2 ounces.
Resin	2 ounces.
Lead plaster	12 ounces.

Melt the oil and resin together, and when cool, add the mercury and rub till the globules disappear; then add the plaster, previously melted, and mix.

Pilulæ Hydrargyri (U. S., Br., Fr.)—Pills of Mercury—Blue Pill.

Take of Mercury	384 grains.
Confection of rose	576 grains.
Liquorice root	192 grains.

Rub the mercury with the confection until globules cease to be visible, add the liquorice root, beat into a mass, and divide into 384 pills.

Dose: 1 to 5 pills.

The British Pharmacopœia directs: mercury, 2 ounces; confection of rose, 3 ounces; liquorice root, 1 ounce, and leaves the mass undivided—a much better plan.

Unguentum Hydrargyri (U. S. et al. Ph.)—Mercurial Ointment.

Take of Mercury	24 ounces.
Lard	12 ounces.
Suet	12 ounces.

Rub the mercury with 1 ounce of the suet and a small portion of lard, until the globules cease to be visible, then add the remainder of the lard and suet, and mix.

Compound Ointment of Mercury (Br.).

Take of Ointment of mercury	6 ounces.
Yellow wax	3 ounces.
Olive oil, (by weight)	3 ounces.
Camphor	1½ ounce.

Melt the wax, add the oil, then, when nearly cold, add the camphor in powder, and the ointment of mercury.

Pills of Mercury with Chalk.

Take of Mercury with chalk	2½ grains.
Dover's powder	2½ grains.
Treacle	sufficient.

Make 1 pill. *St. Bartholomew's Hospital.*

Take of Mercury with chalk	2½ grains.
Rhubarb	2½ grains.
Treacle	sufficient.

Make 1 pill. *St. Bartholomew's Hospital.*

Pill of Mercury with Iron.

Take of Blue pill.................................... 2 grains.
Dried sulphate of iron...................... 1 grain.

Make 1 pill. Dose: 3 to 6 daily. *F. R. Sturgis.*

Pill of Mercury with Quinia.

Take of Blue pill.................................... 2 grains.
Sulphate of quinia .... 1 grain.

Make 1 pill. Dose: 3 to 6 daily.

This, and the preceding have the advantage over blue pill of being not only tonic, but much more easily tolerated. *F. R. Stnrgis.*

Compound Pills of Mercury.

Take of Blue pill.................................... 1 grain.
Compound extract of colocynth .... 4 grains.

Make 1 pill. *Samaritan Hospital.*

Take of Blue pill.................................... 1 grain.
Extract of colocynth....................... 2 grains.
Extract of henbane......................... 2 grains.

Make 1 pill. Dose: 1 or 2 pills. *St. Mary's Hospital.*

Take of Blue pill.................................... 2 grains.
Dover's powder............................. 3 grains.

Make 1 pill. *London Fever Hospital.*

Take of Blue pill.................................... 10 grains.
Powdered aloes............................. 20 grains.
Resin of podophyllum .... 3 grains.
Extract of henbane......................... 10 grains.
Extract of dandelion...................... sufficient.

Mix, and divide into 20 pills. *Hospital Formulary.*

Triplex Pills.

Take of Blue pill.................................... 20 grains.
Resin of scammony .... 20 grains.
Powdered aloes............................. 20 grains.
Oil of caraway............................. sufficient.

Mix, and divide into 20 pills. *Hospital Formulary.*

Compound Powders of Mercury with Chalk.

Take of Mercury with chalk .... 2 grains.
Subnitrate of bismuth .... 2 grains.
Powder of rhubarb and soda................. 6 grains.

Mix. One dose. *Westminster Hospital.*

Take of Mercury with chalk .... 2 grains.
Sugar.................................... 2 grains.
Powdered Belladonna leaves................ 1 grain.

Mix. One dose. *University College Hospital.*

COMPOUND MERCURIAL OINTMENTS.

Take of Extract of belladonna...................... 1 drachm.
Mercurial ointment........................to 1 ounce.

Mix. *London Hospital.*

Take of Extract of conium.................. .... 1 drachm.
Mercurial ointment........................ 1 ounce.

Mix. *Great Northern Hospital.*

Take of Mercurial ointment........................ 1 ounce.
Camphor .... 1 drachm.
Proof spirit .... sufficient.

Powder the camphor with a few drops of the spirit, and mix it with the mercurial ointment. *St. George's Hospital.*

Take of Mercurial ointment........................ 2 ounces.
Soap cerate plaster .... 2 ounces.
Camphor.................................. ½ ounce.

Mix. *London Hospital.*

HYDRARGYRUM AMMONIATUM (U. S., Br., Ger.)—AMMONIATED MERCURY—WHITE PRECIPITATE.

Take of Corrosive chloride of mercury .... 6 ounces.
Water of ammonia .... 8 ounces.
Distilled water........................... 8 pints.

Dissolve the chloride in the distilled water with the aid of heat, and, when cold, add the water of ammonia, frequently stirring, wash the precipitate until the washings become nearly tasteless, and dry it.

Ammoniated mercury is a white powder of a metallic taste, and insoluble in water. It is not used internally, but is employed topically in the form of an ointment in cutaneous affections.

PREPARATIONS.

Unguentum Hydrargyri Ammoniati (U. S., Br., Ger.)—Ointment of Ammoniated Mercury—White Precipitate Ointment.

Take of Ammoniated mercury...................... 40 grains.
Ointment .... 1 ounce.

Mix.

The British Pharmacopœia employs 1 part of ammoniated mercury with 7 of simple ointment, and the German Pharmacopœia 1 with 9 of lard.

14

Take of Ammoniated mercury........................ 4 parts.
Thymol.................................. 1 part.
Vaseline .... 45 parts.

Mix. Use in squamous eczema of the scalp. *G. H. Fox.*

COMPOUND OINTMENT OF AMMONIATED MERCURY.

Take of Ammoniated mercury........................ 40 grains.
Oxide of zinc............................. 40 grains.
Red oxide of mercury...................... 5 grains.
Lard .... 1 ounce.

Mix. *Middlesex Hospital.*

Take of Ammoniated mercury .... 6 grains.
Sublimed sulphur.......................... 30 grains.
Benzoated lard .... 1 ounce.

Mix. *London Ophthalmic Hospital.*

Take of Ammoniated mercury........................ 40 grains.
Sublimed sulphur.......................... 1 ounce.
Nitrate of potassium .... 1 drachm.
Oil of lavender........................... 4 minims.
Prepared lard .... 1 ounce.

Mix. "Itch Ointment." *Hospital for Ruptured and Crippled.*

HYDRARGYRI CHLORIDUM CORROSIVUM (U. S. et al. Ph.)—CORROSIVE CHLORIDE OF MERCURY—CORROSIVE SUBLIMATE—BICHLORIDE OF MERCURY.

Take of Mercury................................. 24 ounces.
Sulphuric acid (by weight)................ 36 ounces.
Chloride of sodium........................ 18 ounces.

Boil the mercury with the acid by means of a sand-bath until a dry, white mass is left. Rub this, when cold, with the chloride of sodium, then sublime with a gradually increasing heat.

It occurs as a white, crystalline powder or mass, of a disagreeable, metallic taste, soluble in 2 parts of boiling, and in 16 of cold water, 3 of alcohol, and 4 of ether.

Taken in small doses it produces the constitutional effects of mercury; in overdoses it is a violent corrosive poison. Locally, solutions are used in various cutaneous affections.

Dose: $\frac{1}{30}$ to $\frac{1}{8}$ grain.

PREPARATIONS.

Compound Pills of Corrosive Sublimate (Fr.)—Dupuytren's Pills.

Take of Corrosive sublimate .... 1 part.
Extract of opium .... 2 parts.
Extract of guaiac........................ 4 parts.

Make a mass, and divide into pills of 1 grain each.

Solution of Corrosive Sublimate (Br.).

Take of Corrosive sublimate........................ 10 grains.
Chloride of ammonium .... 10 grains.
Distilled water .... 20 ounces.

Dissolve.

Dose: ½ to 2 drachms.

Yellow Mercurial Lotion (Br.).

Take of Corrosive sublimate .... 18 grains.
Solution of lime .... 10 ounces.

Mix. Commonly known as "yellow wash." Used as an application to indolent chancres, and other syphilitic ulcers.

Lotions of Corrosive Sublimate.

Take of Corrosive sublimate ....2 to 4 grains.
Alcohol........................ 4 drachms.
Chloride of ammonium........................ ½ drachm.
Rose water........................ to 6 ounces.

Mix. Use in scabies, phtheiriasis, and tinea versicolor.

Tilbury Fox.

Take of Corrosive sublimate .... 1 grain.
Water........................ 8 ounces.

Dissolve. *London Ophthalmic Hospital.*

Take of Corrosive sublimate........................ 1 grain.
Water........................ 1 ounce.

Dissolve. *British Skin Hospital.*

Take of Corrosive sublimate .... ½ grain.
Hydrocyanic acid .... 8 minims.
Glycerin........................ 2 drachms.
Water........................to 1 ounce.

Mix. *St. Bartholomew's Hospital.*

Take of Corrosive sublimate .... 1 grain.
Rose water........................ 2 ounces.
Water........................ 6 ounces.

Mix. Apply every three hours. In purulent ophthalmia of infants.

J. Lewis Smith.

MIXTURES OF CORROSIVE SUBLIMATE.

Take of Corrosive sublimate .... $\frac{1}{12}$ grain.
Tincture of bark........................... 60 minims.
Tincture of rhubarb......................... 30 minims.
Water.................................... 1 ounce.

Mix. One dose. *Guy's Hospital.*

Take of Corrosive sublimate .... $\frac{1}{16}$ grain.
Iodide of potassium......................... 3 grains.
Infusion of quassia.......................... 1 ounce.

Mix. One dose. *Great Northern Hospital.*

Take of Corrosive sublimate.......................... 2 grains.
Iodide of potassium .... 2 drachms.
Compound tincture of cinchona............. 3 ounces.

Mix. Dose: 1 drachm.

Take of Corrosive sublimate.......................... $\frac{1}{2}$ grain.
Iodide of potassium .... 1 drachm.
Citrate of iron and ammonium............... 1 drachm.
Syrup .... 6 ounces.

Mix. Dose: 1 drachm three times a day, for a child three to five years old. In syphilis. *J. Lewis Smith.*

Take of Corrosive sublimate ....1 to 2 grains.
Compound syrup of sarsaparilla............. 2 ounces.
Water...................................... 8 ounces.

Mix. Dose: 1 drachm three times a day. In syphilis of children. *J. Lewis Smith.*

HYDRARGYRI CHLORIDUM MITE (U. S. et al. Ph.)—MILD CHLORIDE OF MERCURY—CALOMEL.

Take of Mercury .... 48 ounces.
Sulphuric acid (by weight).................. 36 ounces.
Chloride of sodium........................... 18 ounces.
Distilled water................................ sufficient.

Boil 24 ounces of mercury with the acid until a dry, white mass is left. Rub this, when cold, with the remainder of the mercury, add the chloride of sodium, and continue to triturate until the globules of mercury cease to be visible; then sublime into a large chamber so that the sublimate may fall in powder. Wash the sublimate with boiling distilled water until the washings are not precipitated by water of ammonia, and dry it.

Calomel is a heavy, white powder, tasteless and insoluble. It is one of the mildest and least irritating preparations of mercury, and is used as an alterative, a cathartic, a sedative, etc. Externally, it is used in powder and in the form of an ointment in cutaneous affections.

Dose: $\frac{1}{16}$ to 1 grain as an alterative; 5 to 20 grains as a purgative (20 to 60 grains for sedative effect—*Leaming*).

PREPARATIONS.

Compound Calomel Pill (Br.).

Take of	
Calomel	1 ounce.
Sulphurated antimony	1 ounce.
Guaiacum resin	2 ounces.
Castor oil	1 ounce, or sufficient.

Triturate the calomel with the antimony, then add the guaiac, and with the oil form a pilular mass.

Dose: 5 to 10 grains.

Known as *Plummer's Pills.* Used in chronic rheumatism and chronic cutaneous affections, especially when there is a syphilitic taint. See also *Compound Pills of Antimony.*

Black Mercurial Lotion (Br., Fr.).

Take of	
Calomel	30 grains.
Solution of lime	10 ounces.

Mix. Generally known as *Black Wash,* and used as an application to syphilitic ulcers.

Calomel Ointment (Br.).

Take of	
Calomel	80 grains.
Prepared lard	1 ounce.

Mix.

Barker's Post-Partum Pills.

Take of	
Calomel	3 drachms.
Compound extract of colocynth	3 drachms.
Extract of henbane	40 grains.
Extract of nux vomica	20 grains.
Aloes	20 grains.
Ipecac	20 grains.

Mix, and divide into 120 pills.

Dose: 1 or 2 pills in the morning before breakfast, as a laxative for puerperal women. *Hospital Formulary.*

White's Gout Pills.

Take of	
Calomel	60 grains.
Aloes	60 grains.
Ipecac	60 grains.
Acetic extract of colchicum	60 grains.

Mix, and divide into 60 pills.

Dose: 1 or 2 pills. *Hospital Formulary.*

Pills of Calomel and Colocynth.

Take of Calomel.................................. 1 grain.
Compound extract of colocynth .... 4 grains.

Make 1 pill. *King's College Hospital.*

Take of Calomel.................................. 1 grain.
Extract of henbane........................ 1 grain.
Compound extract of colocynth.............. 3 grains.

Make 1 pill. *London Ophthalmic Hospital.*

Pills of Calomel and Opium.

Take of Calomel.................................. 2 grains.
Extract of opium.......................... ¼ grain.
Treacle.. sufficient.

Make 1 pill. *London Chest Hospital.*

Take of Calomel.................................. 2 grains.
Opium.. 1 grain.
Confection of roses......................... sufficient.

Make 1 pill. *Westminster Ophthalmic Hospital.*

Take of Calomel.................................. 1 grain.
Dover's powder .... 4 grains.
Mucilage .... sufficient.

Make 1 pill. *London Ophthalmic Hospital.*

Pills of Calomel and Scammony.

Take of Calomel.................................. 1 grain.
Scammony .... 3 grains.
Treacle .... sufficient.

Make 1 pill. *St. Bartholomew's Hospital.*

Powders of Calomel and Jalap.

Take of Calomel.................................. 1 part.
Jalap... 2 parts.

Mix. Dose: 15 grains. *St. Mary's Hospital.*

Take of Calomel.................................. 2 grains.
Ginger... 2 grains.
Jalap.. 8 grains.

Mix. *St. Thomas's Hospital.*

POWDERS OF CALOMEL AND TARTAR EMETIC.

Take of Calomel .... 2 grains.
Tartar emetic .... $\frac{1}{6}$ grain.
Sugar .... 3 grains.

Mix. Place dry on the back of the tongue, and repeat every three hours until free catharsis. Useful in commencing tonsillitis to abort the inflammation. Will also hasten suppuration when too far advanced to be aborted. *J. R. Leaming.*

HYDRARGYRI CYANIDUM (U. S., Fr.)—CYANIDE OF MERCURY

Take of Ferrocyanide of potassium .... 5 ounces.
Sulphuric acid .... 4 ounces and 120 grains.
Red oxide of mercury .... sufficient.
Water .... sufficient.

Dissolve the ferrocyanide in 20 ounces of water, and add the solution to the acid previously diluted with 10 ounces of water, and contained in a glass retort. Distil nearly to dryness into a receiver containing 10 ounces of water and 3 ounces of red oxide of mercury. Set aside 2 ounces of the distillate, and to the remainder add sufficient red oxide to destroy the odor of hydrocyanic acid, filter, add the reserved liquid, and evaporate in a dark place, that crystals may form. Lastly, dry the crystals and preserve them in a well-stopped bottle, protected from the light.

It is in colorless crystals, of a bitter, metallic taste, soluble in water, blackened by exposure to light, and very poisonous. Occasionally used in syphilis and skin diseases.

Dose: $\frac{1}{16}$ to $\frac{1}{8}$ grain.

HYDRARGYRI IODIDUM RUBRUM (U. S. et al. Ph.)—RED IODIDE OF MERCURY—BINIODIDE OF MERCURY.

Take of Corrosive chloride of mercury .... 1 ounce.
Iodide of potassium .... 600 grains.
Distilled water .... sufficient.

Dissolve the corrosive chloride in $1\frac{1}{2}$ pint, and the iodide in $\frac{1}{2}$ pint of distilled water, and mix the solutions. Collect, wash, and dry the precipitate.

It is a powder of a brilliant scarlet color, sparingly soluble in water, but freely soluble in solutions of iodide of potassium, chloride of sodium, etc.

Used both externally and internally in syphilis.

Dose $\frac{1}{16}$ to $\frac{1}{4}$ grain.

PREPARATIONS.

Unguentum Hydrargyri Iodidi Rubri (U. S., Br.)—Ointment of Red Iodide of Mercury.

Take of Red iodide of mercury .... 60 grains.
Ointment .... 420 grains.

Mix thoroughly.

Used as a dressing for obstinate venereal ulcers.

MIXTURES OF RED IODIDE OF MERCURY.

Take of Red iodide of mercury .... $\frac{1}{20}$ grain.
Iodide of potassium .... 5 grains.
Decoction of yellow bark .... 1 ounce.

Mix. One dose. *Westminster Hospital.*

Take of Red iodide of mercury .... $\frac{1}{8}$ grain.
Iodide of potassium .... $\frac{1}{2}$ grain.
Water .... 1 ounce.

Mix. One dose. *British Skin Hospital.*

Take of Red iodide of mercury .... $\frac{1}{2}$ grain.
Iodide of potassium .... 2 drachms.
Syrup of orange peel .... 1 ounce.
Tincture of orange peel .... 1 drachm.
Water .... to 4 ounces.

Mix. Dose: 1 drachm after eating. Used in syphilis.

E. L. Keyes.

PILLS OF RED IODIDE OF MERCURY.

Take of Red iodide of mercury .... 3 grains.
Iodide of potassium .... 150 grains.
Gum tragacanth .... sufficient.
Glycerin .... sufficient.

Mix, and divide into 50 pills.

Dose: 1 pill. In syphilis. *E. L. Keyes.*

HYDRARGYRI IODIDUM VIRIDE (U. S. et al. Ph.)—GREEN IODIDE OF MERCURY—PROTIODIDE OF MERCURY.

Take of Mercury .... 1 ounce.
Iodine .... 300 grains.
Stronger alcohol .... sufficient.

Triturate the mercury and iodine with $\frac{1}{2}$ ounce of the alcohol until thoroughly mixed. Stir occasionally, and after two hours triturate again until nearly dry. Then rub with stronger alcohol into a thin paste, transfer to a filter, and wash with stronger alcohol until the washings cease to produce permanent cloudiness when dropped into a large quantity of water. Lastly, dry in a dark place, and keep in a well-stopped bottle, protected from light.

It is a greenish-yellow powder, almost insoluble in water and entirely so in alcohol and ether. Extensively employed in syphilis.

Dose: $\frac{1}{8}$ to 3 grains.

PREPARATIONS.

Pills of Protiodide of Mercury and Opium (Fr.).

Take of		
	Protiodide of mercury	5 parts.
	Extract of opium	2 parts.
	Confection of rose	10 parts.
	Powdered liquorice root	sufficient.

Mix the extract of opium and confection of rose, add the protiodide, and sufficient liquorice root to make a mass. Divide into pills of 3 grains each.

Dose: 1 pill.

An excellent preparation in secondary syphilis. The small quantity of opium used is generally sufficient to prevent gastro-intestinal irritation.

Take of		
	Protiodide of mercury	$\frac{1}{2}$ grain.
	Powdered opium	$\frac{1}{4}$ grain.
	Extract of gentian	2 grains.

Make 1 pill. *British Skin Hospital.*

Pills of Protiodide of Mercury and Henbane.

Take of		
	Protiodide of mercury	5 grains.
	Extract of henbane	20 grains.
	Powdered liquorice root	sufficient.

Mix, and divide into 20 pills.

Dose: 1 or 2 pills.

HYDRARGYRI NITRAS (Fr.)—NITRATE OF MERCURY.

PREPARATIONS.

Liquor Hydrargyri Nitratis (U. S. et al. Ph.)—Solution of Nitrate of Mercury.

Take of		
	Mercury	3 ounces.
	Nitric acid (by weight)	5 ounces.
	Distilled water	6 drachms.

Dissolve the mercury, with the aid of a gentle heat, in the acid previously mixed with the water. When reddish vapors cease to arise, evaporate to 7½ ounces, by weight, and keep in a well-stopped bottle. It may also be prepared thus:

Take of Red oxide of mercury	26 drachms.
Nitric acid (by weight)	29 drachms.
Distilled water	6 drachms.

Mix the acid and water, dissolve the oxide in the mixture, and evaporate to 7½ ounces, by weight.

It is a transparent, nearly colorless liquid, having the sp. gr. 2,165, and possessing caustic properties. Used as a caustic, and, properly diluted, as a stimulating lotion for indolent ulcers, etc.

Unguentum Hydrargyri Nitratis (U. S., Br., Fr.)—Ointment of Nitrate of Mercury—Citrine Ointment.

Take of Mercury	1½ ounce.
Nitric acid (by weight)	3½ ounces.
Lard	16½ ounces.

Dissolve the mercury in the acid, add the solution to the lard, heated to 200°, stir constantly with a wooden spatula so long as effervescence continues, then occasionally until it cools.

OINTMENT OF NITRATE OF MERCURY WITH ZINC.

Take of Ointment of nitrate of mercury	2 drachms.
Ointment of oxide of zinc	2 ounces.

Mix. Apply to the nostrils by means of a nasal sponge three times a day. In syphilitic coryza of infants. *J. Lewis Smith.*

HYDRARGYRI OXIDUM FLAVUM (U. S., Br.)—YELLOW OXIDE OF MERCURY.

Take of Corrosive chloride of mercury	4 ounces.
Solution of potassa (by weight)	17 ounces.
Distilled water	sufficient.

Dissolve the chloride in 5 pints of distilled water, and mix with the solution of potassa. Allow it to settle, then decant, wash the precipitate with distilled water until the washings cease to be affected by a solution of nitrate of silver, dry it on bibulous paper, in the dark, and preserve in bottles protected from light.

It is a heavy, yellowish powder, which becomes darker by exposure to light. Used externally in skin diseases, syphilitic ulcers, etc.

PREPARATIONS.

Unguentum Hydrargyri Oxidi Flavi (U. S.)—Ointment of Yellow Oxide of Mercury.

Take of Yellow oxide of mercury	1 drachm.
Ointment	7 drachms.

Mix thoroughly.

Oleate of Mercury.

Take of Yellow oxide of mercury .... 10 parts.
Purified oleic acid........................... 90 parts.

Add the oxide gradually to the acid, and triturate frequently until dissolved. *Report of Am. Ph. Ass'n.*

HYDRARGYRI OXIDUM RUBRUM (U. S. et al. Ph.)—RED OXIDE OF MERCURY—RED PRECIPITATE.

Take of Mercury .... 36 ounces.
Nitric acid (by weight)........................ 24 ounces.
Water .... 2 pints.

Dissolve the mercury with the aid of a gentle heat, in the acid and water previously mixed, evaporate to dryness, rub into powder, and heat in a very shallow vessel until red vapors cease to arise.

It is in bright, shining, red scales, which by trituration yield an orange-red powder. Used chiefly as a topical application in skin diseases, syphilitic ulcers, etc. Occasionally employed internally.

Dose: $\frac{1}{16}$ to $\frac{1}{4}$ grain.

PREPARATIONS.

Unguentum Hydrargyri Oxidi Rubri (U. S. et al. Ph.)—Ointment of Red Oxide of Mercury—Ointment of Red Precipitate.

Take of Red oxide of mercury .... 1 drachm.
Ointment .... 7 drachms.

Mix thoroughly.

Eye Ointments.

No. 1, Red oxide of mercury, 1 part to lard.......... 7 parts.
No. 2, Red oxide of mercury, 1 part to lard.......... 16 parts.
No. 3, Red oxide of mercury, 1 part to lard.......... 60 parts.

University College Hospital.

HYDRARGYRI SULPHAS FLAVA—(U. S., Fr.)—YELLOW SULPHATE OF MERCURY—TURPETH MINERAL.

Take of Mercury.................................. 4 ounces.
Sulphuric acid (by weight).................. 6 ounces.

Mix, and boil until a dry, white mass remains. Rub this to powder, throw it into boiling water, decant, wash the precipitate repeatedly with hot water, and dry it.

Turpeth mineral is a heavy, bright yellow powder, of an acrid taste, and sparingly soluble in water. It is powerfully irritant and corrosive, but is occasionally employed as an alterative, and as an emetic, especially in croup. Its action as an emetic is, however, so extremely violent that most physicians discard it.

Dose : $\frac{1}{8}$ to $\frac{1}{2}$ grain, alterative ; 2 to 5 grains, emetic.

HYDRARGYRI SULPHURETUM RUBRUM (U. S., Fr., Ger.)—RED SULPHURET OF MERCURY—VERMILION.

Take of Mercury .... 40 ounces.
Sublimed sulphur .... 8 ounces.

To the sulphur, previously melted, gradually add the mercury, with constant stirring, and continue the heat until the mass begins to swell. Remove from the fire, cool, powder, and sublime.

It is a bright scarlet powder, tasteless, and insoluble. Seldom used except as a fumigation.

HYDRASTIS (U. S.)—HYDRASTIS.

The rhizome and rootlets of Hydrastis Canadensis, L. (*Nat. ord., Ranunculaceæ*) a small, herbaceous perennial, indigenous to North America.

Hydrastis has an intensely bitter taste, and possesses tonic and diuretic properties. Its medicinal effects are due to *berberina*, and a peculiar alkaloid termed *hydrastia*. It is used in atonic dyspepsia, intermittent fever, and catarrhal affections, especially those of the genito-urinary tract, etc.

Dose : $\frac{1}{2}$ to 2 drachms.

PREPARATIONS.

Extractum Hydrastis Fluidum (U. S.)—Fluid Extract of Hydrastis.

Take of Hydrastis .... 16 ounces.
Glycerin .... 2 ounces.
Alcohol .... sufficient.
Water .... sufficient.

Mix the glycerin with 14 ounces of alcohol, moisten the hydrastis with 4 ounces of the mixture, and proceed according to the general formula, page 161. Finish the percolation with a menstruum of 2 parts of alcohol and 1 part of water.

Dose : $\frac{1}{2}$ to 2 drachms.

TINCTURE OF HYDRASTIS.

Take of Hydrastis .... 1 part.
Diluted alcohol .... sufficient.

Moisten, pack, and percolate .... to 5 parts.

Dose : 1 to 2 drachms. *J. P. Remington, Report of Am. Ph. Ass'n.*

Lotion of Hydrastis.

Take of Tincture of hydrastis	1 to 2 drachms.
Water	1 pint.

Mix. Used as an application to ulcers, hæmorrhoids, sore nipples, etc., and as an injection in gonorrhœa. *Phillips.*

Hydrastia—Hydrastine, White Alkaloid.

To the mother liquor from which sulphate of berberine has been crystallized, add an equal bulk of water, and evaporate the alcohol. Permit the residuum to cool and stand twenty-four hours. Filter; add to the filtrate ammonia water until in excess. Wash the precipitate with water, dissolve in cold water acidulated with hydrochloric acid, and precipitate this with ammonia water. Collect, wash, and dry the precipitate. Purify by repeated solutions in boiling alcohol, and crystallization.

It is in white crystals resembling strychnine, tasteless at first, but eventually imparting an acrid sensation to the throat and fauces; *not bitter*, as some authorities state it to be. Forms soluble salts which are, as a rule, uncrystallizable and are acrid to the taste.

Dose: 1 to 3 grains. *J. U. Lloyd.*

This, the *true hydrastia*, should be carefully distinguished from the following, which is erroneously so called.

Hydrochlorate of Berberine—Hydrochlorate of Hydrastine—Muriate of Hydrastine—Hydrastine.

Take of Hydrastis	16 parts.
Alcohol (sp. gr. 0.835)	sufficient.

Moisten, pack, and percolate until 16 parts of tincture are obtained, or the hydrastis is exhausted. To this, add 4 parts of distilled water, and evaporate the alcohol. Mix 12 parts of cold distilled water with the residue, and allow the mixture to stand in a cool place for twenty-four hours, then filter it. Add to the filtrate hydrochloric acid until it ceases to produce a precipitate. Collect the precipitate, and purify it by solution in boiling water, and crystallization. Dry by exposure to the air.

It is of a lemon-yellow color, soluble in about 500 parts of cold water almost insoluble in cold alcohol, insoluble in ether and in chloroform.

Dose: 1 to 3 grains. *J. U. Lloyd.*

This is the article originally employed by the eclectics under the name *hydrastine.* The term *hydrastine alkaloid* is applied to the yellow alkaloid of hydrastis, *berberine.*

It may be substituted for quinia in cases of debility, in convalescence from acute diseases, and in various cachexiæ, especially the paludal. "As a remedy for intermittents, it ranks next to quinia."—(*Bartholow.*)

Injection of Hydrastine.

Take of Hydrastia (hydrochlorate of berberine)........ 1 drachm.
Mucilage of gum arabic.................... 4 ounces.

Use in gonorrhœa after the acute symptoms have subsided, and in gleet.
Bartholow.

Powder of Hydrastine.

Take of hydrastine (hydrochlorate of berberine)...... ½ drachm.
Camphor .... 10 grains.
Subnitrate of bismuth...................... 1½ drachm.
Sugar .... to 1 ounce.

Mix. For insufflation into the larynx or nares. *G. M. Lefferts.*

Take of Hydrastis Canadensis...................... 5 grains.
Indigo .... ½ grain.
Camphor .... 2 grains.
Carbolic acid.............................. 2 grains.
Chloride of sodium....................... 1 drachm.

Mix. "Catarrh snuff." *G. M. Lefferts.*

Sulphate of Berberine—Sulphate of Hydrastine.

To the alcoholic percolate, as obtained in the preceding process, add a considerable excess of sulphuric acid. After twenty-four hours, collect the crystalline precipitate, and purify it by dissolving in the minimum amount of boiling water, and crystallizing. If it is not free from sulphuric acid it will not dry. In this case, repeat the operation of dissolving in boiling water and cooling. Dry the salt by exposure to the air.

It is in orange-yellow, crystalline tufts, quite soluble in cold water, and very soluble in boiling water.

Dose: 1 to 3 grains. *J. U. Lloyd.*

HYOSCYAMUS—HENBANE.

HYOSCYAMI FOLIA (U. S. et al. Ph.)—HYOSCYAMUS LEAVES.

HYOSCYAMI SEMEN (U. S. et al. Ph.)—HYOSCYAMUS SEED.

The leaves and seeds of Hyoscyamus niger, L. (*Nat. ord., Solanaceæ*), an annual or biennial herb, indigenous to the Old World, but naturalized in this country.

The fresh plant has a strong, offensive odor, and a disagreeable, slightly acrid taste. The seeds have something of the odor of the plant, and a bitter taste. The activity of henbane is due to an alkaloid termed *hyoscyamia.*

Henbane affects the system much like belladonna. It is a narcotic and anodyne, less powerful than opium, but producing its effects without constipating the bowels. It is used as an anodyne and hypnotic in a great variety of nervous and painful affections.

Dose: of the leaves, 5 to 10 grains; of the seeds, 3 to 8 grains.

PREPARATIONS.

Extractum Hyòscyami (U. S. et al. Ph.)—Extract of Hyoscyamus.

Take of hyoscyamus leaves, fresh.................... 12 ounces.

Bruise in a mortar, sprinkling with a little water, express the juice, heat to 212°, strain, and evaporate to a proper consistence.

Dose: 2 to 3 grains.

This is an unreliable preparation. The one following is much better.

Extractum Hyoscyami Alcoholicum (U. S., Fr.)—Alcoholic Extract of Hyoscyamus.

Take of Hyoscyamus leaves, recently dried........... 24 ounces.
Alcohol................................ 4 pints.
Water.................................. 2 pints.
Diluted alcohol........................ sufficient.

Mix the alcohol and water, percolate the powder with the mixture, continuing the process with diluted alcohol until 6 pints are obtained, then evaporate on a water-bath to a proper consistence.

Dose: 1 to 2 grains.

Extractum Hyoscyami Fluidum (U. S.)—Fluid Extract of Hyoscyamus.

Take of Hyoscyamus leaves......................... 16 ounces.
Glycerin............................... 4 ounces.
Water.................................. sufficient.
Alcohol................................ sufficient.

Mix 12 ounces of alcohol, 3 of glycerin, and 1 of water, moisten the hyoscyamus with ½ pint of the mixture, and proceed according to the general formula, page 161.

Dose: 5 to 20 minims.

Tinctura Hyoscyami (U. S., Fr.)—Tincture of Hyoscyamus.

Take of Hyoscyamus leaves......................... 4 ounces.
Diluted alcohol........................ sufficient.

Moisten, pack, and percolate to 2 pints.

Dose: 15 to 60 minims.

Extract of Hyoscyamus Seed (Fr.).

Take of Hyoscyamus seed, in powder .... 1 part.
Diluted alcohol .... 6 parts.

Macerate, express, filter, then evaporate to dryness on a water-bath. Dissolve the product in four times its weight of cold distilled water, filter, and evaporate to a solid extract.

Dose: ½ to 1 grain.

Juice of Hyoscyamus (Br.).

Take of Fresh leaves and young branches of hyoscyamus, at will.
Rectified spirit.......................... sufficient.

Bruise the hyoscyamus, press out the juice, and to every 3 measures of juice, add 1 of spirit. After seven days, filter.

Dose: ½ to 1 drachm.

Infused Oil of Hyoscyamus (Ger., Fr.).

Take of Hyoscyamus leaves....................... 2 parts.
Alcohol................................ 1 part.
Olive oil.............................. 20 parts.

Macerate the hyoscyamus in the alcohol for several hours, add the oil, digest until the alcohol is evaporated, express, and filter.

Used as an embrocation.

The French Codex directs 1 part of fresh leaves to be boiled in 2 parts of olive oil until the water of the plant is evaporated.

Ointment of Hyoscyamus (Ger., Fr.).

Take of Extract of hyoscyamus..................... 1 part.
Wax ointment........................... 9 parts.

Mix.

Compound Pills of Hyoscyamus and Valerian (Fr.).

Take of Alcoholic extract of hyoscyamus............. 1 part.
Alcoholic extract of valerian .... 1 part.
Oxide of zinc.......................... 1 part.

Mix, and divide into pills of 2½ grains each.

Hyoscyamus Plaster (Ger.).

Take of Yellow wax............................ 4 parts.
Turpentine............................ 1 part.
Olive oil.............................. 1 part.
Hyoscyamus leaves, powdered............... 2 parts.

Melt together the wax, turpentine, and oil, and when partially cool, stir in the hyoscyamus.

INHALATION OF HYOSCYAMUS.

Take of Fluid extract of hyoscyamus.................. 3 to 10 minims.
Distilled water.......................... 1 ounce.

Mix. Used by inhalation, in spasmodic croup. *Da Costa.*

Take of Extract of hyoscyamus...................... 5 to 10 grains.
Water.............................. 1 ounce.

Mix. Used by means of an atomizer in conjunction with stimulating inhalations, when the latter are attended with pain or irritation.

F. H. Bosworth.

PILLS OF HYOSCYAMUS AND CAMPHOR.

Take of Extract of henbane........................ 2½ grains.
Camphor.............................. 2½ grains.

Make 1 pill. *Great Northern Hospital.*

PILL OF HYOSCYAMUS AND DOVER'S POWDER.

Take of Extract of henbane........................ 2 grains.
Dover's powders .... 2 grains.

Make 1 pill. *St. Thomas' Hospital.*

PILL OF HYOSCYAMUS, SQUILL AND IPECAC.

Take of Extract of henbane........................ 2 grains.
Compound squill pill..................... 2 grains.
Ipecac............................... ½ grain.
Make 1 pill. *London Hospital.*

PILLS OF HYOSCYAMUS AND IRON.

Take of Extract of hyoscyamus..................... ½ drachm.
Valerianate of iron....................... 1 drachm.

Mix, and divide in 40 pills.
Dose: 1 pill thrice daily, in chorea. *Reveil.*

HORDEUM (U. S. et al. Ph.)—BARLEY.

The seed, deprived of their husks, of Hordeum distichon, L. (*Nat. ord., Gramineæ*), common barley, which is supposed to have been derived originally from Tartary, but is now cultivated in most countries.

It is used in medicine, both as a demulcent and as a nutritive food of easy digestion.

PREPARATIONS.

Decoctum Hordei (U. S., Br., Fr.)—Decoction of Barley.

Take of Barley	2 ounces.
Water	sufficient.

Wash the barley, boil it with ½ pint of water for five minutes, pour off the water, and throw it away. Then pour on 4 pints of water, boil down to 2 pints, and strain.

ICHTHYOCOLLA (U. S. et al. Ph.)—ISINGLASS.

A gelatin prepared from the swimming bladders of the sturgeon (*Acipenser huso, L.*), and of other fishes.

It is emollient and nutritive, and is a useful addition to certain liquid foods for the sick. It forms the basis of court-plaster.

IGNATIA (U. S., Fr.)—IGNATIA.

The seed of Strychnos Ignatia, Lindley (*Nat. ord., Loganiaceæ*), a large climbing vine of the Philippine Islands, and of Cochin China where it has been introduced. It contains a large percentage of strychnia, and produces nearly the same medicinal effects as nux vomica.

Dose: 1 to 3 grains.

PREPARATIONS.

Extractum Ignatiæ (U. S.)—Extract of Ignatia.

Take of Ignatia, powdered	12 ounces.
Alcohol	sufficient.

Obtain 3 pints of tincture by maceration and percolation. Distil off the alcohol until the tincture is reduced to ½ pint, then evaporate to a proper consistence.

Dose: ½ to 1½ grain.

Alkaline Tincture of Ignatia (Fr.).

Take of Ignatia, powdered	500 parts.
Carbonate of potassium	5 parts.
Soot	1 part.
Alcohol, 60%	1000 parts.

Macerate ten days, express and filter.

Dose: 2 to 6 minims.

A tincture of ignatia may be made by macerating 1 part of the powdered seed in 5 parts of diluted alcohol, for two weeks.

ILLICIUM (Fr.)—STAR ANISE.

The fruit of Illicium anisatum, Loureiro (*Nat. ord., Magnoliaceæ*), a small tree indigenous to Southwestern China, but early introduced into Japan, and planted near the Buddhist temples.

Though so widely differing, botanically, from the officinal anise, its properties are very similar. It is anodyne, stimulant, and carminative, and may be used with benefit in flatulent colic, indigestion, etc.

Dose: 10 to 30 grains.

PREPARATIONS.

Star Anise Water (Fr.).

Take of Anise .... 1 part.
Water .... sufficient.

Macerate two hours, and distil 4 parts.

Used as a vehicle.

Spirit of Star Anise (Fr.).

Take of Star anise .... 1 part.
Alcohol (80%) .... 8 parts.

Macerate two days, and distil off the alcohol.

Dose: ½ to 1 drachm.

INULA (U. S., Fr., Ger.)—ELECAMPANE.

The root of Inula Helenium, L. (*Nat. ord., Compositæ*), a stout perennial, three to five feet high, introduced into this country from Europe, where it is indigenous. It is tonic, diuretic, diaphoretic, and expectorant, and has been used from remote antiquity in catarrhal affections of the mucous membranes.

Dose: 20 to 60 grains.

PREPARATIONS.

Extract of Elecampane (Fr., Ger.).

Take of Elecampane root .... 1 part.
Cold distilled water .... sufficient.

Exhaust the elecampane by maceration and percolation, heat the infusion to the boiling point, strain, and evaporate.

Dose: 10 to 30 grains.

The German preparation is an alcoholic extract.

Tincture of Elecampane (Fr.).

Take of	Elecampane root	1 part.
	Alcohol (60%)	6 parts.

Macerate ten days, express and filter.

Dose: 1 to 2 drachms.

IODINIUM (U. S. et al. Ph.)—IODINE.

A metalloid obtained from the ashes of sea-weed. It is found also in many marine animals, but not in sufficient quantities to make its extraction from them practicable. It is in crystalline scales, of a bluish black color, a metallic lustre, and has a hot, acrid taste. It is very volatile, slightly soluble in water, but freely so in alcohol, ether, and glycerin.

Iodine is a most valuable alterative and resolvent. It stimulates the absorbent and glandular systems to a remarkable degree, and is of great value in scrofulous and syphilitic affections. It is applied topically to glandular swellings, rheumatic and gouty joints, to the chest in phthisis, pleurisy, etc.

It is seldom, if ever, administered in substance.

PREPARATIONS.

Liquor Iodinii Compositus (U. S., Br.)—Compound Solution of Iodine.

Take of	Iodine	360 grains.
	Iodide of potassium	1½ ounce.
	Distilled water	1 pint.

Dissolve.

Dose: 2 to 6 minims.

Tinctura Iodinii (U. S., Fr., Ger.)—Tincture of Iodine.

Take of	Iodine	1 ounce.
	Alcohol	1 pint.

Dissolve.

Dose: 1 to 10 minims. Chiefly used externally.

The French Codex and German Pharmacopœia direct 1 part to 12, and 1 to 10, respectively.

Tinctura Iodinii Composita (U. S.)—Compound Tincture of Iodine.

Take of	Iodine	½ ounce.
	Iodide of potassium	1 ounce.
	Alcohol	1 pint.

Dissolve.

Dose: 5 to 15 minims. *Tincture of Iodine, Br.*

Unguentum Iodinii (U. S.)—Iodine Ointment.

Take of Iodine	20 grains.
Iodide of potassium	4 grains.
Water	6 minims.
Lard	1 ounce.

Rub the powders first with the water, then with the lard.

Unguentum Iodinii Compositum (U. S., Br., Fr.)—Compound Iodine Ointment.

Take of Iodine	15 grains.
Iodide of potassium	30 grains.
Water	30 minims.
Lard	1 ounce.

Rub the powders first with the water, then with the lard.

The British Pharmacopœia directs: 16 grains each of iodine and iodide of potassium, ½ drachm of proof spirit, and 1 ounce of lard, and terms the preparation *Ointment of Iodine.* The French Codex directs the proportion of 1, 5, and 40, with sufficient water to dissolve the iodide.

Colorless Tincture of Iodine (Ger.).

Take of Iodine	10 parts.
Hyposulphite of sodium	10 parts.
Water	10 parts.
Spirit of ammonia	16 parts.
Alcohol	75 parts.

Digest the iodine and hyposulphite in the water until dissolved, then add the spirit of ammonia, and lastly, the alcohol. After three days, filter.

Used externally as a counter-irritant in cases where the color of the ordinary tincture is objectionable.

Liniment of Iodine (Br.).

Take of Iodine	1¼ ounce.
Iodide of potassium	½ ounce.
Camphor	¼ ounce.
Rectified spirit	10 ounces.

Dissolve.

Inhalation of Iodine (Br.).

Take of Tincture of iodine	1 drachm.
Water	1 ounce.

Mix in a suitable apparatus, and, having applied a gentle heat, let the vapor that arises be inhaled.

Take of Tincture of iodine	2 to 10 minims.
Iodide of potassium	3 to 20 grains.
Water	to 1 ounce.

Stimulant application to the throat. Used by means of a spray apparatus. *G. M. Lefferts.*

Churchill's Tincture of Iodine.

Take of Iodine.................................. 2½ ounces.
Iodide of potassium........................ ½ ounce.
Alcohol (75%)................................ 16 ounces.

Mix.

Churchill's Iodine Caustic.

Take of Iodine.................................. 1 drachm.
Iodide of potassium .... 2 drachms.
Water.. ½ ounce.

Mix.

Gargles of Iodine.

Take of Tincture of Iodine......................... 2 drachms.
Water sufficient..........................to 5 ounces.

Mix. *St. Thomas's Hospital.*

Take of Tincture of iodine.......................... ½ drachm.
Tincture of bark .... 1½ drachm.
Water .... 4 ounces.

Mix. *St. Mary's Hospital.*

Hypodermic Injection of Iodine.

Take of Iodine.................................. 40 grains.
Absolute alcohol .... 1 ounce.

Dose: 10 to 15 minims. *London Throat Hospital.*

Liniments of Iodine and Belladonna.

Take of Compound solution of iodine .... 1 ounce.
Tincture of belladonna..................... 1 ounce.
Soap liniment.............................. 6 ounces.

Mix.

Take of Compound solution of iodine .... 1 ounce.
Tincture of belladonna..................... 1 ounce.
Oil of cajuput.............................. 1 drachm.
Soap liniment.............................. 6 ounces.

Mix.

This and the preceding are used in cases of interstitial effusion, as in synovitis, the inflammation having been subdued, and where little torpidity exists. *Hospital for Ruptured and Crippled.*

Take of Compound solution of iodine .... 1 part.
Glycerin .... 1 part.

Mix. Apply by inunction. In strumous adenitis. *J. Lewis Smith.*

Mixtures of Iodine.

Take of	Tincture of iodine	3 minims.
	Water	1 ounce.

Mix. One dose. *Women's Hospital, London.*

Take of	Iodine	$\frac{1}{8}$ grain.
	Iodide of potassium	3 grains.
	Compound tincture of lavender	30 minims.
	Water	to 1 ounce.

Mix. One dose. *Guy's Hospital.*

Compound Iodine Plaster.

Take of	Iodine	2 drachms.
	Iodide of potassium	3 drachms.
	Lead plaster	16 ounces.
	Opium plaster	6 ounces.

Mix. *St. George's Hospital.*

Solutions of Iodine for the Throat.

Take of	Tincture of iodine	3 to 12 minims.
	Iodide of potassium	5 to 20 grains.
	Glycerin	$\frac{1}{2}$ ounce.
	Water	$\frac{1}{2}$ ounce.

Mix. Stimulant application to the throat. Applied with the laryngeal brush. *G. M. Lefferts.*

Take of	Iodine	15 grains.
	Oil of sweet almonds	1 ounce.

Mix. Used in the same manner as the preceding. *G. M. Lefferts.*

IODOFORMUM (U. S., Ger.)—IODOFORM.

Iodoform is produced by the action of iodine upon alcohol in the presence of an alkali. It occurs in small, yellow crystals, having a peculiar, penetrating odor, and a sweet taste. It is insoluble in water, but soluble in ether, chloroform, alcohol, and oils.

It is stimulant, alterative, anæsthetic, and, in overdoses, poisonous. It is used internally in syphilis, gastralgia, neuralgia, gastric ulcer, etc. Externally it is applied to a great variety of ulcers and sores, frequently relieving pain and promoting the healing process.

Dose: 1 to 3 grains.

Ointment of Iodoform.

Take of Iodoform.................................. 1 part.
Balsam of Peru .... 5 parts.
Cosmoline.. 14 parts.

Mix. A stimulating application to foul or indolent ulcerations.

G. H. Fox.

Take of Iodoform .... 1 drachm.
Cerate.................................. 1 ounce.

Mix. Apply twice a week; also daily to the abdominal wall. For peri- and para-metritis. *R. Tauszky.*

Pills of Iodoform.

Take of Iodoform.................................. 2 grains.
Sugar of milk.............................. 1 grain.
Glycerin of tragacanth .... sufficient.

Triturate the iodoform with the sugar to a fine powder, then with glycerin of tragacanth make 1 pill.

Dose: 1 pill two or three times a day. *London Throat Hospital.*

Take of Iodoform .... 20 grains.
Corrosive sublimate .... 1 grain.
Reduced iron.............................. 20 grains.

Mix, and make 20 pills.

Dose: 1 pill three times a day. In constitutional syphilis.

Bartholow.

Take of Iodoform .... 20 grains.
Extract of gentian .... sufficient.

Mix, and divide into 20 pills.

Dose: 1 pill three times a day.

Powder of Iodoform and Camphor.

Take of Iodoform ....1 to 4 drachms.
Powdered camphor......................... 20 grains.
Powdered gum arabic .... to 1 ounce.

Mix. Used by insufflation into the larynx or nares.

G. M. Lefferts.

Powder of Iodoform and Tannic Acid.

Take of Iodoform .... 1 part.
Tannic acid.............................. 1 part.

Mix. Used by insufflation into the larynx or nares.

G. M. Lefferts.

Powder of Iodoform and Morphia.

Take of Iodoform .... 1 ounce.
Sulphate of morphia .... 10 to 20 grains.

Mix. Used by insufflation into the larynx or nares.

This, and the two preceding, are stimulating, alterative, and sedative.

G. M. Lefferts.

Powder of Iodoform and Calamine.

Take of Iodoform .... 30 grains.
Calamine .... 1 drachm.
Starch .... to 1 ounce.

Mix. *University College Hospital.*

Solution of Iodoform.

Take of Iodoform .... 1 part.
Ether .... 4 to 10 parts.

Dissolve. Painted over the surface of an ulcer, it acts as a protective, allays pain and promotes healing.

Suppositories of Iodoform.

Take of Iodoform .... 15 to 18 grains.
Oil of theobroma .... 6 drachms.

Rub the iodoform with a small portion of the oil, then add the remainder, previously melted, and make 6 suppositories.

Used in painful affections of the uterus, vagina, and rectum.

IPECACUANHA (U. S. et al. Ph.)—IPECACUANHA.

The root of Cephælis Ipecacuanha, A. Richard (*Nat. ord., Rubiaceæ*), a small shrub growing in South America. It owes its medicinal effects to an alkaloid, called *emetia or emetina,* which exists in it in combination with ipecacuanhic acid.

Ipecacuanha is diaphoretic, emetic, expectorant, and purgative, these different effects depending chiefly upon the size of the dose employed. Moreover, it seems to have certain specific effects, notably in the case of dysentery, which are not well understood, or easily explained.

Dose: $\frac{1}{5}$ to 2 grains, diaphoretic and expectorant; 15 to 30 grains, emetic.

PREPARATIONS.

Extractum Ipecacuanhæ Fluidum (U. S.)—Fluid Extract of Ipecacuanha.

Take of	Ipecacuanha	16 ounces.
	Glycerin	½ pint.
	Stronger alcohol	1½ pint.
	Water	12 ounces.
	Diluted alcohol	sufficient.

Mix the stronger alcohol and water, macerate the ipecacuanha in 18 ounces of the mixture for four days, then percolate with the remainder, continuing the process with diluted alcohol until 2 pints are obtained. Mix this with the glycerin, and evaporate below 140° to 1 pint.

Dose: 1 to 30 minims.

Pulvis Ipecacuanhæ Compositus (U. S. et al. Ph.)—Compound Powder of Ipecacuanha—Dover's Powder.

Take of	Ipecacuanha, in powder	1 drachm.
	Opium, in powder	1 drachm.
	Sulphate of potassium	1 ounce.

Rub together into a very fine powder.

Dose: 5 to 15 grains.

Dr. Piffard recommends the substitution of sugar of milk for sulphate of potassium in this preparation, thereby improving its taste without diminishing its therapeutic value.

Syrupus Ipecacuanhæ (U. S., Fr., Ger.)—Syrup of Ipecacuanha.

Take of	Fluid extract of ipecacuanha	2 ounces.
	Syrup	30 ounces.

Mix.

Dose: 30 to 60 minims as an expectorant; 4 to 8 drachms as an emetic.

Trochisi Ipecacuanhæ (U. S. et al. Ph.)—Troches of Ipecacuanha.

Take of	Ipecacuanha	120 grains.
	Tragacanth	120 grains.
	Arrow-root	2 ounces.
	Sugar	8 ounces.
	Syrup of orange peel	sufficient.

Rub the powders together, then with the syrup form a mass, to be divided into 480 troches.

Dose: 1 or 2 troches.

Vinum Ipecacuanhæ (U. S., Br., Ger.)—Wine of Ipecacuanha.

Take of Fluid extract of ipecacuanha .... 2 ounces.
Sherry wine .... 30 ounces.

Mix and filter.

Dose: 10 to 30 minims as an expectorant; ½ to 1 ounce as an emetic.

Very small doses—1 to 5 drops—are sometimes used with excellent effect in allaying vomiting.

Alcoholic Extract of Ipecacuanha (Fr.).

Take of Ipecacuanha .... 1 part.
Diluted alcohol .... 6 parts.

Percolate the ipecacuanha with the alcohol, then distil off the alcohol, and evaporate to the consistence of a soft extract.

Used by the French Codex in making the syrup of ipecacuanha.

Tincture of Ipecacuanha (Fr., Ger.).

Take of Ipecacuanha .... 1 part.
Diluted alcohol .... 5 parts.

Macerate ten days, express and filter.

Dose: 5 to 20 minims as an expectorant; 2 to 3 drachms as an emetic.

Mixtures of Ipecacuanha.

Take of Wine of ipecacuanha .... 15 minims.
Tincture of squill .... 7½ minims.
Tincture of opium .... 4 minims.
Glycerin .... 18 minims.
Water .... to 1 ounce.

Mix. One dose. *University College Hospital.*

Take of Wine of ipecacuanha .... 20 minims.
Bicarbonate of sodium .... 10 grains.
Spirit of nitrous ether .... 15 minims.
Tincture of henbane .... 20 minims.
Treacle .... 1 drachm.
Water .... to 1 ounce.

Mix. One dose. *Great Northern Hospital.*

IRIS FLORENTINA (U. S., Fr., Ger.)—FLORENTINE ORRIS.

The rhizome of Iris Florentina, L. (*Nat. ord., Iridaceæ*) an herbaceous perennial indigenous to Southern Europe.

It is seldom used in this country except as a dentifrice.

Dose: 1 to 6 drachms.

IRIS VERSICOLOR (U. S.)—BLUE FLAG.

The rhizome of Iris versicolor, L. (*Nat. ord.*, *Iridaceæ*), our common blue flag which grows in wet places and blooms in May and June.

When fresh, the root has a slightly nauseous odor, and a bitter, acrid taste. It is emetic and cathartic, and in overdoses may produce great prostration. The fresh root, or a tincture made from it, has been found useful in sick headache. It should be administered in very small doses, insufficient to provoke vomiting.

Dose: 5 to 15 grains.

PREPARATIONS.

Tincture of Iris.

Take of Iris versicolor, fresh	1 part.
Alcohol	2 parts.

Macerate one week, and filter.

Dose: 5 to 30 minims.

Oleoresin of Iris—Irisin.

Take of Iris versicolor, fresh	16 parts.
Alcohol (sp. gr. 0.835)	sufficient.

Moisten, pack, and percolate until 16 parts of tincture are obtained. Evaporate this to a thick, syrupy consistence, pour into ten times its bulk of cold water, stir slightly, and allow the mixture to stand for twenty-four hours. Draw off the aqueous solution, transfer the oily precipitate to an evaporating basin, and evaporate on a water-bath, stirring well, until the water is expelled.

It is an oily substance, liquid at ordinary temperatures, having the odor of fresh blue flag root, and a disagreeable, oily taste.

Dose: 2 to 4 grains. *J. U. Lloyd.*

JALAPA (U. S. et al. Ph.)—JALAP.

The tuber of Ipomæa Purga, Hayne (*Nat. ord.*, *Convolvulaceæ*), an herbaceous, perennial, twining vine, indigenous to the mountainous regions of Mexico.

Jalap has a peculiar, coffee-like odor, a nauseous taste, and is an active, hydragogue cathartic. Its action is generally attended with some pain, and, in overdoses, it may produce dangerous hypercatharsis.

It is used chiefly in dropsy, and as a revulsive in cerebral affections.

Dose: 15 to 30 grains.

PREPARATIONS.

Extractum Jalapæ (U. S., Br.)—Extract of Jalap.

Take of Jalap	12 ounces.
Alcohol	4 pints.
Water	sufficient.

Percolate the jalap with the alcohol, continuing the process with water until 4 pints are obtained; then percolate with water until 6 pints of infusion are obtained. Evaporate the two liquids separately to the consistence of thin honey, then mix them, and evaporate to a proper consistence.

Dose: 10 to 20 grains.

Pulvis Jalapæ Compositus (U. S., Br.)—Compound Powder of Jalap.

Take of Jalap	1 ounce.
Bitartrate of potassium	2 ounces.

Rub together thoroughly.

Dose: ½ to 1 drachm. An excellent hydragogue cathartic.

Resina Jalapæ (U. S. et al. Ph.)—Resin of Jalap.

Take of Jalap	16 ounces.
Alcohol	sufficient.
Water	sufficient.

Macerate the jalap in 16 ounces of alcohol for four days, then percolate until 24 ounces are obtained. Reduce this to 6 ounces by distilling off the alcohol, and mix it with 7 pints of water. Decant the supernatant liquid, wash the precipitate, and dry it with a gentle heat.

Dose: 2 to 5 grains.

Tinctura Jalapæ (U. S., Br., Fr.)—Tincture of Jalap.

Take of Jalap	6 ounces.
Alcohol, 2 parts—Water, 1 part	sufficient.

Moisten, pack, and percolate to 2 pints.

Dose: ½ to 2 drachms.

Compound Tincture of Jalap (Fr.).

Take of Jalap	8 parts.
Turpeth	1 part.
Scammony	2 parts.
Alcohol (60%)	96 parts.

Macerate ten days, and filter.

Dose: ½ to 2 drachms.

Jalap Soap (Ger.).

Take of Resin of jalap	4 parts.
Soap	4 parts.
Diluted alcohol	8 parts.

Dissolve, and evaporate on a water-bath until the whole weighs 9 parts.

Dose: 5 to 10 grains.

JALAP PILLS (Ger.).

Take of Jalap soap .... 3 parts.
Jalap root, powdered .... 1 part.

Beat into a mass, and divide into pills of 1½ grain each.

Dose: 3 to 6 pills.

CONFECTION OF JALAP.

Take of Powdered jalap .... ¼ ounce.
Powdered senna .... 2 ounces.
Ginger .... 60 grains.
Treacle .... 8 ounces.

Mix. Dose: 1 to 2 drachms. *St. George's Hospital.*

CONFECTION OF JALAP AND SULPHUR.

Take of Confection of jalap .... 5 ounces.
Precipitated sulphur .... 1 ounce.

Mix. Dose: 1 to 2 drachms. *St. George's Hospital.*

POWDER OF JALAP AND SCAMMONY.

Take of Jalap .... 3 grains.
Scammony .... 7 grains.

Mix. One dose. *Westminster Ophthalmic Hospital.*

JUGLANS (U. S.)—BUTTERNUT.

The inner bark of Juglans cinerea, L. (*Nat. ord., Juglandaceæ*), our common butternut. It possesses mild cathartic properties, resembling those of rhubarb, and is well suited to cases of habitual constipation.

Dose: ½ to 1 drachm.

PREPARATION.

Extractum Juglandis (U. S.)—Extract of Butternut.

Take of Butternut .... 12 ounces.
Water .... sufficient.

Exhaust the butternut by percolation with water, boil the infusion to three-fourths of its bulk, strain, and evaporate on a water-bath to a proper consistence.

Dose: 4 to 30 grains.

JUNIPERUS (U. S. et al. Ph.)—JUNIPER.

The fruit of Juniperus communis, L. (*Nat. ord., Coniferæ*), a common evergreen shrub, often growing with Juniperus Virginiana, L. (*Red cedar*), from which it may be distinguished by having its leaves in threes, with a slender prickly point, while the latter has scale-shaped leaves.

Juniper berries have a sweetish, terebinthinate taste, and possess diuretic properties. Used in urinary diseases.

Dose: 1 to 2 drachms.

PREPARATIONS.

Infusum Juniperi (U. S.)—Infusion of Juniper.

Take of Juniper, bruised	1 ounce.
Boiling water	1 pint.

Macerate for an hour, and strain.

Dose: 2 to 3 ounces.

Oleum Juniperi (U. S. et al. Ph.)—Oil of Juniper.

The oil obtained from juniper by distillation.

Dose: 5 to 10 minims.

Spiritus Juniperi (U. S., Br., Ger.)—Spirit of Juniper.

Take of Oil of juniper	1 ounce.
Stronger alcohol	3 pints.

Dissolve.

Dose: ½ to 1 drachm.

Spiritus Juniperi Compositus (U. S.)—Compound Spirit of Juniper.

Take of Oil of juniper	1½ drachm.
Oil of caraway	10 minims.
Oil of fennel	10 minims.
Alcohol	5 pints.
Water	3 pints.

Dissolve the oils in the alcohol, add the water, and mix.

Dose: 2 to 4 drachms.

EXTRACT OF JUNIPER (Ger., Fr.).

Take of Juniper berries	1 part.
Hot water	4 parts.

Infuse, strain, and evaporate to a thin extract.

Dose: 1 to 2 drachms.

ROTTLERA (U. S., Br., Ger.)—KAMEELA—KAMALA.

The glandular powder and hairs obtained from the fruit of Rottlera tinctoria, Roxb. (*Nat. ord., Euphorbiaceæ*), a shrub or small tree indigenous to Abyssinia and Southern Arabia.

Kameela is used almost exclusively for the expulsion of tape-worm.

Dose: 1 to 3 drachms.

PREPARATION.

TINCTURE OF KAMEELA.

Take of Kameela .... 3 ounces.
Alcohol .... 10 ounces.

Digest, and filter.
Dose 3 to 8 drachms. *Anderson.*

KINO (U. S. et al. Ph.)—KINO.

The inspissated juice of Pterocarpus Marsupium, Roxb. (*Nat. ord., Leguminosæ*), a tree indigenous to the Indian Peninsula and Ceylon.

Kino occurs in small, angular pieces, of a blackish-red color, and an astringent taste. Employed internally as an astringent in diarrhœa and dysentery, and topically in leucorrhœa, etc.

Several other varieties of kino occur in market, the products of as many different species of trees. All are astringents, and are used for the same purposes as the East Indian drug.

Dose: 10 to 30 grains.

Tinctura Kino (U. S. et al. Ph.)—Tincture of Kino.

Take of Kino .... 360 grains.
Alcohol .... sufficient.
Water .... sufficient.

Mix 2 measures of alcohol with 1 of water, and percolate the kino, previously mixed with an equal bulk of dry sand, with the mixture, until ½ pint of tincture is obtained.

Dose: ½ to 2 drachms.

COMPOUND POWDER OF KINO (Br.).

Take of Kino .... 3¾ ounces.
Opium .... ¼ ounce.
Cinnamon .... 1 ounce.

Mix. Dose: 5 to 20 grains.

INFUSION OF KINO.

Take of Kino .... 2 drachms.
Boiling water .... 8 ounces.

Infuse and strain.

Dose: 1 ounce. *Wood.*

Used as an injection in leucorrhœa and gleet, and as a gargle in various forms of sore throat.

Troches of Kino.

Take of Kino	700 grains.
Tragacanth	70 grains.
Refined sugar	280 grains.
Red currant paste	sufficient.

Mix the powders, then add the paste, and divide into 350 troches of 20 grains each. *Mackenzie.*

KRAMERIA (U. S. et al. Ph.)—RHATANY.

The root of Krameria triandra, Ruiz et Pav. (*Nat. ord., Polygalaceæ*), a small shrub indigenous to Peru.

Krameria has a very astringent, and slightly bitter taste. It is somewhat tonic, and powerfully astringent. Used in diarrhœa, dysentery, hemorrhages, etc.

Dose: 10 to 40 grains.

PREPARATIONS.

Extractum Krameriæ (U. S. et al. Ph.)—Extract of Rhatany.

Take of Rhatany	12 ounces.
Water	sufficient.

Exhaust the rhatany by percolation with water, heat the infusion to the boiling point, strain, and evaporate, at a temperature not exceeding 160°, to a proper consistence.

Dose: 5 to 20 grains.

Extractum Krameriæ Fluidum (U. S.)—Fluid Extract of Rhatany.

Take of Rhatany	16 ounces.
Glycerin	4 ounces.
Alcohol	sufficient.
Water	sufficient.

Mix 8 ounces of alcohol, 3 of glycerin, and 5 of water, moisten the rhatany with 4 ounces of the mixture, and proceed according to the general formula, page 161. Finish the percolation with diluted alcohol, reserve 14 ounces, and add 1 ounce of glycerin to the remainder, before evaporation.

Dose: 10 to 40 minims.

Infusum Krameriæ (U. S., Br., Fr.)—Infusion of Rhatany.

Take of Rhatany	1 ounce.
Water	sufficient.

Moisten, pack, and percolate to 1 pint.

Dose: 1 to 2 ounces. Often used as a gargle in sore throat.

Syrupus Krameriæ (U. S., Fr.)—Syrup of Rhatany.

Take of Rhatany	12 ounces.
Sugar	30 ounces.
Water	sufficient.

Obtain 4 pints of infusion by percolation with water, and evaporate it on a water-bath to 17 ounces; then add the sugar, dissolve, and strain while hot. Or, mix 12 ounces of fluid extract of rhatany with 24 ounces of syrup.

Dose: 1 to 4 drachms.

Tinctura Krameriæ (U. S. et al. Ph.)—Tincture of Rhatany.

Take of Rhatany	6 ounces.
Diluted alcohol	sufficient.

Moisten, pack, and percolate to 2 pints.

Dose: ½ to 2 drachms.

Suppositories of Extract of Rhatany (Fr.).

Take of Extract of rhatany	1 part.
Oil of theobroma	4 parts.

Melt the oil, and when cooling add the extract, mix thoroughly, and pour into moulds having the capacity of 1 drachm each.

Used in fissure of the anus, etc.

Injection of Rhatany.

Take of Extract of rhatany	1 to 2 drachms.
Tincture of rhatany	1 drachm.
Water	5 ounces.

Mix. Used in fissure of the anus. *Trousseau.*

Mixture of Rhatany.

Take of Tincture of rhatany	1 drachm.
Tincture of opium	6 drops.
Bicarbonate of sodium	20 grains.
Syrup of ginger	7 drachms.
Water	2 ounces.

Mix. Dose: 1 drachm two or three times a day, for children one or two years old. In diarrhœa. *Meigs and Pepper.*

Troches of Rhatany.

Take of Extract of rhatany	1,050 grains.
Tragacanth	70 grains.
Refined sugar	280 grains.
Red currant paste	sufficient.

Mix the powders, add the paste, and divide into 350 troches, each containing 3 grains of extract of rhatany. *Mackenzie.*

LACTUCARIUM (U. S., Fr., Ger.)—LACTUCARIUM.

The concrete juice of garden lettuce, Lactuca sativa, L., Lactuca virosa, L., and other species of Lactuca (*Nat. ord., Compositæ*).

Lactucarium has an unpleasant, opium-like odor, and a very bitter taste. It possesses some of the anodyne and soporific properties of opium, but does not produce such unpleasant after-effects. It is, however, an unreliable drug, and should not be substituted for opium except when the latter disagrees.

Dose: 8 to 20 grains.

PREPARATIONS.

Syrupus Lactucarii (U. S.)—Syrup of Lactucarium.

Take of Lactucarium	1 ounce.
Syrup	14 ounces.
Diluted alcohol	sufficient.

Rub the lactucarium with enough diluted alcohol to bring it to a syrupy consistence, then percolate with more until ½ pint of tincture is obtained. Evaporate, at or below 160°, to 2 ounces, mix with the syrup previously heated, and strain while hot.

Dose: 2 to 3 drachms.

In this connection may be considered

Extract of Lettuce (Br., Fr., Ger.).

Take of flowering herb of lettuce, a convenient quantity.

Bruise, express the juice, heat to 130°, and separate the green coloring matter by a calico filter. Heat the strained liquid to 200°, filter, evaporate to the consistence of thin syrup, add the coloring matter previously separated, continue the evaporation, at or below 140°, to a proper consistence for forming pills.

Dose: 5 to 15 grains.

Considered inferior to lactucarium, though there is not very much difference between them.

Syrup of Extract of Lettuce (Fr.).

Take of Extract of lettuce	2 parts.
Syrup	98 parts.
Distilled water	sufficient.

Dissolve the extract in 8 times its weight of water, filter, mix with the syrup, and evaporate to the sp. gr. 1.26.

Water of Lettuce (Fr.).

Take of	Flowering herb of lettuce	1 part.
	Water	2 parts.

Distil 1 part.

Pills of Lettuce and Henbane.

Take of	Extract of lettuce	2 grains.
	Extract of henbane	2 grains.

Make 1 pill. *Brompton Consumption Hospital.*

LAVANDULA (U. S. et al. Ph.)—LAVENDER.

The flowers of Lavandula vera, DC. (*Nat. ord., Labiatæ*), a small shrub indigenous to Southern Europe, but cultivated in many temperate regions.

Lavender has a strong, agreeable odor, and an aromatic, bitterish taste. It is much used as a stimulant in hysterical and other nervous affections.

PREPARATIONS.

Oleum Lavandulæ (U. S. et al. Ph.)—Oil of Lavender.

The oil obtained from lavender by distillation.

Dose: 1 to 5 minims.

Spiritus Lavandulæ (U. S. et al. Ph.)—Spirit of Lavender.

Take of	Oil of lavender	1 ounce.
	Stronger alcohol	3 pints.

Dissolve.

Dose: ½ to 1 drachm.

The French and German preparations are prepared by distillation.

Spiritus Lavandulæ Compositus (U. S., Br.)—Compound Spirit of Lavender.

Take of	Oil of lavender	1 ounce.
	Oil of rosemary	2 drachms.
	Cinnamon	2 ounces.
	Cloves	½ ounce.
	Nutmeg	1 ounce.
	Red saunders	6 drachms.
	Alcohol	6 pints.
	Water	2 pints.
	Diluted alcohol	sufficient.

Dissolve the oils in the alcohol, and add the water. Mix the powders, and percolate them with the alcoholic solution, continuing the process with diluted alcohol until 8 pints are obtained.

Dose: 10 to 60 minims.

Compound Tincture of Lavender, Br.

LEPTANDRA (U. S.)—LEPTANDRA.

The rhizome and rootlets of Veronica Virginica, L. (*Leptandra Virginica, Nutt., Nat. ord., Scrophulariaceæ*), a tall, herbaceous perennial indigenous to North America, growing in rich woodlands.

Leptandra is an excellent cholagogue cathartic. Its effects are due to a peculiar principle, termed *leptandrin*, which, however, is not as yet separated in sufficient quantities to be of commercial importance. The *leptandrin* of the shops is a resinoid substance precipitated from a concentrated tincture of leptandra.

Dose: ½ to 1 drachm.

PREPARATIONS.

Fluid Extract of Leptandra.

Take of Leptandra .... 16 parts.
Diluted alcohol, 5 parts—glycerin, 1 part..... sufficient.

Prepare according to the general formula, page 161.

This is the menstruum proposed in the *Report of the Am. Ph. Ass'n.*

Dose: ½ to 1 drachm.

Resin of Leptandra—Leptandrin.

Take of Leptandra .... 16 parts.
Alcohol (sp. gr. 0.835)..................... sufficient.

Moisten, pack, and percolate until 16 parts of tincture are obtained. Evaporate this to a thick, syrupy consistence, pour into ten times its bulk of cold water, stir well, and allow the mixture to stand for twenty-four hours. Then decant the supernatant liquid, wash the precipitate with hot water, and permit it to cool. Then transfer to an evaporating basin, and evaporate almost to dryness on a water-bath. Crush the mass in a mortar, expose it to the air until perfectly dry, then powder it.

A dark, almost black powder having the odor of Leptandra. As above made it is almost tasteless, but the commercial article is usually bitter, owing to the presence of extractive matters.[1]

Dose: 2 to 5 grains.

LIMON—LEMON.
LIMONIS CORTEX (U. S. et al. Ph.)—LEMON PEEL.
LIMONIS SUCCUS (U. S. et al. Ph.)—LEMON JUICE.

The fruit of Citrus limonum, Risso (*Nat. ord., Aurantiaceæ*), a tree cultivated in many subtropical countries. The lemons of commerce come chiefly from the Mediterranean coast.

---

[1] For remarks upon commercial leptandrin, see article by J. U. Lloyd, in Proceedings of Am. Phar. Ass'n, 1880.

All parts of the lemon are used medicinally. The rind has a fragrant odor, a bitter, aromatic taste, and is used as a flavoring agent. The juice abounds in citric acid, and is used in preparing cooling and refreshing drinks.

PREPARATIONS.

Oleum Limonis (U. S. et al. Ph.)—Oil of Lemon.

The volatile oil obtained from lemon peel.

Used for flavoring.

Spiritus Limonis (U. S., Fr.)—Spirit of Lemon.

Take of		
	Oil of lemon	2 ounces.
	Lemon peel (freshly grated)	1 ounce.
	Stronger alcohol	2 pints.

Dissolve the oil in the alcohol, add the peel, macerate for twenty-four hours, and filter.

Used for flavoring.

The French preparation is made by distillation.

Syrupus Limonis (U. S. et al. Ph.)—Syrup of Lemon.

Take of		
	Lemon juice	1 pint.
	Sugar	48 ounces.
	Water	1 pint.

Mix the lemon juice and water, add the sugar, dissolve with a gentle heat, and strain while hot.

Used as a vehicle.

The British Pharmacopœia adds also lemon peel.

Tincture of Lemon Peel (Br., Fr.).

Take of		
	Fresh lemon peel	2½ ounces.
	Proof spirit	20 ounces.

Macerate seven days, express, filter, and add sufficient proof spirit to make 20 ounces.

Dose: ½ to 2 drachms. Used chiefly for flavoring.

The French Codex directs: lemon peel, 1 part; alcohol (80%), 2 parts.

LINUM (U. S. et al. Ph.)—FLAXSEED—LINSEED.

The seed of Linum usitatissimum, L. (*Nat. ord., Linaceæ*), common flax, a plant which has been cultivated from the remotest antiquity.

Flaxseed is an excellent demulcent, and is largely used in febrile and inflammatory affections. It also possesses important nutrient properties, and has been used as a substitute for cod-liver oil. It is employed externally as an emollient.

PREPARATIONS.

Infusum Lini Compositum (U. S., Br.)—Compound Infusion of Flaxseed.

Take of Flaxseed	½ ounce.
Liquorice root	2 drachms.
Boiling water	1 pint.

Macerate two hours, and strain.

May be used *ad libitum* as a drink in febrile affections.

Lini Farina (U. S. et al. Ph.)—Flaxseed Meal.

The meal prepared by grinding flaxseed. The British and German Pharmacopœias employ the meal from which the oil has been expressed, while the United States Pharmacopœia and French Codex direct the freshly ground seed. Used in poultices.

Oleum Lini (U. S. et al. Ph.)—Flaxseed Oil—Linseed Oil.

The fixed oil obtained by expression from ground flaxseed.

Dose: ½ to 2 ounces. Chiefly used externally.

Sulphurated Flaxseed Oil (Ger.).

Take of Flaxseed Oil	6 parts.
Sublimed sulphur	1 part.

Boil them, stirring constantly, until they have united into a homogeneous mass.

Linseed Poultice (Br.).

Take of Linseed meal	4 ounces.
Olive oil	½ ounce.
Boiling water	10 ounces.

Mix the meal gradually with the water, then add the oil, with constant stirring.

LITHIUM—LITHIUM.

A very rare metal found in combination in a few minerals, as *spodumene* and *lepidolite,* and in some mineral waters. Its protoxide, lithia, is a powerful alkali like potassa or soda, though less soluble.

The salts of lithium are chiefly employed as alkaline diuretics in gout, rheumatism, etc.

LITHII BENZOAS—BENZOATE OF LITHIUM.

Prepared by adding benzoic acid to a solution of carbonate of lithium, and evaporating to dryness.

Dose: 2 to 5 grains.

LITHII BROMIDUM—BROMIDE OF LITHIUM.

Prepared by adding carbonate of lithium to hydrobromic acid, and evaporating. It is a very deliquescent salt.

It produces the general effects of the bromides, and is, by some physicians, esteemed most highly of them all. Used in epilepsy and other nervous affections, and in rheumatism.

Dose: 1 to 3 grains.

Mixture of Bromide of Lithium.

Take of Bromide of lithium	3 drachms.
Syrup of ginger	½ ounce.
Water	1½ ounce.

Mix. Dose: 1 drachm three times a day, in rheumatism, when the smaller joints are swollen and tender after the subsidence of acute symptoms. *Bartholow.*

LITHII CARBONAS (U. S., Br., Ger.)—CARBONATE OF LITHIUM.

Obtained chiefly from the mineral lepidolite. It is a white powder, sparingly soluble in water, and of a feeble, alkaline reaction.

Dose: 2 to 6 grains.

PREPARATION.

Effervescing Solution of Lithium (Br.).

Take of Carbonate of lithium	10 grains.
Water	20 ounces.

Dissolve, and, by means of a suitable apparatus, impregnate the solution with carbonic acid gas.

Dose: 5 to 10 ounces.

Mixture of Carbonate of Lithium.

Take of Carbonate of lithium	1 drachm.
Citric acid	2 drachms.
Water	2 ounces.

Mix. Dose: 1 drachm every four hours. *Bartholow.*

LITHII CITRAS (U. S., Br.)—CITRATE OF LITHIUM.

Take of	Carbonate of lithium	100 grains.
	Citric acid, in crystals	200 grains.
	Distilled water	2 ounces.

Dissolve the acid in the water, add the carbonate, and evaporate to dryness.

Dose: 2 to 6 grains.

LOBELIA (U. S. et al. Ph.)—LOBELIA.

The leaves and tops of Lobelia inflata, L. (*Nat. ord., Lobeliaceæ*), a small annual indigenous to North America, growing abundantly in dry pastures and by roadsides.

Lobelia has an unpleasant odor, and an acrid, nauseous taste. It produces effects analogous to those of tobacco, and is used in small doses as a diuretic, diaphoretic, and sedative. In large doses it acts as a violent emetic, and may produce fatal prostration. It is employed chiefly in chest affections, especially those of a spasmodic character, as asthma, etc.

Dose: 5 to 20 grains.

PREPARATIONS.

Acetum Lobeliæ (U. S.)—Vinegar of Lobelia.

Take of	Lobelia	4 ounces.
	Diluted acetic acid	sufficient.

Moisten, pack, and percolate to 2 pints.

Dose: 10 to 60 minims.

Tinctura Lobeliæ (U. S. et al. Ph.)—Tincture of Lobelia.

Take of	Lobelia	4 ounces.
	Diluted alcohol	sufficient.

Moisten, pack, and percolate to 2 pints.

Dose: 10 to 60 minims.

Ethereal Tincture of Lobelia (Br.).

Take of	Lobelia	$2\frac{1}{2}$ ounces.
	Spirit of ether	20 ounces.

Macerate seven days, express, filter, and add sufficient menstruum to make 20 ounces.

Dose: 10 to 30 minims.

MIXTURES OF LOBELIA.

Take of	Tincture of lobelia	1 ounce.
	Iodide of ammonium	2 drachms.
	Bromide of ammonium	3 drachms.
	Syrup of tolu	2 ounces.

Mix. Dose: 1 drachm every one, two, three, or four hours, in the paroxysm of asthma. *Bartholow.*

Take of	Tincture of lobelia	1 ounce.
	Tincture of hyoscyamus	1 ounce.
	Compound spirit of ether	1 ounce.
	Syrup of tolu	1 ounce.

Mix. 1 drachm every half-hour during the paroxysm of asthma, afterward at longer intervals. *Da Costa.*

Take of	Ethereal mixture of lobelia	2 drachms.
	Mixture of chloride of iron	2½ drachms.
	Camphor water	4 ounces.

M. Dose: ½ ounce three times a day, in emphysema. *Chambers.*

Take of	Ethereal tincture of lobelia	15 minims.
	Camphor water	1 ounce.

Mix. One dose. *London Hospital.*

LUPULINA (U. S., Ger.)—LUPULIN.

The yellow powder separated from the strobiles of Humulus Lupulus, L. Dose: 5 to 10 grains. *See Hops.*

PREPARATIONS.

Extractum Lupulinæ Fluidum (U. S.)—Fluid Extract of Lupulin.

Take of	Lupulin	16 ounces.
	Stronger alcohol	sufficient.

Moisten the lupulin with 6 ounces of the stronger alcohol, and proceed according to the general formula, page 161.

Dose: 10 to 30 minims.

Oleoresina Lupulinæ (U. S.)—Oleoresin of Lupulin.

Take of	Lupulin	12 ounces.
	Ether	sufficient.

Obtain 20 ounces of tincture by percolation, distil off the greater portion of the ether, and allow the remainder to evaporate spontaneously.

Dose: 2 to 5 grains.

Tinctura Lupulinæ (U. S.)—Tincture of Lupulin.

Take of Lupulin	4 ounces.
Alcohol	sufficient.

Moisten, pack, and percolate to 2 pints.

Dose: ½ to 2 drachms.

MIXTURE OF LUPULIN AND CAPSICUM.

Take of Fluid extract of lupulin	1 ounce.
Tincture of capsicum	1 ounce.

Mix. Dose: 1 or 2 drachms when necessary, as a substitute for alcoholic stimulants, and when delirium tremens is threatened. *Bartholow.*

LYCOPODIUM (U. S., Fr., Ger.)—LYCOPODIUM.

The sporules of Lycopodium clavatum, L., and of other species of Lycopodium (*Nat. ord., Lycopodiaceæ*), the common club-moss of this country and Europe.

Lycopodium is a fine powder, odorless and tasteless. It is used as a protective coating in intertrigo, eczema, and other cutaneous affections; as a diluent of medicinal powders, etc.

MAGNESIUM—MAGNESIUM.

Magnesium, though not employed medicinally in the metallic form, furnishes several compounds which are largely used. Its protoxide, magnesia, is a mild alkali, as is also the carbonate, while the citrate and sulphate are extensively used as cathartics.

MAGNESIA (U. S. et al. Ph.)—CALCINED MAGNESIA.

Take of carbonate of magnesium, at will.

Expose it in an earthen crucible to a red heat for two hours, or until the carbonic acid is expelled.

Dose: 10 to 30 grains, as an antacid. In larger doses it is mildly laxative.

PREPARATION.

Trochisi Magnesiæ (U. S., Fr., Ger.)—Troches of Magnesia.

Take of Magnesia	3 ounces.
Nutmeg	60 grains.
Sugar	9 ounces.
Mucilage of tragacanth	sufficient.

Rub the powders together, then with the mucilage form a mass, to be divided into 480 troches.

Dose: 1 to 3 troches.

MIXTURE OF MAGNESIA.

Take of	Magnesia	12 grains.
	Tincture of opium	3 drops.
	White sugar	sufficient.
	Water	1 ounce.

Mix. Dose: 1 drachm every two hours until the bowels are tranquil. In thrush. *Dewees.*

MAGNESII CARBONAS (U. S. et al. Ph.)—CARBONATE OF MAGNESIUM.

May be prepared by mixing solutions of sulphate of magnesium and carbonate of sodium, washing and drying the precipitate, but is prepared on a large scale from the bittern of salt works. It occurs in two varieties, called *light* and *heavy*, respectively, the former of which is directed by the U. S. Pharmacopœia. The difference between the two depends upon the manner of preparation, the one containing more water than the other.

Carbonate of magnesium is used for the same purposes as magnesia.

Dose: ½ to 2 drachms.

PREPARATIONS.

SOLUTION OF CARBONATE OF MAGNESIUM (Br.).

Take of	Sulphate of magnesium	2 ounces.
	Carbonate of sodium	2½ ounces.
	Distilled water	sufficient.

Dissolve the salts separately, each in 10 ounces of water, heat the magnesium solution to the boiling point, add the sodium solution, and boil until gas ceases to be evolved. Wash the precipitate, mix it with 20 ounces of distilled water, impregnate it with carbonic acid, and bottle.

Dose: 1 to 2 ounces.

TROCHES OF CARBONATE OF MAGNESIUM (Fr.).

Take of	Carbonate of magnesium	20 parts.
	White sugar	80 parts.
	Mucilage of tragacanth	12 parts.

Form a mass, and divide into troches of 15 grains each.

TROCHES OF CARBONATE OF MAGNESIUM AND CATECHU (Fr.).

Take of	Carbonate of magnesium	10 parts.
	Catechu	5 parts.
	White sugar	85 parts.
	Mucilage of tragacanth	12 parts.

Form a mass, and divide into troches of 15 grains each.

POWDER OF MAGNESIA AND RHUBARB (Ger.).

Take of	Carbonate of magnesium	60 parts.
	Oleosaccharate of fennel	40 parts.
	Rhubarb	15 parts.

Mix.

The *oleosaccharates* of the German Pharmacopœia are formed by rubbing 1 drop of any essential oil with ½ drachm of white sugar.

MIXTURES OF CARBONATE OF MAGNESIUM.

Take of	Carbonate of magnesium	½ drachm.
	Tincture of asafetida	40 drops.
	Tincture of opium	20 drops.
	Sugar	1 drachm.
	Distilled water	1 ounce.

Mix. Dose: 20 drops to 1 drachm, according to age, in flatulent colic and diarrhœa of infants. *Dewees.*

Take of	Carbonate of magnesium	15 grains.
	Rhubarb	5 grains.
	Cinnamon water	1 ounce.

Mix. One dose. *London Ophthalmic Hospital.*

Take of	Carbonate of magnesium	½ drachm.
	Sulphate of magnesium	3 drachms.
	Aromatic spirit of ammonia	1 drachm.
	Tincture of rhubarb	½ ounce.
	Tincture of hyoscyamus	½ drachm.
	Peppermint water	4 ounces.

Mix. Dose: ½ ounce two or three times a day, as a mild cathartic. *Meigs.*

Dr. Bartholow substitutes tincture of gelsemium for tincture of hyoscyamus in this mixture.

MAGNESII CITRAS—CITRATE OF MAGNESIUM.

PREPARATIONS.

Liquor Magnesii Citratis (U. S., Br., Fr.).—Solution of Citrate of Magnesium.

Take of	Carbonate of magnesium	200 grains.
	Citric acid	400 grains.
	Syrup of citric acid	2 ounces.
	Bicarbonate of potassium	40 grains.
	Water	sufficient.

Dissolve the acid in 4 ounces of water, add the carbonate, and stir until dissolved. Filter into a twelve-ounce bottle containing the syrup, add the bicarbonate, and sufficient water to nearly fill the bottle; cork, and fasten with wire or twine.

Dose: 4 to 6 ounces, laxative; 6 to 12 ounces, purgative.

A pleasant, but rather unreliable preparation.

EFFERVESCING CITRATE OF MAGNESIUM (Ger.).

Take of	Carbonate of magnesium	25 parts.
	Citric acid	75 parts.

Mix, form into a thick paste with distilled water, and dry, at or below 86°.

Take of	This mass	14 parts.
	Bicarbonate of sodium	13 parts.
	Citric acid	6 parts.
	White sugar	3 parts.

Mix, moisten with sufficient alcohol, pass through a sieve to form a coarse granular powder, and dry.

MAGNESII SULPHAS (U. S. et al. Ph.)—SULPHATE OF MAGNESIUM—EPSOM SALT.

Sulphate of magnesium is a constituent of sea-water, and of the waters of many mineral springs, and is prepared on a large scale from the mineral *magnesite.*

It occurs in transparent, colorless crystals, resembling those of oxalic acid, and has a bitter, saline taste. It is a mild, but effective, cooling purgative.

Dose: 2 drachms to 1 ounce.

MIXTURES OF SULPHATE OF MAGNESIUM.

Take of	Sulphate of magnesium	1 drachm.
	Diluted sulphuric acid	10 minims.
	Syrup of red poppies	30 minims.
	Spearmint water	to 1 ounce.

Mix. One dose. *Westminster Hospital.*

Take of	Sulphate of magnesium	80 grains.
	Carbonate of magnesium	10 grains.
	Peppermint water	1 ounce.

Mix. One dose. *Guy's Hospital.*

Take of	Sulphate of magnesium	1 drachm.
	Gallic acid	10 grains.
	Diluted sulphuric acid	5 minims.
	Water	to 1 ounce.

Mix. One dose. *Royal Chest Hospital.*

Take of Sulphate of magnesium .... 20 grains.
Sulphate of iron .... 1 grain.
Diluted sulphuric acid .... 10 minims.
Peppermint water ....to 1 ounce.

Mix. One dose. *Middlesex Hospital.*

Take of Sulphate of magnesium .... 1 drachm.
Sulphate of quinia .... 1 grain.
Diluted sulphuric acid .... 4 minims.
Syrup of ginger .... 1 drachm.
Dill water ....to 1 ounce.

Mix. One dose. *Samaritan Hospital.*

Take of Sulphate of magnesium .... 1 drachm.
Tincture of rhubarb .... 2 drachms.
Syrup of ginger .... 1 drachm.
Caraway water .... 10 drachms.

Mix. Dose: 1 drachm for a child one year old. In diarrhœa.

West.

Take of Sulphate of magnesium .... 20 grains.
Nitrate of potassium .... 10 grains.
Solution of acetate of ammonium .... 2 drachms.
Syrup of lemon .... ½ ounce.
Water ....to 4 ounces.

Mix. Dose: 2 drachms for a child two years old.

Samaritan Hospital.

Take of Sulphate of magnesium .... ½ ounce.
Syrup of seneka .... ½ ounce.
Syrup of squill .... ½ ounce.
Tincture of hyoscyamus .... 1 ounce.
Compound mixture of liquorice .... 2 ounces.

Mix. Dose: 2 drachms every four hours, in subacute and chronic bronchitis of adults. *H. S. Dessau.*

MAGNOLIA (U. S.)—MAGNOLIA.

The bark of Magnolia acuminata, L. (*cucumber tree*); M. glauca, L. (*sweet-bay*); and M. Umbrella, Lam. (*Umbrella tree*); (*Nat. ord., Magnoliaceæ*), trees indigenous to the United States, and growing most abundantly in the southern portions.

Magnolia bark has an aromatic, bitter taste, and possesses aromatic, stimulant, and tonic properties. In hot decoction it is employed to produce diaphoresis in acute inflammatory and febrile affections.

Dose: ½ to 1 drachm.

DECOCTION OF MAGNOLIA.

Take of Magnolia .... 1 ounce.
Water .... 1½ pint.

Boil to 1 pint, and strain.

Dose : 1 to 2 ounces.

MALTUM—MALT.

Barley is generally employed in the preparation of malt, though not exclusively. The grain is soaked in water, spread out until it has germinated, and then dried. It has an agreeable odor, a sweet taste, and yields to water a brown infusion, which, properly evaporated, constitutes the extract of malt so well known.

Malt is nutritious and easily digested, and has been very much employed the past few years in chronic and wasting diseases.

PREPARATIONS.

EXTRACT OF MALT (Ger.).

Take of Barley malt .... 1 part.
Water .... 5 parts.

Mix the malt with 1 part of water, macerate three hours, add the remainder, and digest, at or below 140°, for an hour ; heat to 212°, express, strain, and evaporate to a thick extract.

FERRATED EXTRACT OF MALT (Ger.).

Take of Extract of malt .... 95 parts.
Pyrophosphate of iron with citrate of ammonium .... 2 parts.
Water .... 3 parts.

Dissolve the iron in the water, and mix with the extract.

MANGANESIUM—MANGANESE.

MANGANESII OXIDUM NIGRUM (U. S. et al. Ph.)—BLACK OXIDE OF MANGANESE.

Occurs as a bog ore in many sections of the country. It is generally an amorphous powder, of a dull black color, and without odor or taste. It is used chemically in the preparation of chlorine, and therapeutically, with excellent results, in painful indigestion, gastralgia, gastric ulcer, etc.

Dose : 2 to 10 grains.

OINTMENT OF OXIDE OF MANGANESE.

Take of Oxide of manganese .... 1 ounce.
Sulphur........................ 1 ounce.
Hard soap........................ 1 ounce.
Lard........................ 3 drachms.

Mix. Used in porrigo. *Bartholow.*

MANGANESII SULPHAS (U. S., Fr.)—SULPHATE OF MANGANESE.

Prepared by treating black oxide of manganese with sulphuric acid, evaporating to dryness, and heating to redness, in order to decompose the sulphate of iron present as an impurity. The mass is then mixed with water, filtered, evaporated, and crystallized.

It occurs in pale red crystals, of a styptic taste, and freely soluble in water. Used as a cholagogue.

Dose: 2 to 10 grains.

PILLS OF SULPHATE OF MANGANESE AND OX BILE.

Take of Sulphate of manganese, dried........................ 40 grains.
Purified ox bile........................ 1 drachm.
Resin of podophyllum........................ 5 grains.

Mix, and divide into 20 pills.

Dose: 1 pill three times a day, in catarrhal jaundice. *Bartholow.*

MANNA (U. S. et al. Ph.)—MANNA.

A concrete exudation from Fraxinus Ornus, L. (*Nat. ord.*, *Oleaceæ*), a tree growing in Southern Italy. Manna occurs in irregular shaped pieces, from one to six inches long, of a white or yellowish-white color, a honey-like odor, and has a sweetish, and afterward, acrid taste. It contains a peculiar saccharine principle, termed mannite, a little sugar, and a resin to which it owes its cathartic properties.

It is nutritive and laxative. On account of its agreeable taste it is easily administered to children.

Dose: 1 to 4 drachms for children; 1 to 2 ounces for adults.

PREPARATIONS.

SYRUP OF MANNA (Ger.).

Take of Manna........................ 3 parts.
Distilled water........................ 12 parts.
White sugar........................ 16 parts.

Dissolve the manna in the water, filter, add the sugar, and heat to the boiling point.

Troches of Manna (Fr.).

Take of Manna	150 parts.
Sugar	800 parts.
Gum arabic	50 parts.
Orange flower water	75 parts.

Dissolve the manna in the water with a gentle heat, and strain; add the gum, previously mixed with twice its weight of sugar, then the remainder of the sugar, and divide into troches of 15 grains each.

MASTICHE (U. S. et al. Ph.)—MASTIC.

A concrete resinous exudation from the bark of Pistacia Lentiscus, L. (*Nat. ord., Anacardiaceæ*), a shrub or small tree indigenous to the coasts of the Mediterranean.

Mastic occurs in roundish tears about the size of peas, of a yellow or greenish color, and a balsamic odor. Whatever medicinal properties it possesses are of a terebinthinate character.

Dose: 2 to 5 grains.

PREPARATION.

Ethereal Tincture of Mastic (Fr.).

Take of Mastic	at will.

Prepare a saturated tincture by maceration in a menstruum composed of 89 parts of pure ether and 36 parts of alcohol (90%).

Used to fill the cavities of aching teeth.

See *Pilulæ Aloes et Mastiches.*

MATICO (U. S., Br., Fr.)—MATICO.

The leaves of Piper angustifolium, Ruiz and Pav. (*Nat. ord., Piperaceæ*), a shrub growing in South America.

Matico is an agreeable aromatic tonic, and appears to exert its influence mainly upon the mucous membranes, particularly those of the genito-urinary organs. It is used, both internally and locally, as a hæmostatic.

Dose: 20 to 60 grains.

PREPARATIONS.

Extractum Matico Fluidum (U. S.)—Fluid Extract of Matico.

Take of Matico	16 ounces.
Glycerin	4 ounces.
Alcohol	sufficient.
Water	sufficient.

Mix 12 ounces of alcohol, 3 of glycerin, and 1 of water, moisten the matico with ½ pint of the mixture and proceed according to the general formula, page 161. Finish the percolation with diluted alcohol, reserve 14 ounces, and add 1 ounce of glycerin to the remainder, before evaporation.

Dose: 20 to 60 minims.

Infusion of Matico (Br.).

Take of Matico leaves	½ ounce.
Boiling distilled water	10 ounces.

Infuse for half an hour, and strain.

Dose: 1 to 4 ounces.

Either of the above preparations fairly represents the leaves, and may be used instead of them, except when a styptic effect is desired; in the latter case the leaves are applied in substance to the bleeding part.

The so-called "*injections*" and "*capsules of matico*" contain, according to Bumstead, no matico whatever.

MATRICARIA (U. S., Fr., Ger.)—GERMAN CHAMOMILE.

The flower heads of Matricaria Chamomilla, L. (*Nat. ord., Compositæ*), a common European plant. The flowers are smaller than those of chamomile, but possess similar properties, and are often substituted for them.

Dose: 10 to 30 grains.

PREPARATIONS.

Extract of German Chamomile (Ger.).

Take of German chamomile	2 parts.

Exhaust by maceration in diluted alcohol, and evaporate to a thick extract.

Dose: 2 to 10 grains.

Infusion of German Chamomile (Ger.).

Take of German chamomile	1 part.
Boiling water	5 parts.

Infuse until cold, then express and strain.

Dose: ½ to 1 ounce.

Made according to the general formula for infusions of the German Pharmacopœia.

Oil of German Chamomile (Ger.).

A thick oil obtained from German chamomile by distillation.

Dose: 3 to 10 minims.

INFUSED OIL OF GERMAN CHAMOMILE (Ger.).

Take of German chamomile........................ 2 parts.
Alcohol.............................. 1 part.
Olive oil.............................. 20 parts.

Mix the alcohol with the chamomile, set aside in a covered vessel for a few hours, then add the oil, digest until the alcohol is evaporated, express and filter. Used as an embrocation.

SYRUP OF GERMAN CHAMOMILE (Ger.).

Take of German chamomile........................ 3 parts.
Boiling distilled water...................... 15 parts.

Macerate a few hours, and, in 10 parts of the filtered liquid, dissolve 18 parts of sugar. Used as a vehicle.

CHAMOMILE WATER (Ger.).

Take of German chamomile........................ 1 part.
Water.................................... sufficient.

Distil 10 parts. Used as a vehicle.

MEL (U. S. et al. Ph.)—HONEY.

A saccharine liquid collected from flowers by the common honey-bee *(Apis mellifica, L.)*, and deposited in the honey-comb.

Used chiefly as a vehicle.

PREPARATIONS.

Mel Despumatum—Clarified Honey.

Take of Honey............................a convenient quantity.

Melt by means of a water-bath, and remove the scum.

HYDROMEL (Fr.).

Take of Clarified honey.......................... 1 part.
Warm water............................ 10 parts.

Dissolve, and strain. A slightly laxative drink.

OXYMEL (Br., Ger.).

Take of Clarified honey (by weight).................. 40 ounces.
Acetic acid.............................. 5 ounces.
Distilled water.......................... 5 ounces.

Liquefy the honey by heat, then mix with the acid and water.

Dose: 1 to 2 drachms.

The German Pharmacopœia employs 1 part of acid with 40 parts of honey.

Syrup of Honey (Fr.).

Take of Honey.. 4 parts.
Water.. 1 part.

Mix, skim, clarify, and strain. Used as a vehicle.

Suppositories of Honey (Fr.).

Take of honey................................a convenient quantity.

Boil until it solidifies on cooling, then pour into moulds having the capacity of 1 drachm each.

Honey Cough Mixture.

Take of Honey.. 1 part.
Jamaica rum.............................. 1 part.
Lemon juice.............................. 1 part.

Mix. Dose : ½ ounce every two or three hours, in the chronic bronchitis of old people.

MELISSA (U. S., Fr., Ger.)—BALM.

The leaves and tops of Melissa officinalis, L. (*Nat. ord.*, *Labiatæ*), an herbaceous perennial indigenous to the old world, but naturalized in this country.

Balm possesses the stimulant and aromatic properties common to many of the labiatæ. In hot decoction it is used as a diaphoretic.

Dose : 2 to 4 drachms.

PREPARATIONS.

Balm Water (Fr., Ger.).

Take of Balm tops, fresh.............................. 1 part.
Water.. sufficient.

Distil 1 part. Used as a vehicle.

Compound Spirit of Balm (Ger., Fr.).

Take of Balm leaves.................................. 14 parts.
Lemon peel.................................. 12 parts.
Coriander seeds.............................. 6 parts.
Nutmeg.. 6 parts.
Cassia bark.................................. 3 parts.
Cloves.. 3 parts.
Mix, bruise, and add of alcohol......................150 parts.
Water..250 parts.
Distil 200 parts.

Dose : 1 to 2 drachms, as a stomachic and carminative.

MENTHA PIPERITA (U. S. et al. Ph.)—PEPPERMINT.

The leaves and tops of Mentha piperita, L. (*Nat. ord.*, *Labiatæ*), a perennial herb which grows freely in all temperate regions. It is extensively cultivated in this country for the oil which it yields by distillation.

Peppermint has a pungent, biting taste, followed by a cooling and refreshing sensation. It is a stimulant and carminative, besides having feeble anodyne powers. It is frequently used to relieve the pains of flatulent colic, to expel flatus, to arrest vomiting, etc.

Dose: 5 to 20 grains.

PREPARATIONS.

Aqua Menthæ Piperitæ (U. S. et al. Ph.)—Peppermint Water.

Take of Oil of peppermint	½ drachm.
Carbonate of magnesium	1 drachm.
Distilled water	2 pints.

Rub the oil first with the carbonate, then with the water added gradually, and filter. Or, mix 18 ounces of peppermint with 16 pints of water, and distil 8 pints. Used as a vehicle.

The British Pharmacopœia directs: oil of peppermint, 1½ drachm; water, 1½ gallon; distil 1 gallon. The French Codex and German Pharmacopœia direct to distil from the fresh and dried plant, respectively.

Oleum Menthæ Piperitæ (U. S. et al. Ph.)—Oil of Peppermint.

The oil obtained from peppermint by distillation.

Dose: 1 to 3 minims.

Spiritus Menthæ Piperitæ (U. S. et al. Ph.)—Spirit of Peppermint—Essence of Peppermint.

Take of Oil of peppermint	1 ounce.
Peppermint, coarsely powdered	2 drachms.
Stronger alcohol	15 ounces.

Dissolve the oil in the alcohol, add the peppermint, macerate twenty-four hours, and filter.

Dose: 5 to 30 minims.

The British preparation of the same name has the strength of 1 volume of oil to 49 of alcohol, while its preparation known as *essence of peppermint* has the strength of 1 to 4. The German spirit has the strength of 1 to 9, and the French is prepared by distillation from the fresh plant.

Trochisci Menthæ Piperitæ (U. S., Fr.)—Troches of Peppermint.

Take of Oil of peppermint	1 drachm.
Sugar	12 ounces.
Mucilage of tragacanth	sufficient.

Rub the oil and sugar together, then with the mucilage form a mass, to be divided into 480 troches.

Dose: 1 or 2 troches, as required.

SYRUP OF PEPPERMINT (Fr., Ger.).

Take of Peppermint water	50 parts.
White sugar	95 parts.

Dissolve without heat, and filter. Used as a vehicle.

The German preparation is made in the same manner as syrup of spearmint, which see.

MENTHA VIRIDIS (U. S. et al. Ph.)—SPEARMINT.

The leaves and tops of Mentha viridis, L. (*Nat. ord., Labiatæ*), a perennial herb indigenous to the Old World, but long since naturalized here.

Spearmint, like peppermint, is an aromatic stimulant and carminative, but is less efficient, and, to most persons, less agreeable. It is used for the same purposes. The French Codex and German Pharmacopœia recognize Mentha crispa, L., a closely allied species.

Dose: 5 to 20 grains.

PREPARATIONS.

Aqua Menthæ Viridis (U. S., Br., Ger.)—Spearmint Water.

Take of Oil of spearmint	½ drachm.
Carbonate of magnesium	1 drachm.
Distilled water	2 pints.

Rub the oil first with the carbonate, then with the water added gradually, and filter. Or mix 18 ounces of spearmint with 16 pints of water, and distil 8 pints. Used as a vehicle.

The British Pharmacopœia directs: oil of spearmint, 1½ drachm; water, 1½ gallon; distil 1 gallon. The German Pharmacopœia directs to distil from the dried plant.

Oleum Menthæ Viridis (U. S., Br., Ger.)—Oil of Spearmint.

The oil obtained from spearmint by distillation.

Dose: 1 to 5 minims.

Spiritus Menthæ Viridis (U. S., Ger.)—Spirit of Spearmint—Essence of Spearmint.

Take of	Oil of spearmint	1 ounce.
	Spearmint	2 drachms.
	Stronger alcohol	15 ounces.

Dissolve the oil in the alcohol, add the spearmint, macerate for twenty-four hours, and filter.

Dose : 5 to 30 minims.

The German preparation has the strength of 1 to 9.

Syrup of Spearmint (*Curled Mint*, Ger.).

Take of	Spearmint	3 parts.
	Boiling distilled water	15 parts.

Macerate a few hours, and, in 10 parts of the filtered liquid, dissolve 18 parts of sugar. Used as a vehicle.

Compound Infusion of Spearmint.

Take of	Spearmint	2 ounces.
	Red rose petals	80 grains.
	Diluted sulphuric acid	2 drachms.
	Sugar	1½ ounce.
	Boiling water	20 ounces.

Macerate half an hour, strain, and add the sugar.

Dose : 1 to 2 ounces. *Guy's Hospital.*

MEZEREUM (U. S. et al. Ph.)—MEZEREON.

The bark of Daphne Mezereum, L. (*Nat. ord.*, *Thymelaceæ*), a shrub indigenous to Northern Europe.

Formerly very popular as a remedy for syphilis, chronic rheumatism, and other chronic diseases, but now seldom used. Applied locally it is an irritant, and, in the form of an ointment is used to maintain the discharge from blistered surfaces. It is a constituent of compound decoction of sarsaparilla.

Dose : 5 to 10 grains.

PREPARATIONS.

Extractum Mezerei Fluidum (U. S.)—Fluid Extract of Mezereon.

Take of	Mezereon	16 ounces.
	Stronger alcohol	sufficient.

Moisten the mezereon with 6 ounces of the alcohol, and proceed according to the general formula, page 161.

Dose : 1 to 10 minims.

Unguentum Mezerei (U. S., Ger.)—Mezereon Ointment.

Take of	Fluid extract of mezereon	4 ounces.
	Lard	14 ounces.
	Yellow wax	2 ounces.

Melt the lard and wax together, add the mezereon, and stir until the alcohol has evaporated and the mass has become cool.

Extract of Mezereon (Ger.).

Take of	Mezereon	1 part.
	Alcohol	7 parts.

Exhaust the mezereon by maceration with the alcohol, and evaporate to a thin extract.

Ethereal Extract of Mezereon (Br.).

Take of	Mezereon	1 pound (avoir.).
	Rectified spirit	8 pints (imp.).
	Ether	1 pint (imp.).

Exhaust the mezereon by maceration in the spirit, and evaporate to a soft extract. Put this into a bottle with the ether, macerate twenty-four hours, shaking frequently, decant, and evaporate the ethereal solution to a soft extract.

Used in *Compound Liniment of Mustard, Br.*

MORPHIA (U. S., Fr., Ger.) MORPHIA—MORPHINE.

Take of	Opium, sliced	12 ounces.
	Water of ammonia	6 ounces.
	Animal charcoal	sufficient.
	Alcohol	sufficient.
	Distilled water	sufficient.

Macerate the opium with 4 pints of distilled water for twenty-four hours, work it with the hands, macerate twenty-four hours longer, and strain. Macerate the residue twice successively in like manner, and strain; mix the infusions, evaporate to 6 pints, and filter. Then add 5 pints of alcohol, and afterward 3 ounces of water of ammonia, previously mixed with ½ pint of alcohol. After twenty-four hours, pour on the remainder of the water of ammonia, previously mixed with ½ pint of alcohol, and set aside for twenty-four hours to crystallize. Boil the crystals with 2 pints of alcohol, filter, while hot, through animal charcoal, and set aside to crystallize.

Morphia, thus prepared, is in colorless crystals, almost insoluble in water, but freely soluble in boiling alcohol.

Dose: ⅛ to ½ grain. Seldom used except in the form of its salts.

PREPARATIONS.

Morphiæ Acetas (U. S., Br., Ger.)—Acetate of Morphia.

Take of	Morphia	1 ounce.
	Distilled water	½ pint.
	Acetic acid	sufficient.

Mix the morphia with the water, carefully add acetic acid until the morphia is neutralized and dissolved, evaporate to a syrupy consistence, then set aside till it concretes. Lastly, dry with a gentle heat, and rub into powder.

Dose: ⅛ to ½ grain.

Morphiæ Hydrochloras (U. S. et al. Ph.)—Hydrochlorate of Morphia —Muriate of Morphia.

Take of	Morphia	1 ounce.
	Distilled water	4 ounces.
	Hydrochloric acid	sufficient.

Mix the morphia with the water, carefully add hydrochloric acid until the morphia is neutralized and dissolved, evaporate, and crystallize. Lastly, drain the crystals, and dry them on bibulous paper.

Dose: ⅛ to ½ grain.

Morphiæ Sulphas (U. S., Fr., Ger.).—Sulphate of Morphia.

Take of	Morphia	1 ounce.
	Distilled water	½ pint.
	Diluted sulphuric acid	sufficient.

Mix the morphia with the water, carefully add diluted sulphuric acid until the morphia is neutralized and dissolved, evaporate, and crystallize. Lastly, drain the crystals, and dry them on bibulous paper.

Dose: ⅛ to ½ grain.

Liquor Morphiæ Sulphatis (U. S.)—Solution of Sulphate of Morphia.

Take of	Sulphate of morphia	8 grains.
	Distilled water	½ pint.

Dissolve.

Dose: 1 to 4 drachms.

There is no good reason why this preparation should be officinal. It has no advantage over extemporaneous solutions, and, besides, it is liable to be mistaken for *Magendie's Solution of Morphia*, which contains 16 grains of sulphate of morphia to 1 ounce of water. This latter solution is a very convenient one for hypodermic use.

Suppositoria Morphiæ (U. S., Br.)—Suppositories of Morphia.

Take of	Sulphate of morphia	6 grains.
	Oil of theobroma	354 grains.

Mix the morphia with 1 drachm of the oil, then, having melted the remainder and cooled it to 95°, mix all together, and pour into suitable moulds, making 12 suppositories.

An eligible form in which to administer morphia by the rectum or vagina.

The British Pharmacopœia employs hydrochlorate of morphia, and a mixture of oil of theobroma, white wax, and benzoated lard.

Trochisci Morphiæ et Ipecacuanhæ (U. S., Br.)—Troches of Morphia and Ipecacuanha.

Take of	Sulphate of morphia	12 grains.
	Ipecacuanha	40 grains.
	Sugar	10 ounces.
	Oil of gaultheria	5 minims.
	Mucilage of tragacanth	sufficient.

Rub the powders together, add the oil, and with the mucilage form a mass, to be divided into 480 troches.

Dose: 1 to 6 troches.

Used to allay cough and to promote expectoration.

The British Pharmacopœia employs hydrochlorate of morphia in nearly the same proportion. There are also officinal in the British Pharmacopœia *Troches of Morphia*, made in the same manner, and of the same strength, but without ipecacuanha.

Solution of Acetate of Morphia (Br.).

Take of	Acetate of morphia	4 grains.
	Diluted acetic acid	8 minims.
	Rectified spirit	2 drachms.
	Distilled water	6 drachms.

Mix the acid, spirit, and water, and dissolve the morphia in the mixture.

Dose: 10 to 60 minims.

Solution of Hydrochlorate of Morphia (Br.).

Take of	Hydrochlorate of morphia	4 grains.
	Diluted hydrochloric acid	8 minims.
	Rectified spirit	2 drachms.
	Distilled water	6 drachms.

Mix the acid, spirit, and water, and dissolve the morphia in the mixture.

Dose: 10 to 60 minims.

Pills of Hydrochlorate of Morphia (Fr.).

Take of Hydrochlorate of morphia	15 grains.
Sugar of milk	15 grains.
Honey	sufficient.

Mix, and form a mass, to be divided into 100 pills.

Dose: 1 to 3 pills.

Suppositories of Morphia with Soap (Br.).

Take of Hydrochlorate of morphia	6 grains.
Glycerin of starch	50 grains.
Curd soap	100 grains.
Starch	sufficient.

Mix the morphia with the glycerin of starch and the soap; add starch to form a paste of suitable consistence, and divide into 12 equal parts, each of which is to be made into a conical form.

Hypodermic Injection of Morphia (Br.).

A slightly acid solution of acetate of morphia, containing 1 grain of the salt in 12 minims.

Dose: 1 to 6 minims.

Hypodermic Injection of Morphia and Atropia.

Take of Sulphate of atropia	$\frac{1}{6}$ grain.
Hypodermic injection of morphia (Br.)	1 drachm.

Mix. Dose: 1 to 4 minims. *London Hospital.*

Collyrium of Morphia.

Take of Sulphate of morphia	4 to 8 grains.
Distilled water	1 ounce.

Mix. A few drops to be put in the eyes, when necessary, in conjunctivitis. *Bartholow.*

Collyrium of Morphia, Zinc, and Atropia.

Take of Sulphate of morphia	4 grains.
Sulphate of zinc	2 to 8 grains.
Sulphate of atropia	1 grain.
Distilled water	1 ounce.

Mix. Used like the preceding, in iritis, and other inflammatory affections of the eyes. *Bartholow.*

Mixture of Morphia and Hydrocyanic Acid.

Take of Solution of hydrochlorate of morphia	10 minims.
Diluted hydrocyanic acid	3 minims.
Spirit of chloroform	10 minims.
Water	to 1 ounce.

Mix. One dose. *Westminster Hospital.*

Mixture of Morphia and Iron.

Take of Hydrochlorate of morphia .... $\frac{1}{12}$ grain.
Solution of perchloride of iron .... 15 minims.
Diluted hydrocholoric acid .... 3 minims.
Spirit of chloroform .... 10 minims.
Water .... to 1 ounce.

Mix. One dose. *St. Bartholomew's Hospital.*

Morphia Cough Mixtures.

Take of Sulphate of morphia .... 3 grains.
Diluted hydrocyanic acid .... $\frac{1}{2}$ drachm.
Syrup of seneka .... 1 ounce.
Syrup of tolu .... 1 ounce.
Water .... 1 ounce.

Mix. Dose: 1 drachm three or four times a day. *F. H. Bosworth.*

Take of Solution of sulphate of morphia (U. S.) ...... 1 ounce.
Compound spirit of ether .... 1 ounce.

Mix. Dose: 1 drachm three or four times a day. *F. H. Bosworth.*

Take of Sulphate (or other salt) of morphia .... $\frac{1}{40}$ grain.
Spirit of chloroform .... 3 minims.
Glycerin or syrup .... 1 drachm.

Mix. One dose. To be taken frequently, when cough is troublesome, until the paroxysm is subdued. *Ringer.*

Pills of Morphia and Zinc.

Take of Sulphate of morphia .... 1 to 2 grains.
Zinc oxidi .... $\frac{1}{2}$ drachm.

Mix, and make 10 pills.

Dose: 1 pill three times a day, before each meal, in painful diseases of the stomach. *Bartholow.*

Powder of Morphia and Bismuth.

Take of Sulphate of morphia .... 1 to 2 grains.
Subnitrate or subcarbonate of bismuth .... 3 drachms.
Aromatic powder .... 1 drachm.

Mix, and divide into 12 powders.

Dose: a powder in milk before each meal, in painful diseases of the stomach. *Bartholow.*

Tully's Powder.

Take of Sulphate of morphia .... 1 grain.
Camphor .... 20 grains.
Powdered liquorice root .... 20 grains.
Prepared chalk .... 20 grains.

Mix. Dose: 10 grains, containing $\frac{1}{6}$ grain of morphia.

Powder of Morphia and Quinia.

Take of Sulphate of morphia........................ 1 grain.
Sulphate of quinia........................ 6 grains.
Capsicum................................ 6 grains.

Mix, and divide into 6 powders.

Dose: 1 powder every two or three hours, in sick headache.

Lotion of Morphia and Hydrocyanic Acid.

Take of Hydrochlorate of morphia.................. 16 grains.
Diluted hydrocyanic acid.................. ½ ounce.
Borax................................ 2 drachms.
Glycerin.............................. 2 ounces.
Water ....to 8 ounces.

Mix. Used in pruritus ani. *J. W. Wright.*

MOSCHUS (U. S. et al. Ph.)—MUSK.

A peculiar concrete substance obtained from the male musk deer (*Moschus moschiferus*, L.), which inhabits the mountains of Central Asia. The musk-bag, situated between the prepuce and umbilicus, is about two and a half inches long, and one and a half broad.

Musk occurs in grains cemented together, more or less mixed with hairs; is of a reddish-brown color, and has a peculiar, aromatic odor, and a bitter taste. It is a powerful nervous stimulant and antispasmodic. Used in low fevers, and in nervous diseases.

Dose: 5 to 10 grains.

PREPARATIONS.

Tincture of Musk (Fr., Ger.).

Take of Musk.. 1 part.
Alcohol (80%)................................ 10 parts.

Macerate ten days, express, and filter.

Dose: ½ to 1 drachm.

The German Pharmacopœia employs, as a menstruum, diluted alcohol and water, in equal parts, and makes the strength 1 to 50.

Ethereal Tincture of Musk (Fr.).

Take of Musk.... .... 1 part.

Macerate ten days in 10 parts of a menstruum composed of 89 parts of pure ether, and 36 parts of alcohol (90%), then filter.

Dose: ½ to 1 drachm.

MIXTURES OF MUSK.

Take of Musk	15 to 60 grains.
Infusion of valerian	4 ounces.
Spirit of orange flowers	1 ounce.

Mix. Dose: ½ ounce every two hours, in ataxic fevers, and typhoid pneumonia. *Guibourt.*

Take of Musk	1 drachm.
Ether	1½ drachm.
Tincture of opium	1½ drachm.
Cinnamon water	2½ ounces.
Syrup	1½ ounce.

Mix. Dose: ½ ounce three times a day, in the subsultus of typhoid, and other low and malignant fevers. *Horace Green.*

MYRISTICA (U. S. et al. Ph.)—NUTMEG.
MACIS (U. S., Fr.)—MACE.

These are obtained from the fruit of Myristica fragrans, Houttuyn (*Nat. ord.*, *Myristiceæ*), an evergreen tree indigenous to the East India Islands, but cultivated in numerous tropical countries. Nutmeg is the seed, while mace is its immediate covering, both being covered by an outer rind, or pericarp.

Nutmeg and mace are pleasant aromatics, and are used mainly as flavoring agents. Both yield a volatile oil by distillation, and nutmeg contains a large percentage of fixed oil, or fat.

Dose: Of either, 5 to 20 grains.

PREPARATIONS.

Oleum Myristicæ (U. S., Br., Ger.)—Oil of Nutmeg.

The oil obtained from nutmeg by distillation.

Dose: 1 to 3 minims.

Spiritus Myristicæ (U. S.)—Spirit of Nutmeg.

Take of Oil of nutmeg	1 ounce.
Stronger alcohol	3 pints.

Dissolve.

Dose: ½ to 1 drachm.

EXPRESSED OIL OF NUTMEG (Br., Fr., Ger.).

A concrete oil obtained from nutmeg by means of heat and expression. Used as an embrocation, and in making plasters and cerates.

NUTMEG CERATE (Ger.).

Take of Expressed oil of nutmeg .... 6 parts.
Olive oil .... 2 parts.
Yellow wax.................................. 1 part.

Melt together.

MYRRHA (U. S. et al. Ph.)—MYRRH.

A gum-resin which exudes from Balsamodendron Myrrha, Nees. (*Nat. ord., Amyridaceæ*), a small tree of Arabia Felix. It occurs in small, semi-transparent, reddish-yellow tears, of a peculiar, agreeable odor, and a bitter, aromatic taste.

Myrrh is a stimulant and tonic, which exerts an influence chiefly upon the mucous membranes, hence its use in bronchitis, leucorrhœa, and cystitis. It acts as an emmenagogue by virtue of its tonic properties. Locally it is used as an astringent for relaxed or spongy gums.

Dose: 5 to 30 grains.

PREPARATIONS.

Tinctura Myrrhæ (U. S. et al. Ph.)—Tincture of Myrrh.

Take of Myrrh.................................. 3 ounces.
Alcohol.................................. sufficient.

Moisten, pack, and percolate to 2 pints.
Dose: ½ to 1 drachm.

EXTRACT OF MYRRH (Ger.).

Take of Myrrh.................................. 1 part.
Distilled water.......................... 5 parts.

Macerate two days, decant, filter, and evaporate to a dry extract.

GARGLES OF MYRRH.

Take of Tincture of myrrh .... ½ ounce.
Glycerite of borax...................... ½ to 1½ ounce.
Water ....to 10 ounces.

Mix. Astringent. *G. M. Lefferts.*

Take of Borax .... 2 drachms.
Tincture of myrrh........................ 1 ounce.
Glycerin .... 2 ounces.
Water .... 8 ounces.

Mix. *N. W. Dispensary.*

MIXTURES OF MYRRH.

Take of Myrrh .... 40 grains.
Decoction of liquorice ....to 1 ounce.

Mix. One dose. *Guy's Hospital.*

Take of Tincture of myrrh........................ 4 ounces.
Tincture of saffron .... 3 ounces.
Tincture of aloes.......................... 3 ounces.

Mix. Dose: 1 to 3 drachms twice a day, as an emmenagogue, in chlorosis. Known as *Elixir of Paracelsus.*

NUX VOMICA (U. S. et al. Ph.)—NUX VOMICA.

The seeds of Strychnos Nux-vomica, L. (*Nat. ord., Loganiaceæ*), a tree growing in various parts of India, which bears fruit of the size, shape, and color of an orange, and containing from one to five seeds imbedded in a juicy pulp. The seeds are nearly round, less than one inch in diameter, about a quarter of an inch thick, convex on one side, concave on the other, and have an extremely bitter taste. They contain two active principles, *strychnia* and *brucia*, to which the medicinal effects of the drug are due.

Nux vomica is a powerful excitant of the cerebro-spinal system, and in overdoses produces tetanic convulsions and death. In small doses it is an excellent tonic, especially of the nervous system, and is both diuretic and laxative. It is much used in paralysis when not caused by structural lesions, in chorea, incontinence of urine, neuralgia, dyspepsia, etc.

Strychnia is more generally employed than nux vomica, since, clinically, there seems to be little, if any, difference in their therapeutic effects, while the former is much more certain than the latter.

Dose: 1 to 5 grains.

PREPARATIONS.

Extractum Nucis Vomicæ (U. S. et al. Ph.)—Extract of Nux Vomica.

Take of Nux vomica.............................. 12 ounces.
Alcohol.......... .... sufficient.

Exhaust the nux vomica by percolation with alcohol, and evaporate to a proper consistence.

Dose: $\frac{1}{4}$ to $\frac{1}{2}$ grain.

The German Pharmacopœia prepares also an aqueous extract.

Tinctura Nucis Vomicæ (U. S. et al. Ph.)—Tincture of Nux Vomica.

Take of Nux vomica .... 8 ounces.
Alcohol .... sufficient.

By digestion and percolation obtain 2 pints of tincture.

Dose: 10 to 30 minims.

The British and German Pharmacopœias direct 1 part of nux vomica to 10 of alcohol; the French Codex, 1 to 5.

ETHEREAL TINCTURE OF NUX VOMICA (Ger.).

Take of Nux vomica .... 1 part.
Spirit of ether .... 10 parts.

Prepare the tincture by maceration.

Dose: 10 to 30 minims.

MIXTURES OF NUX VOMICA.

Take of Tincture of nux vomica .... 2 drachms.
Tincture of capsicum .... 6 drachms.

Mix. Dose: 20 drops in water every four hours. Used to diminish the craving for stimulants when they are withdrawn, and to sustain the nervous system. *Bartholow.*

Take of Tincture of nux vomica .... 2 drachms.
Tincture of aloes and myrrh .... 6 drachms.

Mix. Dose: 15 to 30 drops two or three times a day, in habitual constipation. *Bartholow.*

Take of Tincture of nux vomica .... 7½ minims.
Diluted nitric acid .... 10 minims.
Water .... to 1 ounce.

Mix. One dose. *University College Hospital.*

Take of Tincture of nux vomica .... 10 minims.
Tincture of chloride of iron .... 10 minims.
Water .... to 1 drachm.

Mix. One dose. To be taken thrice daily, after meals. *Hospital Formulary.*

Take of Tincture of nux vomica .... 2 drachms.
Tincture of chloride of iron .... 6 drachms.

Mix. Dose: 10 to 15 drops. *N. W. Dispensary.*

PILLS OF NUX VOMICA.

Take of Extract of nux vomica .... 5 grains.
Reduced iron .... 20 grains.
Sulphate of quinia .... 10 grains.
Syrup .... sufficient.

Mix, and divide into 20 pills. One to be taken three times a day, after meals, in nervous headache. *Hammond.*

Take of Extract of nux vomica	¼ grain.
Compound extract of colocynth	1 grain.
Extract of henbane	1 grain.
Compound rhubarb pill	1 grain.

Make 1 pill. *London Hospital.*

WALKER'S PILLS.

Take of Extract of nux vomica	5 grains.
Extract of belladonna	5 grains.
Dried sulphate of iron	10 grains.
Extract of aloes	10 grains.

Mix, and divide into 20 pills. *Hospital Formulary.*

Take of Extract of nux vomica	5 grains.
Extract of aloes	20 grains.
Powdered rhubarb	10 grains.
Extract of taraxacum	30 grains.

Mix, and divide into 20 pills. *W. T. Lusk.*

OLEUM CADINUM (Fr., Ger.)—OIL OF CADÉ.

A tar originally obtained by the destructive distillation of the wood of Juniperus Oxycedrus, L. (*Nat. ord., Coniferæ*), a shrub or small tree indigenous to the countries bordering the Mediterranean. That now in use is imported from Europe, but where made, or from what wood, is uncertain. It is used locally in skin diseases.

OINTMENTS OF OIL OF CADE.

Take of Oil of cade	15 parts.
Oil of sweet almonds	45 parts.

Mix. For chronic eczema of the scalp. *G. H. Fox.*

Take of Oil of cade	2 drachms.
Ointment of oxide of zinc	1 ounce.

Mix. *N. Y. Dispensary.*

Take of Oil of cade	6 drachms.
Ointment of oxide of zinc	1½ ounce.
Powdered camphor	1 drachm.

Mix. *N. W. Dispensary.*

OLEUM CAJUPUTI (U. S. et al. Ph.)—OIL OF CAJUPUT.

An oil obtained by distillation from the leaves of Melaleuca Leucadendron, L. (*Nat. ord., Myrtaceæ*), a tree growing in the Indian Archipelago and Australia.

Oil of cajuput is a transparent, mobile fluid, of a pale bluish-green color, a camphor-like odor, and an aromatic, bitterish taste.

It is used internally in flatulent colic, cholera morbus, nervous vomiting, etc., and externally in scaly diseases of the skin, and rheumatism.

Dose: 1 to 5 minims.

PREPARATIONS.

Refined Oil of Cajuput (Ger.).

Take of Oil of cajuput............................ 1 part.
Water.................................. 6 parts.

Distil as long as a colorless, or slightly yellow oil passes over.

Spirit of Cajuput (Br.).

Take of Oil of cajuput............................ 1 ounce.
Rectified spirit........................... 49 ounces.

Dissolve.

Dose: ½ to 1 drachm.

Mixture of Oil of Cajuput.

Take of Oil of cajuput............................ 1 drachm.
Spirit of chloroform..................... 1 ounce.
Tincture of cinnamon.................... 1 ounce.

Mix. Dose: 1 drachm every half hour in glycerin or syrup, and water. In cholera morbus, nervous vomiting, etc. *Bartholow.*

Liniment of Oil of Cajuput.

Take of Oil of cajuput............................ 1 part.
Olive oil................................ 1 part.

Mix. Used as an embrocation in rheumatic and gouty pains.

Wood.

OLEUM MORRHUÆ (U. S. et al. Ph.)—COD-LIVER OIL.

The fixed oil obtained from the liver of the common cod (*Gadus morrhua*), and other species of Gadus.

When pure, cod-liver oil is of a pale yellow color, and has a fishy odor and taste. There are, however, varieties in market ranging in color from pale yellow to dark, reddish-brown, but the lighter colored oils are generally the least disagreeable, while containing, it is believed, all the virtues of the darker varieties.

Cod-liver oil deserves to rank rather as a food than a medicine. Its tonic and restorative effects are most strongly marked in wasting diseases, as consumption, necrosis, chronic dysentery and diarrhœa, etc.

Dose: ½ to 2 ounces.

Emulsions of Cod-Liver Oil.

Take of Cod-liver oil.............................. 8 ounces.
Lime water .... 8 ounces.
Oil of cinnamon.............................. 10 drops.

Mix. Dose: ½ ounce. *Hospital Formulary.*

Take of Cod-liver oil.............................. 2 ounces.
Lime water .... 1½ ounce.
Syrup of lactophosphate of lime .... 1½ ounce.

Mix. Dose: 1 drachm. *Bosley.*

Take of Cod-liver oil.............................. 28 parts.
Glyconin .... 9 parts.
Aromatic spirit of ammonia.................. 1 part.
Sherry wine.............................. 20 parts.
Spirit of bitter almond..................... 2 parts.

Mix. Dose: ½ ounce.

All by weight. The spirit of bitter almonds is made by mixing 1 part of oil of bitter almonds with 64 parts of alcohol. Glyconin or *Glycerite of Yolk of Eggs*, is made thus:

Take of Yolk of eggs.............................. 4 parts.
Glycerin.............................. 5 parts.

Beat the yolks in the usual manner, pour into a bottle, add the glycerin, and shake well together. About 4 ounces of glyconin are required to emulsionize a pint of cod-liver oil. *Charles Rice.*

Phosphorated Emulsion of Cod-Liver Oil.

Take of Cod-liver oil.............................. 20 parts.
Phosphorated oil (1%) .... 2 parts.
Glyconin.............................. 7 parts.
Aromatic spirit of ammonia.................. 1 part.
Syrup .... 10 parts.
Diluted phosphoric acid..................... 4 parts.
Spirit of bitter almonds..................... 2 parts.

Rub the glyconin with the oils, added gradually, then add the other ingredients in the order in which they are named.

Dose: ½ ounce. *Charles Rice.*

Mixtures of Cod-Liver Oil.

Take of Cod-liver oil.............................. 3 ounces.
Syrup of wild cherry..................... 1 ounce.
Lime water.............................. 1 ounce.

Mix. *N. Y. and N. W. Dispensaries.*

Take of Cod-liver oil.............................. 1 ounce.
Carbonate of potassium..................... ¼ grain.
Wine of iron .... 1 ounce.

Mix. Dose: 2 drachms. *St. Mary's Hospital.*

Take of Cod-liver oil .... 6 drachms.
Solution of potassa .... 40 minims.
Stronger solution of ammonia .... 2 minims.
Oil of cassia .... 1 minim.
Syrup .... 2 drachms.

Mix. One dose. *Brompton Consumption Hospital.*

PHOSPHORATED COD-LIVER OIL.

Take of Phosphorated oil (1%) .... 100 grains.
Ether .... 2 drachms.
Cod-liver oil .... to 16 ounces.

Mix. Dose: ½ ounce. *Bellevue Hospital.*

OLEUM OLIVÆ (U. S. et al. Ph.)—OLIVE OIL—SWEET OIL.

The fixed oil obtained from the fruit of Olea Europæa, L. (*Nat. ord., Oleaceæ*). The olive tree is extensively cultivated in the countries bordering on the Mediterranean, where the fruit and oil are important articles of diet and commerce.

Olive oil is nutritious, emollient, demulcent, and laxative. It is employed both externally and internally.

Dose: ½ to 2 ounces.

ENEMA OF OLIVE OIL.

Take of Olive oil .... 4 ounces.
Decoction of barley .... to 20 ounces.

Mix. *Middlesex Hospital.*

MIXTURE OF OLIVE OIL.

Take of Olive oil .... 1 drachm.
Carbonate of potassium .... 4 grains.
Water .... to 1 ounce.

Mix. One dose. *Guy's Hospital.*

OLEUM RICINI (U. S. et al. Ph.)—CASTOR OIL.

The fixed oil obtained from the seeds of Ricinus communis, L. (*Nat. ord., Euphorbiaceæ*), a native of India, but cultivated in many countries. It is grown extensively in Southern Illinois, and that State now furnishes a large percentage of the oil used in this country.

In warmer countries the castor oil plant is a tree of twenty to forty feet in height, but in temperate regions it is an annual herb, five or ten feet high.

Castor oil is a thick, viscous liquid, of a faint, unpleasant odor, and a disagreeable, nauseous taste. It is a mild but certain cathartic, operating with little pain, and well suited to cases where free evacuation is desired without abdominal irritation, as in pregnancy, after childbirth, in typhoid fever, dysentery, diarrhœa from indigestible food, and for children.

Dose: 1 to 8 drachms.

PREPARATIONS.

Emulsion of Castor Oil (Fr.).

Take of Castor oil	30 parts.
Gum arabic	8 parts.
Peppermint water	15 parts.
Water	60 parts.
Syrup	30 parts.

Rub the gum with its weight of water, then add the oil and other ingredients gradually.

Dose: 2 to 4 ounces.

Mixture of Castor Oil.

Take of Castor oil	1 ounce.
Mucilage of gum arabic	½ ounce.
Syrup	½ ounce.
Cinnamon water	2 ounces.

Mix. Dose: ½ ounce every four to six hours, in sporadic dysentery, after the more acute symptoms have subsided. When there are much pain, tenesmus, and frequent passages, 10 to 20 drops of laudanum may be added to each dose; when there are much depression, a low state of the arterial tension, and a dry, glazed tongue, 5 drops of turpentine may also be added. *Bartholow.*

OLEUM TEREBINTHINÆ (U. S. et al. Ph.)—OIL OF TURPENTINE—SPIRIT OF TURPENTINE.

The volatile oil obtained by distilling the oleoresinous exudation from Pinus palustris, L. (*Nat. ord., Coniferæ*), and other species of pine.

It is a clear, colorless, volatile liquid, of a pungent, terebinthinate odor and taste. In small doses it is a stimulant to the skin, mucous membranes, and kidneys. Very large doses pass off by the bowels, and may produce strangury, bloody urine, etc. In large doses it is employed as an anthelmintic; in small doses, to relieve the tympanites of typhoid fever, to arrest hemorrhage from mucous membranes, in chronic catarrhs, etc. Externally it is employed as a rubefacient.

Dose: 5 to 30 minims. As an anthelmintic, ½ to 2 ounces.

PREPARATIONS.

Linimentum Terebinthinæ (U. S., Br., Fr.)—Liniment of Turpentine.

Take of Resin cerate	12 ounces.
Oil of turpentine	½ pint.

Add the oil to the cerate, previously melted, and mix.

Used as an application to burns and scalds.

The British Pharmacopœia directs: soft soap, 2 ounces; camphor, 1 ounce; oil of turpentine, 16 ounces. The French Codex employs equal parts of oil of turpentine and infused oil of camomile. These latter preparations are therefore much stronger than that of the United States Pharmacopœia, and are used for rubefacient effect.

Liniment of Turpentine and Acetic Acid (Br.).

Take of Oil of turpentine	1 ounce.
Acetic acid	1 ounce.
Liniment of camphor	1 ounce.

Mix.

Confection of Turpentine (Br.).

Take of Oil of turpentine	1 ounce.
Liquorice root	1 ounce.
Clarified honey (by weight)	2 ounces.

Rub the turpentine with the liquorice, then add the honey, and mix.

Dose: 1 to 2 drachms.

Enema of Turpentine (Br.).

Take of Oil of turpentine	1 ounce.
Mucilage of starch	15 ounces.

Mix.

Ointment of Turpentine (Br., Ger.).

Take of Oil of turpentine	1 drachm.
Resin	60 grains.
Yellow wax	½ ounce.
Prepared lard	½ ounce.

Melt together on a water-bath, and stir until cool.

This closely resembles *Liniment of Turpentine, U. S.* The German Pharmacopœia employs equal parts of turpentine, oil of turpentine, and yellow wax.

Emulsions of Turpentine.

Take of Oil of turpentine	1 drachm.
Expressed oil of almonds	½ ounce.
Tincture of opium	2 drachms.
Mucilage of gum arabic	5 drachms.
Cherry-laurel water	½ ounce.

Mix. Dose: 1 drachm every three, four, or six hours, in chronic intestinal catarrh. *Bartholow.*

Take of Oil of turpentine	3 drachms.
Fluid extract of digitalis	1 drachm.
Mucilage of gum arabic	½ ounce.
Peppermint water	1 ounce.

Mix. Dose: 1 drachm every three hours, in passive hemorrhages.

Bartholow.

Take of Oil of turpentine	3 drachms.
Tincture of opium	3 drachms.
Gum arabic	3 drachms.
White sugar	3 drachms.
Peppermint water	3 ounces.

Mix. Dose: 1 drachm every four hours, in acute dysentery.

N. S. Davis.

MIXTURE OF TURPENTINE AND ETHER.

Take of Oil of turpentine	3 drachms.
Sulphuric ether	2 drachms.

Mix. Dose: 30 minims night and morning, in biliary calculus.

Durand.

OLEUM THEOBROMÆ (U. S. et al. Ph.)—OIL OF THEOBROMA—BUTTER OF CACAO.

Obtained from the kernels of the seeds of Theobromæ Cacao, L. (*Nat. ord., Sterculiaceæ*), a tree indigenous to Central and South America, where it is extensively cultivated for the manufacture of chocolate.

Butter of cacao is a yellowish-white fat, solid at ordinary temperatures, but melting readily at the temperature of the human body, and possessing the agreeable odor of chocolate. It is used as an emollient dressing for cracked nipples, chafed and excoriated surfaces, etc., and in the preparation of suppositories.

OLEUM TIGLII (U. S. et al. Ph.)—CROTON OIL.

The fixed oil obtained from the seeds of Croton Tiglium, L. (*Nat. ord., Euphorbiaceæ*), a small tree indigenous to the Malabar Coast and Tavoy, but cultivated in many Eastern countries.

Croton oil is a transparent, amber-colored liquid, of a slightly rancid odor, and an acrid, oily taste. It is a drastic purgative, and in over-doses may produce death. It is chiefly used in cases of great urgency. Externally it is employed as a counter-irritant in cases where it is desired to keep up irritation for some time, as in pleurisy with effusion, phthisis, etc.

Dose: ¼ to 2 minims.

Liniment of Croton Oil (Br.).

Take of Croton oil	1 ounce.
Oil of cajuput	3½ ounces.
Rectified spirit	3½ ounces.

Mix.

Pills of Croton Oil.

Take of Croton oil	2 drops.
Crumb of bread	sufficient.
Alcohol	sufficient.

Dissolve the oil in a little alcohol, form a mass with bread, and divide into 4 pills.

Dose: 1 every hour or two. *Stillé.*

Pills of Croton Oil and Colocynth.

Take of Croton oil	3 drops.
Compound extract of colocynth	20 grains.
Extract of belladonna	3 grains.

Mix, and divide into 6 pills. *Bartholow.*

Compound Liniment of Croton Oil.

Take of Croton oil	2 drachms.
Olive oil,	
Oil of turpentine,	
Water of ammonia,	
Spirit of camphor, each equal parts, sufficient to make	2 ounces.

Mix. *Hospital Formulary.*

Croton Oil Paint.

Take of Croton oil	½ drachm.
Ether	1 drachm.
Compound tincture of iodine	2½ drachms.

Mix.

Used in pleurisy with effusion. *Corson.*

OPIUM (U. S. et al. Ph.)—OPIUM.

The concrete juice of the unripe capsules of Papaver somniferum, L. (*Nat. ord., Papaveraceæ*), an annual plant which has been cultivated in the East from the earliest times. Some varieties of the opium poppy are cultivated in the United States, and have yielded a fair article of opium.

Opium occurs in commerce as a dark chestnut-colored mass, of a peculiar and not disagreeable odor, and of a bitter, slightly acrid taste. It contains a variety of chemical constituents, but the most important one is the alkaloid *morphia*. This, when isolated, is in colorless crystals, without odor, but having a bitter taste.

Other alkaloids, as *narceia, codeia, narcotina*, etc., are separated from the drug, and have some commercial and medicinal importance.

Opium is our most valuable narcotic. It is used for the relief of pain in nearly all circumstances and in all diseases. Its primary action is stimulant, quickly followed, however, by a sedative effect. Depending upon the indications, it may be used as a diaphoretic, a sedative, an anodyne, a hypnotic, to check mucous discharges, etc.

The salts of morphia are largely used instead of opium, producing essentially the same effects.

Dose: ¼ to 2 grains.

PREPARATIONS.

Acetum Opii (U. S.)—Vinegar of Opium—Black Drop.

Take of		
	Opium, dried	5 ounces.
	Nutmeg	1 ounce.
	Sugar	8 ounces.
	Diluted acetic acid	sufficient.

Macerate the opium and nutmeg in 1 pint of diluted acetic acid for twenty-four hours, then percolate until 26 ounces are obtained. In this, dissolve the sugar, strain, and add sufficient diluted acid to make the finished product 2 pints.

Dose: 5 to 10 minims.

Confectio Opii (U. S., Br.)—Confection of Opium.

Take of		
	Opium	270 grains.
	Aromatic powder	6 ounces.
	Clarified honey (by weight)	14 ounces.

Rub the opium with the aromatic powder, then add the honey, and beat the whole together.

Dose: 5 to 36 grains. The British preparation is about one-fourth stronger.

Emplastrum Opii (U. S., Br., Ger.)—Opium Plaster.

Take of		
	Extract of opium	1 ounce.
	Burgundy pitch	3 ounces.
	Lead plaster	12 ounces.
	Water	sufficient.

Mix the extract with 3 ounces of water, and evaporate on a water-bath to 1½ ounce; add the pitch and plaster, and continue the heat, stirring constantly, until the moisture is evaporated.

Extractum Opii (U. S. et al. Ph.)—Extract of Opium.

Take of	Opium	12 ounces.
	Water	5 pints.

Macerate the opium in 1 pint of water for twenty-four hours, and express the liquid. Treat it in like manner with each of the 4 pints of water remaining, mix the liquids, filter, and evaporate to a proper consistence.

Dose: ¼ to ½ grain.

Pilulæ Opii (U. S.)—Pills of Opium.

Take of	Opium	24 grains.
	Soap	6 grains.

Beat together with water, and divide into 24 pills.

Dose: 1 pill.

See also *Pilula Saponis Composita.*

Suppositoria Opii (U. S.)—Suppositories of Opium.

Take of	Extract of opium	12 grains.
	Oil of theobroma	348 grains.
	Water	sufficient.

Rub the extract into a smooth paste with water, then mix with 1 drachm of the oil, and having melted the remainder and cooled it to 95°, mix all together, and pour into moulds, forming 12 suppositories.

Tinctura Opii (U. S. et al. Ph.)—Tincture of Opium—Laudanum.

Take of	Opium, dried and in powder	2½ ounces.
	Water	1 pint.
	Alcohol	1 pint.
	Diluted alcohol	sufficient.

Macerate the opium with the water for three days, then add the alcohol and continue the maceration three days longer. Introduce into a percolator, and pour on diluted alcohol until 2 pints are obtained.

Dose: 5 to 20 minims. The French Codex employs 1 part of extract of opium with 12 parts of alcohol (60 per cent.).

Tinctura Opii Acetata (U. S.)—Acetated Tincture of Opium.

Take of	Opium, dried and in powder	2 ounces.
	Distilled vinegar	12 ounces.
	Alcohol	½ pint.

Rub the opium with the vinegar, add the alcohol, macerate seven days, and filter.

Dose: 5 to 20 minims.

Tinctura Opii Camphorata (U. S. et al. Ph.)—Camphorated Tincture of Opium—Paregoric Elixir.

Take of	Opium, dried and in powder	60 grains.
	Benzoic acid	60 grains.
	Camphor	40 grains.
	Oil of anise	60 grains.
	Clarified honey (by weight)	2 ounces.
	Diluted alcohol	2 pints.

Mix the ingredients, macerate seven days, and filter.

Dose: 1 to 2 drachms. *Compound Tincture of Camphor, Br.*

Tinctura Opii Deodorata (U. S.)—Deodorized Tincture of Opium.

Take of	Opium, dried and in powder	2½ ounces.
	Ether	½ pint.
	Alcohol	½ pint.
	Water	sufficient.

Macerate the opium with ½ pint of water for twenty-four hours, and express; repeat the operation twice with a like quantity of water. Mix the liquids, evaporate to 4 ounces, and shake in a bottle with the ether. Decant the ethereal solution, evaporate the remaining liquid until free from ether, then mix with 20 ounces of water, filter, and add sufficient water through the filter to make 1½ pint. Lastly, add the alcohol.

Dose: 5 to 25 minims. A substitute for *McMunn's Elixir,* which it very much resembles.

Vinum Opii (U. S., Br.)—Wine of Opium.

Take of	Opium, dried and in powder	2 ounces.
	Cinnamon	60 grains.
	Cloves	60 grains.
	Sherry wine	sufficient.

Macerate the powders in 15 ounces of the wine for seven days, transfer to a percolator, pour on wine until 1 pint is obtained.

Dose: 4 to 16 minims. The British preparation contains 1 ounce of extract of opium in 20 fluid ounces of sherry.

Enema of Opium (Br.).

Take of	Tincture of opium	½ drachm.
	Mucilage of starch	2 ounces.

Mix.

Liquid Extract of Opium (Br.).

Take of	Extract of opium	1 ounce.
	Distilled water	16 ounces.
	Rectified spirit	4 ounces.

Macerate the extract in the water for an hour, add the spirit, and filter.

Dose: 10 to 40 minims.

Glycerite of Opium (Fr.).

Take of Extract of opium........................... 10 parts.
Glycerite of starch........................100 parts.

Mix.

Liniment of Opium (Br.).

Take of Tincture of opium........................... 2 ounces.
Liniment of soap........................... 2 ounces.

Mix.

Opium Lozenges (Br.).

Take of Extract of opium........................... 72 grains.
Tincture of tolu........................... ½ ounce.
Refined sugar........................... 16 ounces.
Gum arabic........................... 2 ounces.
Extract of liquorice........................... 6 ounces.
Distilled water........................... sufficient.

Soften the extract with a little water, then add to the tincture of tolu and extract of liquorice heated on a water-bath. When reduced to a proper consistence, add the gum and sugar, and divide into 720 lozenges.

Dose: 1 to 6 lozenges.

Compound Powder of Opium (Br.).

Take of Opium........................... 1½ ounce.
Black pepper........................... 2 ounces.
Ginger........................... 5 ounces.
Caraway........................... 6 ounces.
Tragacanth........................... ½ ounce.

Mix thoroughly. Dose: 2 to 5 grains.

Ammoniated Tincture of Opium (Br.).

Take of Opium...........................100 grains.
Saffron...........................180 grains.
Benzoic acid...........................180 grains.
Oil of anise........................... 1 drachm.
Strong solution of ammonia........................... 4 ounces.
Rectified spirit........................... 16 ounces.

Macerate seven days, strain, press, filter, and add sufficient spirit to make 20 ounces.

Dose: ½ to 1 drachm.

Tincture of Opium and Saffron (Ger., Fr.)—Sydenham's Laudanum.

Take of Opium........................... 16 parts.
Saffron........................... 6 parts.
Cloves........................... 1 part.
Cassia bark........................... 1 part.
Sherry wine...........................152 parts.

Prepare the tincture by digestion.

Dose: 5 to 20 minims. *Compound Wine of Opium, Fr.*

Opium Water (Ger.).

Take of Opium .... 1 parts.
Water .... 10 parts.

Distil 5 parts.

Syrup of Opium (Fr., Ger.).

Take of Extract of opium .... 1 part.
Distilled water .... 8 parts.
Syrup ....990 parts.

Dissolve the extract in the water, filter, and mix with the syrup. The German Pharmacopœia directs to dissolve the extract in white wine.

Opium Ointment (Ger.).

Take of Extract of opium .... 1 part.
Distilled water .... 1 part.
Wax ointment .... 18 parts.

Rub the extract with the water, then mix with the ointment.

Collyrium of Opium (Fr.).

Take of Extract of opium .... 1 part.
Distilled rose water ....500 parts.

Mix.

Collyria of Opium and Zinc.

Take of Sulphate of zinc .... 2 grains.
Tincture of opium .... 20 minims.
Water .... 1 ounce.

Mix. *King's College Hospital.*

Take of Sulphate of zinc .... 4 grains.
Acetate of lead .... 4 grains.
Wine of opium .... ½ ounce.
Rose water .... 4 ounces.

Mix, and filter. Used in conjunctivitis.

Hospital for Ruptured and Crippled.

Enema of Opium and Lead.

Take of Tincture of opium .... 20 minims.
Acetate of lead .... 9 grains.
Diluted acetic acid .... 15 minims.
Distilled water .... 3 ounces.

Mix. *London Fever Hospital.*

Lotion of Opium.

Take of Extract of opium .... 3 grains.
Water .... 1 ounce.

Mix. *Guy's Hospital.*

Lotions of Lead and Opium.

Take of Acetate of lead .... 4 grains.
Opium, in powder .... 4 grains.
Warm water .... 1 ounce.

Mix. *London Fever Hospital.*

Take of Extract of opium .... 1 grain.
Diluted solution of subacetate of lead .... 1 ounce.

Mix. *St. Bartholomew's Hospital.*

Infusion of Opium.

Take of Opium .... 1 drachm.
Boiling water .... 1 pint.

Mix. Applied hot, is an excellent application to inflamed joints, inflamed testicle, etc. *Bartholow.*

Mixtures of Opium, Rhubarb, and Camphor.

Take of Tincture of opium .... ½ drachm.
Aromatic tincture of rhubarb .... ½ drachm.
Spirit of camphor .... ½ drachm.
Compound tincture of cardamom .... 2 drachms.
Anise water .... 4 ounces.

Mix. Dose: 1 drachm, for children, in diarrhœa.

G. H. Swezey.

Take of Tincture of opium .... 1 part.
Tincture of capsicum .... 1 part.
Aromatic tincture of rhubarb .... 1 part.
Spirit of peppermint .... 1 part.
Spirit of camphor .... 1 part.

Mix. Dose: 20 to 40 minims, in diarrhœa.

Ruschenberger.

Squibb's Cholera Mixture.

Take of Tincture of opium .... 1 ounce.
Tincture of capsicum .... 1 ounce.
Spirit of camphor .... 1 ounce.
Chloroform .... 3 drachms.
Alcohol .... to 5 ounces.

Mix. Dose: 20 to 40 minims.

Mixture of Opium, Catechu, and Chalk.

Take of Camphorated tincture of opium .... 2 drachms.
Tincture of catechu .... 2 drachms.
Chalk mixture .... 3½ ounces.

Mix. Dose: 1 drachm, for children. *New York Dispensary.*

PAREIRA (U. S., Br., Fr.)—PAREIRA BRAVA.

The root of Chonodendron tomentosum, Ruiz et Pav. (*Nat. ord., Menispermaceæ*), a climbing shrub indigenous to Brazil and Peru.

Pareira is tonic and diuretic. Used in chronic diseases of the urinary organs.

Dose: ½ to 2 drachms.

PREPARATIONS.

Extractum Pareiræ Fluidum (U. S., Br.)—Fluid Extract of Pareira Brava.

Take of Pareira brava	16 ounces.
Glycerin	4 ounces.
Alcohol	sufficient.
Water	sufficient.

Mix 8 ounces of alcohol, 3 of glycerin, and 5 of water, moisten the pareira with 4 ounces of the mixture, and proceed according to the general formula, page 161.

Dose: ½ to 2 drachms.

Infusum Pareiræ (U. S., Br.)—Infusion of Pareira Brava.

Take of Pareira brava	1 ounce.
Boiling water	1 pint.

Macerate for two hours in a covered vessel, and strain.

Dose: 1 to 2 ounces. *Decoction of Pareira, Br.*

PEPO (U. S.)—PUMPKIN SEED.

The seeds of Cucurbita pepo, L. (*Nat. ord., Cucurbitaceæ*), the common pumpkin.

Pumpkin seeds are used for the expulsion of tape-worm, and are among the most efficient agents for this purpose.

They may be administered in the following manner: From 1 to 2 ounces of pumpkin seed, deprived of their outer envelope, are beaten into a paste with sugar, and, diluted with water or milk, are taken after a fast of twenty-four hours. After three or four hours, a dose of castor-oil should be administered. *Stillé.*

PEPSINUM (Br., Fr.)—PEPSIN.

A peculiar digestive principle obtained from the mucous lining of the stomach of the pig, sheep, calf, or other warm-blooded animals.

Used as an aid to digestion in a great variety of disorders, as convalescence from febrile diseases, dyspepsia, ulcer of the stomach, infantile diarrhœa, etc.

Dose: ½ to 5 grains.

PREPARATIONS.

Saccharated Pepsin.

Take of	Pepsin, fresh and moist	at will.
	Sugar of milk	sufficient.

Mix in such proportions that 10 parts of it, dissolved in 150 parts of water and 3 parts of hydrochloric acid, will dissolve at least 120 parts of egg-albumen, at a temperature of 104°, in five or six hours.

Dose: 5 to 15 grains. *Report of Am. Ph. Ass'n.*

Liquid Pepsin.

Take of	Pepsin	64 grains.
	Water	2½ ounces.
	Hydrochloric acid	½ drachm.
	Glycerin	1½ ounce.

Mix and filter. *Hospital Formulary.*

Take of	Pepsin	2 drachms.
	Diluted hydrochloric acid	1 drachm.
	Mint water	3 ounces.

Mix, filter, and add of syrup of orange peel 1 ounce.

Dose: ½ ounce in an equal quantity of water directly after eating. For indigestion, sense of oppression and flatulence after eating.

" One of the most successful and most agreeable prescriptions that I have ever devised." *Fordyce Barker.*

PHOSPHORUS (U. S. et al. Ph.)—PHOSPHORUS.

A non-metallic element which exists in many minerals, and in the tissues of animals and plants. It is extracted on a large scale from bones, in which it exists as phosphate of calcium.

Phosphorus is a stimulant to the nervous system, and is used in nervous exhaustion, neuralgia, etc.

Dose: $\frac{1}{100}$ to $\frac{1}{20}$ grain. It is never administered in substance.

PREPARATIONS.

Phosphorated Oil.

Take of Phosphorus .... 1 part.
Cod-liver oil .... 99 parts.

Introduce the oil into a bottle fitted with a cork perforated with two glass tubes, one reaching nearly to the surface of the oil. Pass a current of dry carbonic acid through the bottle until all air is expelled, then quickly introduce the phosphorus, cork tight, and gently heat on a water-bath until solution is effected. Transfer to 1-ounce bottles, which have been rinsed with ether, and not dried, and preserve in a cool, dark place.

Squibb's formula, abbreviated.

Dose: 1 to 5 grains, administered in cod-liver oil.

Phosphorus Pill (Br.).

Take of Phosphorus .... 2 grains.
Balsam of tolu .... 120 grains.
Yellow wax .... 60 grains.

Put the phosphorus and balsam into a wedgewood mortar half full of hot water, rub together until no particles of phosphorus are visible, the temperature being kept at 140°. Add the wax, and as it softens, mix with the other ingredients. Cool without exposure to the air, and keep in a bottle immersed in cold water. It may be softened with a few drops of alcohol when made into pills.

Dose: 3 to 6 grains.

PHYSOSTIGMA (U. S. et al. Ph.)—CALABAR BEAN.

The seed of Physostigma venenosum, Balfour (*Nat. ord., Leguminosæ*), a woody, climbing vine indigenous to Western Africa.

In full doses Calabar bean produces giddiness and drowsiness, with pallor and coolness of the skin, weak pulse, relaxation of the muscles, and contraction of the pupil. It has been used with success in chorea, traumatic tetanus, poisoning with strychnia, etc.

Its active principle, *physostigmia*, or *eserina*, is used in ophthalmic practice to contract the pupils.

Dose: 1 to 3 grains.

PREPARATION.

Extractum Physostigmatis (U. S. et al. Ph.)—Extract of Calabar Bean.

Take of Calabar bean, in powder .... 12 ounces.
Alcohol .... sufficient.

Macerate the powder in 12 ounces of alcohol for four days, then percolate until 2 pints are obtained, or the bean is exhausted. Distil off most of the alcohol, then evaporate to a proper consistence.

Dose: $\frac{1}{16}$ to $\frac{1}{4}$ grain.

PILOCARPUS—JABORANDI.

The leaves of Pilocarpus pennatifolius, Lemaire (*Nat. ord., Rutaceæ*), a shrub growing in the eastern provinces of Brazil.

Jaborandi is a very powerful diaphoretic and sialagogue. It has been employed in a great variety of cases, but its most beneficial effects have been obtained in the dropsy of Bright's disease, pleuritic effusion, etc. Its active principle, termed *pilocarpina*, or its salts, may be used hypodermically.

Dose: 5 to 60 grains.

PREPARATION.

Tincture of Jaborandi.

Take of Jaborandi .... 1 part.
Diluted alcohol .... sufficient.

Moisten, pack, and percolate to 5 parts.

Dose: ½ to 2 drachms. *J. P. Remington, Report Am. Ph. Ass'n.*

PIMENTA (U. S., Br., Fr.)—PIMENTO—ALLSPICE.

The unripe berries of Eugenia pimenta, DC. (*Nat. ord., Myrtaceæ*), an evergreen tree indigenous to the West Indies, Central and South America, but cultivated in other tropical regions.

Allspice is aromatic and stimulant, but is rarely used except as a condiment and flavoring agent.

PREPARATIONS.

Oleum Pimentæ (U. S., Br.)—Oil of Pimento.

The oil obtained from pimento by distillation.

Dose: 1 to 5 minims.

Pimento Water (Br.).

Take of Pimento .... 14 ounces.
Water .... 2 gallons.

Distil 1 gallon. Used as a vehicle.

PIPER (U. S., Br.)—BLACK PEPPER.

The unripe berries of Piper nigrum, L. (*Nat. ord., Piperaceæ*), a climbing shrub indigenous to India, but cultivated in both the East and West Indies.

Pepper is a carminative and stimulant. Used as an adjunct to other remedies, and as a condiment.

Dose: 5 to 20 grains.

PREPARATIONS.

Oleoresina Piperis (U. S.)—Oleoresin of Black Pepper.

Take of Black pepper	12 ounces.
Ether	sufficient.

Percolate the pepper with ether until 20 ounces have passed, distil off most of the ether, evaporate the remainder, and when the deposition of piperin has ceased, strain and express.

Dose: 1 to 2 minims.

Confection of Pepper (Br.).

Take of Black pepper	2 ounces.
Caraway	3 ounces.
Clarified honey (by weight)	15 ounces.

Rub well together.

Dose: 60 to 120 grains.

PIX BURGUNDICA (U. S. et al. Ph.)—BURGUNDY PITCH.

The prepared concrete exudation of Abies excelsa, DC. (*Nat. ord., Coniferæ*), the spruce fir tree of Northern Europe.

Burgundy pitch is a gentle rubefacient, rarely producing more than slight inflammation and serous effusion. It is used in the form of plasters, in chronic and subacute rheumatism, lumbago, etc.

PREPARATIONS.

Emplastrum Picis Burgundicæ (U. S., Fr.)—Burgundy Pitch Plaster.

Take of Burgundy pitch	72 ounces.
Yellow wax	6 ounces.

Melt together, strain, and stir while cooling.

The French Codex employs 1 part of wax with 3 of pitch.

Emplastrum Picis cum Cantharide (U. S.)—Plaster of Pitch with Cantharides.

Take of Burgundy pitch	48 ounces.
Cerate of cantharides	4 ounces.

Heat the cerate to 212°, strain, add the pitch, melt, and stir while cooling.

Pitch Plaster (Br.).

Take of Burgundy pitch	26 ounces.
Common frankincense	13 ounces.
Resin	4½ ounces.
Yellow wax	4½ ounces.
Expressed oil of nutmeg	1 ounce.
Olive oil	2 ounces.
Water	2 ounces.

Add the oils and the water to the other ingredients, previously melted together, then, constantly stirring, evaporate to a proper consistence.

PIX LIQUIDA (U. S. et al. Ph.)—TAR.

An impure turpentine obtained from Pinus palustris, L. and other species of pine, (*Nat. ord.*, *Coniferæ*), by burning the wood in pits covered with earth to prevent the access of air. It is a complex body, containing resin, pyroligneous acid, oil of tar, etc., and produces, besides the general effects of the turpentine, others due to these latter principles. It is employed internally in chronic catarrhal affections, and externally in a variety of cutaneous diseases.

Dose: 10 to 60 grains.

PREPARATIONS.

Glyceritum Picis Liquidæ (U. S.)—Glycerite of Tar.

Take of Tar	1 ounce.
Carbonate of magnesium	2 ounces.
Glycerin	4 ounces.
Alcohol	2 ounces.
Water	10 ounces.

Mix the glycerin, alcohol, and water; rub the tar first with the carbonate of magnesium, then with 6 ounces of the mixed liquids added gradually, and strain. Rub the residue with half the remaining liquid, and strain as before. Repeat the process with the remaining liquid. Finally, percolate the residue with the expressed liquids previously mixed, and add sufficient water to make 1 pint.

Dose: ½ to 2 drachms.

Infusum Picis Liquidæ (U. S., Ger.)—Infusion of Tar—Tar Water.

Take of Tar	1 pint.
Water	4 pints.

Mix, and shake the mixture frequently during twenty-four hours, then decant and filter.

Dose: 2 to 4 ounces.

Unguentum Picis Liquidæ (U. S., Fr.)—Tar Ointment.

Take of Tar	12 ounces.
Suet	12 ounces.

Melt the suet, add the tar, strain, and stir while cooling.

Oil of Tar.

An empyreumatic volatile oil obtained in the distillation of tar.

Used externally.

INHALATION OF TAR.

Take of Infusion of tar	1 to 4 ounces.
Water	8 ounces.

Mix.

Stimulant. Used by means of an atomizer. *G. M. Lefferts.*

LOTION OF TAR.

Take of Tar	1 ounce.
Alcohol	1 ounce.
Soft soap	1 ounce.

Mix. Used in eczema. *Tilbury Fox.*

PLUMBUM—LEAD.

PLUMBI ACETAS (U. S. et al. Ph.)—ACETATE OF LEAD.

Prepared by dissolving oxide of lead in acetic acid, evaporating, and crystallizing.

Acetate of lead is in colorless crystals, which effloresce in the air, and have an acetous odor, and a sweetish, astringent taste. It is used as an astringent in hæmoptysis and other hemorrhages, and in dysentery and diarrhœa. Topically it is employed as a styptic, as an injection in gonorrhœa and leucorrhœa, as a lotion in bruises, sprains, etc.

Dose: ½ to 3 grains.

PREPARATIONS.

Ceratum Plumbi Subacetatis (U. S. et al. Ph.)—Cerate of Subacetate of Lead—Goulard's Cerate.

Take of Solution of subacetate of lead	2½ ounces.
White wax	4 ounces.
Olive oil (by weight)	8 ounces.
Camphor	30 grains.

Mix the wax, previously melted, with 7 ounces of the oil, and, while cooling, stir in the solution of lead. Then add the camphor, previously dissolved in the remainder of the oil, and mix. Or,

Take of Cerate	350 grains.
Olive oil	50 grains.
Solution of subacetate of lead	1½ drachm.
Liniment of camphor	12 grains.

Mix.

Linimentum Plumbi Subacetatis (U. S.)—Liniment of Subacetate of Lead.

Take of	Solution of subacetate of lead (by weight)	2 ounces.
	Olive oil (by weight)........................	3 ounces.

Mix.

Liquor Plumbi Subacetatis (U. S. et al. Ph.)—Solution of Subacetate of Lead—Goulard's Extract.

Take of	Acetate of lead..............................	16 ounces.
	Oxide of lead...............................	9½ ounces.
	Boiling water...............................	sufficient.

Put the acetate and oxide into 4 pints of boiling water in a glass or porcelain vessel, and boil for half an hour, adding boiling water to preserve the measure, then filter. Used externally.

Liquor Plumbi Subacetatis Dilutus (U. S. et al. Ph.)—Diluted Solution of Subacetate of Lead—Lead Water.

Take of	Solution of subacetate of lead (by weight).....	3 drachms.
	Distilled water..............................	1 pint.

Mix. Used externally.

Suppositoria Plumbi (U. S.)—Suppositories of Lead.

Take of	Acetate of lead..............................	36 grains.
	Oil of theobroma............................	324 grains.

Mix the acetate with 60 grains of the oil, and having melted the remainder and cooled it to 95°, mix all together, and pour into suitable moulds, forming 12 suppositories.

Suppositoria Plumbi et Opii (U. S., Br.)—Suppositories of Lead and Opium.

Take of	Acetate of lead..............................	36 grains.
	Extract of opium............................	6 grains.
	Oil of theobroma............................	318 grains.
	Water	sufficient.

Rub the acetate and extract into a smooth paste with a few drops of water, then mix with 60 grains of the oil, and proceed as in the preceding preparation, forming 12 suppositories.

The British Pharmacopœia employs a mixture of benzoated lard, white wax, and oil of theobroma as an excipient, and terms them *Compound Lead Suppositories.*

Pill of Lead and Opium (Br.).

Take of	Acetate of lead..............................	36 grains.
	Opium	6 grains.
	Confection of roses.........................	6 grains.

Beat them into a uniform mass.

Dose : 3 to 5 grains.

Ointment of Acetate of Lead (Br.).

Take of Acetate of lead.............................. 12 grains.
Benzoated lard.............................. 1 ounce.

Mix thoroughly.

Injections of Lead.

Take of Solution of subacetate of lead...............½ to 1 ounce.
Water..................................4 to 6 ounces.

Mix. In gonorrhœa. *Bumstead.*

Take of Acetate of lead.............................. 30 grains.
Sulphate of zinc.............................. 25 grains.
Tincture of catechu.............................. 1 drachm.
Wine of opium.............................. 1 drachm.
Rose water.............................. 6 ounces.

Mix. In gonorrhœa. *Ricord.*

Take of Diluted solution of subacetate of lead........ 4 ounces.
Sulphate of zinc.............................. 8 grains.

Mix. In gonorrhœa. *Bartholow.*

Lotions of Lead.

Take of Acetate of lead..............................5 to 10 grains.
Diluted hydrochloric acid..................½ to 2 drachms.
Water.............................. 6 ounces.

Mix. In eczematous and lichenous affections. *Tilbury Fox.*

Take of Acetate of lead.............................. 15 grains.
Diluted hydrocyanic acid.............................. 20 minims.
Alcohol.............................. ½ ounce.
Water..............................to 6 ounces.

Mix. In impetigo. *Tilbury Fox.*

Mixture of Lead.

Take of Acetate of lead.............................. 8 grains.
Acetic acid.............................. 6 drops.
Deodorized tincture of opium.............................. 4 drops.
Distilled water.............................. 1 ounce.

Mix. Dose: 1 drachm every two, three, or four hours, for a child of two years of age. In summer diarrhœa. *Bartholow.*

Powder of Lead, Opium, and Camphor.

Take of Acetate of lead.............................. 24 grains.
Opium.............................. 12 grains.
Camphor.............................. 30 grains.
Sugar.............................. sufficient.

Mix, and divide into 12 powders.

Dose: One powder every hour or two, in choleraic diarrhœa.

Bartholow.

PLUMBI CARBONAS (U. S. et al. Ph.) CARBONATE OF LEAD—WHITE LEAD.

Prepared on a large scale by exposing sheets of lead to the vapor of vinegar or pyroligneous acid.

A heavy, white powder, insoluble in water. Used externally in skin diseases, as erythema, erysipelas, and intertrigo, and as an application to superficial burns and scalds.

PREPARATIONS.

Unguentum Plumbi Carbonatis (U. S. et al. Ph.)—Ointment of Carbonate of Lead.

Take of Carbonate of lead .... 60 grains.
Ointment.................................420 grains.

Mix.

The British preparation is almost identical with this; the French is made in the proportion of 1 part of the carbonate to 5 of lard, and the German 1 to 2.

CAMPHORATED OINTMENT OF WHITE LEAD (Ger.).

Take of Camphor.................................. 5 parts.
Ointment of white lead ....100 parts.

Mix.

OINTMENT OF CARBONATE OF LEAD, WITH GLYCERIN.

Take of Carbonate of lead.......................... 4 grains.
Glycerin.............................. 1 drachm.
Cerate................................. 1 ounce.

Mix. Used in erythema. *Tilbury Fox.*

PLUMBI IODIDUM (U. S. et al. Ph.)—IODIDE OF LEAD.

Take of Nitrate of lead............................ 4 ounces.
Iodide of potassium........................ 4 ounces.
Distilled water............................ sufficient.

With the aid of heat dissolve the nitrate in 1½ pint of distilled water, and the iodide in ½ pint, and mix the solutions. Decant the supernatant liquid, wash, and dry the precipitate with a gentle heat.

A bright yellow, heavy, inodorous powder, sparingly soluble in cold water.

Occasionally used internally as an alterative, but chiefly employed externally in skin diseases.

Dose: ½ to 4 grains.

PREPARATIONS.

Unguentum Plumbi Iodidi (U. S., Br., Fr.)—Ointment of Iodide of Lead.

Take of	Iodide of lead	60 grains.
	Ointment	420 grains.

Mix.

The French Codex employs 1 part of iodide of lead with 9 parts of benzoated lard.

Iodide of Lead Plaster (Br.).

Take of	Iodide of lead	1 ounce.
	Soap plaster	4 ounces.
	Resin plaster	4 ounces.

Add the iodide to the plasters, previously melted, and mix.

PLUMBI NITRAS (U. S., Br.)—NITRATE OF LEAD.

Prepared by dissolving lead in warm nitric acid, evaporating and crystallizing. Used externally as an application to cracks and excoriations of the nipples, ulcers, etc.

Glycerite of Nitrate of Lead.

Take of	Nitrate of lead	10 to 20 grains.
	Glycerin	1 ounce.

Dissolve. For sore nipples.

After nursing, the nipple should be carefully wiped, and the solution applied freely. It should be washed off before the child is again put to the breast. *Fordyce Barker.*

PLUMBI OXIDUM (U. S. et al. Ph.)—OXIDE OF LEAD—LITHARGE.

Prepared by passing a current of air over melted lead heated to dull redness.

Litharge occurs as a yellowish or pale red powder, insoluble in water. Used externally in the form of lead plaster, and occasionally as an application to burns, etc.

PREPARATIONS.

Emplastrum Plumbi (U. S. et al. Ph.)—Lead Plaster.

Take of	Oxide of lead	30 ounces.
	Olive oil (by weight)	56 ounces.
	Water	sufficient.

Rub the oxide with half its weight of the oil; add the mixture to the remainder of the oil, contained in a vessel of a capacity equal to twice the bulk of the ingredients. Add ½ pint of boiling water, and boil until a plaster is formed, adding from time to time a little boiling water, to preserve the measure.

The French Codex and German Pharmacopœia employ equal parts of litharge, olive oil, and lard.

COMPOUND LEAD PLASTER (Ger.).

Take of	Lead plaster	24 parts.
	Yellow wax	3 parts.
	Ammoniac	2 parts.
	Galbanum	2 parts.
	Turpentine	2 parts.

Melt the plaster and wax together, and, when partially cooled, add the other ingredients, previously melted.

SOFT LEAD PLASTER (Ger.).

Take of	Lead plaster	3 parts.
	Lard	2 parts.
	Suet	1 part.
	Yellow wax	1 part.

Melt together, and strain.

HEBRA'S OINTMENT OF LEAD (Ger.).

Take of	Lead plaster	1 part.
	Linseed oil	1 part.

Mix.

PODOPHYLLUM (U. S., Br.)—MAY-APPLE—MANDRAKE.

The rhizome and rootlets of Podophyllum peltatum, L. (*Nat. ord., Berberidaceæ*), an herbaceous perennial indigenous to North America.

Podophyllum is a drastic cathartic, acting particularly upon the upper portion of the alimentary canal, increasing the biliary secretion, and causing considerable pain.

Its active principles reside in a resin (*Resina Podophylli*) which is generally employed instead of the crude drug.

Dose: 5 to 30 grains.

PREPARATIONS.

Extractum Podophylli ((U. S.)—Extract of May-Apple.

Take of	May-apple	12 ounces.
	Alcohol	2 pints.
	Diluted alcohol	sufficient.

Percolate the May-apple with the alcohol, continuing the process with diluted alcohol until 2 pints have passed. Set this aside, and continue the percolation until 2 pints more are obtained. Distil off the alcohol from the tinctures until they have been brought to the consistence of honey, then mix them, and evaporate to a proper consistence.

Dose: 5 to 15 grains.

Resina Podophylli (U. S., Br.)—Resin of May-Apple—Podophyllin.

Take of May-apple	16 ounces.
Hydrochloric acid	2 drachms.
Alcohol	sufficient.
Water	sufficient.

Macerate the May-apple with 1 pint of alcohol for four days, then percolate until 24 ounces have passed. Distil off the alcohol until the tincture is reduced to 6 ounces, then add the residue to 7 pints of water, previously mixed with the acid, collect, wash, and dry the precipitate.

Dose: $\frac{1}{8}$ to $\frac{1}{2}$ grain.

See also *Formula in Report of Am. Ph. Ass'n.*

COMPOUND PILLS OF PODOPHYLLIN (*Res. Podophyl.*).

Take of Podophyllin	10 grains.
Aloes	20 grains.
Extract of belladonna	5 grains.
Extract of nux vomica	5 grains.

Mix, and divide into 20 pills.

Dose: 1 pill at bedtime, in habitual constipation. *E. G. Janeway.*

Take of Podophyllin	6 grains.
Extract of belladonna	3 grains.
Extract of Calabar bean	3 grains.

Mix, and divide into 12 pills.

Dose: 1 pill each night, in habitual constipation. *Bartholow.*

POTASSIUM—POTASSIUM.

POTASSA (U. S. et al. Ph.)—POTASSA—CAUSTIC POTASH.

Take of solution of potassa, 8 pints. Evaporate rapidly in an iron vessel over the fire until ebullition ceases and the potassa melts. Pour into suitable moulds, and when cold, keep in a well-stopped bottle.

Potassa occurs in white, cylindrical sticks, which are very deliquescent. It is a most energetic caustic, and is employed to destroy morbid growths, the virus of poisoned wounds, as the bites of venomous reptiles and rabid animals, etc.

PREPARATIONS.

Potassa cum Calce (U. S.)—Potassa with Lime.

Take of Potassa .... 1 ounce
Lime .... 1 ounce.

Rub together into a powder, and keep it in a well-stopped bottle.

Used as a caustic; it is milder and more manageable than potassa. Made into a paste with alcohol, when required for use, it constitutes the well-known *Vienna Paste.*

Liquor Potassæ (U. S., Br., Ger.)—Solution of Potassa.

Take of Bicarbonate of potassium .... 15 ounces.
Lime .... 9 ounces.
Distilled water .... sufficient.

Dissolve the bicarbonate in 4 pints of distilled water, and heat until effervescence ceases, adding water to make up the loss by evaporation. Mix the lime with 4 pints of distilled water, heat to the boiling-point, and mix with the potassium solution at the same temperature. Boil ten minutes, strain, and add sufficient distilled water to make 7 pints. Or,

Take of Potassa .... 1 ounce.
Distilled water .... 1 pint.

Dissolve, allow the sediment to subside, then decant the clear liquid, and preserve in a well-stopped bottle.

Dose: 5 to 20 minims.

Used as an antacid, antilithic, and diuretic, but is, in general, less eligible than the alkaline carbonates, on account of its irritant properties.

The German preparation contains one-third of its weight of potassa.

LOTION OF POTASSA.

Take of Solution of potassa .... 1 drachm.
Rose water .... 4 ounces.

Mix. Apply with a soft sponge twice a day, in acne occurring in persons with a greasy skin, and prominent and black sebaceous follicles.

Bartholow.

MIXTURES OF POTASSA.

Take of Solution of potassa .... ½ drachm.
Infusion of columbo .... 4 ounces.

Mix. Dose: 2 drachms three times a day before meals, in atonic dyspepsia. *Bartholow.*

Take of Solution of potassa .... 8 parts.
Extract of belladonna .... 1 part.
Camphor water .... 120 parts.

Mix. Dose: 1 drachm in a wineglass of water, three times a day. For irritable bladder. *Daniel Lewis.*

POTASSII ACETAS (U. S. et al. Ph.)—ACETATE OF POTASSIUM.

Take of Acetic acid.............................. 1 pint.
Bicarbonate of potassium.................. sufficient.

Add the bicarbonate gradually to the acid until it is neutralized, then evaporate cautiously on a sand-bath to dryness.

It is a white, very deliquescent salt, of a pungent, saline taste, and wholly soluble in water and in alcohol. It is diuretic, and, in large doses, cathartic. Used in gout, rheumatism, dropsy, etc.

Dose: 20 to 60 grains, diuretic; 1 to 3 drachms, laxative.

MIXTURES OF ACETATE OF POTASSIUM.

Take of Acetate of potassium...................... 1 drachm.
Tincture of digitalis...................... ½ drachm.
Syrup of squill...........................1 to 2 drachms.
Syrup of ginger.......................... 5 drachms.
Water .... to 3 ounces.

Mix. Dose: 1 drachm every two or three hours, for children two or three years old. As a diuretic and febrifuge in scarlatinous dropsy.

Meigs and Pepper.

Take of Acetate of potassium...................... 6 drachms.
Wine of colchicum seed.................... 3 drachms.
Water.....................................to 4 ounces.

Mix. Dose: 1 drachm. In rheumatism. *Hospital Formulary.*

POTASSII ARSENITIS—ARSENITE OF POTASSIUM.

PREPARATION.

Liquor Potassii Arsenitis (U. S. et al. Ph.)—Solution of Arsenite of Potassium—Fowler's Solution.

Take of Arsenious acid........................... 64 grains.
Bicarbonate of potassium.................. 64 grains.
Compound spirit of lavender.. .... ½ ounce.
Distilled water.......................... sufficient.

Boil the acid and bicarbonate with ½ ounce of distilled water until the acid is dissolved, then add 12 ounces of distilled water, and afterward the spirit of lavender, and sufficient distilled water to make 1 pint.

One ounce contains 4 grains of arsenic.

Dose: 3 to 10 minims. Best administered simply diluted with water. It is one of the best of the arsenical preparations, and is employed in almost all cases to which arsenic is in any way applicable. See *Arsenic.*

POTASSII BICARBONAS (U. S. et al. Ph.)—BICARBONATE OF POTASSIUM.

Take of Carbonate of potassium.................... 48 ounces.
Distilled water............................ 10 pints.

Dissolve the carbonate in the water, and pass carbonic acid through the solution until it is saturated. Then filter, and evaporate, at or below 160°, until crystals form. Lastly, pour off the supernatant liquid, and dry the crystals on bibulous paper.

Bicarbonate of potassium occurs in white, transparent crystals, of a feeble alkaline taste, and freely soluble in water. It is much pleasanter to the taste than carbonate of potassium, and is used for like purposes.

Dose: 10 to 60 grains.

PREPARATION.

Effervescing Solution of Potash (Bicarbonate) (Br.).

Take of Bicarbonate of potassium.................. 30 grains.
Water.................................. 20 ounces.

Dissolve, filter, and pass as much carbonic acid into the solution as can be introduced with a pressure of seven atmospheres. Keep in bottles tightly closed.

POTASSII BITARTRAS (U. S. et al. Ph.)—BITARTRATE OF POTASSIUM—CREAM OF TARTAR.

Crude tartar is deposited by grape juice during fermentation; purified by recrystallization, it is the cream of tartar of commerce.

It occurs as a crystalline powder, of a pleasant, acidulous taste, and requiring 200 parts of cold water for solution.

Cream of tartar is diuretic, refrigerant, and laxative. Used in febrile diseases, for diuretic and refrigerant effect, and as a laxative in a variety of affections.

Dose: 1 to 2 drachms, diuretic and laxative; 2 to 8 drachms, purgative.

See *Pulvis Jalapæ Compositus, U. S.*, and *Confection of Sulphur*, Br.

POTASSII BROMIDUM (U. S. et al. Ph.)—BROMIDE OF POTASSIUM.

Take of Bromine (by weight)...................... 2 ounces.
Iron, in filings.............................. 1 ounce.
Pure carbonate of potassium................1,020 grains.
Distilled water............................ 4 pints.

Add the iron, and afterward the bromine, to 1½ pint of the water; stir frequently for half an hour. Heat gently, and, when the liquid becomes greenish, add gradually the carbonate dissolved in 1½ pint of the water, until it ceases to produce a precipitate. Continue the heat for half an hour, then filter. Wash the precipitate with the remainder of the water, boiling hot, and again filter. Mix the filtered liquids, evaporate and crystallize. Lastly, pour off the mother-water, and dry the crystals on bibulous paper.

Bromide of potassium occurs in white crystals, of a pungent, saline taste, and freely soluble in water. It is antispasmodic, hypnotic, and sedative, and has an extremely wide range of usefulness, being employed in a great variety of nervous affections, as convulsions, epilepsy, whooping-cough, sleeplessness, headaches, vomiting from cerebral disturbance, tetanus, etc.

Dose: 10 to 60 grains.

Gargle of Bromide of Potassium.

Take of Bromide of potassium...................... 1½ drachm.
Glycerin...................................... 2 drachms.
Water.......................................to 10 ounces.

Mix. Sedative. *G. M. Lefferts.*

Mixtures of Bromide of Potassium.

Take of Bromide of potassium...................... ½ drachm.
Fluid extract of conium.................... 15 minims.
Water.......................................to 1 drachm.

Mix. One dose, to be taken thrice daily, in epilepsy.

Hospital Formulary.

Take of Bromide of potassium...................... 1 ounce.
Bromide of ammonium....................... ½ ounce.
Water...................................... 7 ounces.

Mix. Dose: 1 drachm. In epilepsy. *E. C. Seguin.*

Take of Bromide of potassium...................... 1 ounce.
Chloral.................................... ½ ounce.
Water...................................... 7 ounces.

Mix. Dose: 1 drachm. In epilepsy. *J. C. Shaw.*

Take of Bromide of potassium...................... 6 drachms.
Compound tincture of cinchona.............. 2 ounces.
Tincture of cinnamon....................... 1 ounce.
Syrup of orange peel....................... 1 ounce.

Mix. Dose: 1 drachm in a wineglass of sweetened water three times a day, commencing eight days before the expected appearance of the menses. For metrorrhagia at the climacteric, when not the result of organic disease requiring surgical treatment. *Fordyce Barker.*

The patient should also use suppositories of extract of ergot, which see.

Take of	Bromide of potassium	1 ounce.
	Bromide of iron	6 grains.
	Water	6 ounces.

Mix. Dose : ½ ounce three times a day. Used in cases requiring both iron and a bromide. *Bartholow.*

Take of	Bromide of potassium	1 ounce.
	Iodide of potassium	½ ounce.
	Water	4 ounces.

Mix. Dose : 1 drachm in water, every half-hour or hour, in spasmodic asthma. *Bartholow.*

POTASSII CARBONAS (U. S. et al. Ph.)—CARBONATE OF POTASSIUM.

Take of	Impure carbonate of potassium	36 ounces.
	Water	2½ pints.

Dissolve the carbonate in the water, filter, evaporate in an iron vessel over a gentle fire until it thickens, then remove from the fire and stir so as to form a granular salt.

The impure carbonate (*Potassii Carbonas Impura, U. S.*) is obtained by lixiviating wood ashes with water, and evaporating to dryness.

Carbonate of potassium has a disagreeable alkaline taste, and is very deliquescent. Used as an antacid, antilithic, diuretic, etc.

Dose : 10 to 30 grains.

PREPARATION.

Potassii Carbonas Pura (U. S., Ger.)—Pure Carbonate of Potassium.

Take of	Bicarbonate of potassium	12 ounces.
	Distilled water	12 ounces.

Heat the bicarbonate gradually in an iron crucible until the water of crystallization is driven off, then raise the heat to redness, and maintain it for half an hour. Remove from the fire, cool, dissolve in the distilled water, filter and evaporate, as in the preceding preparation.

Used for the same purposes as carbonate of potassium.

Dose : 10 to 30 grains.

POTASSII CHLORAS (U. S. et al. Ph.)—CHLORATE OF POTASSIUM.

Prepared by passing chlorine through a solution of caustic potassa mixed with lime, filtering, evaporating, and crystallizing.

Chlorate of potassium occurs in crystalline plates, of a cooling, saline taste, and soluble in 16 parts of cold water.

Used internally and topically in nearly all inflammatory and ulcerative affections of the mouth and throat, scarlatina, diphtheria, etc.

Dose: 10 to 30 grains.

PREPARATION.

Trochisci Potassii Chloratis (U. S., Br., Fr.)—Troches of Chlorate of Potassium.

Take of Chlorate of potassium	5 ounces.
Sugar	18 ounces.
Tragacanth	2 ounces.
Vanilla	30 grains.

Rub the vanilla with a small quantity of the sugar into a uniform powder, then mix with the other powders, avoiding pressure. Then with water form a mass, to be divided into 480 troches.

Each contains 5 grains of chlorate of potassium.

GARGLES OF CHLORATE OF POTASSIUM.

Take of Chlorate of potassium	½ to 2 drachms.
Glycerin	2 drachms.
Water	to 10 ounces.

Mix. *G. M. Lefferts.*

Take of Chlorate of potassium	1 drachm.
Carbolic acid	½ drachm.
Distilled water	4 ounces.

Mix. In ulcerous disease of the mouth, follicular pharyngitis, etc. *Bartholow.*

MIXTURE OF CHLORATE OF POTASSIUM.

Take of Chlorate of potassium	1 to 2 drachms.
Tincture of chloride of iron	2 drachms.
Syrup	1 ounce.
Water	to 4 ounces.

Mix. Dose: 1 drachm, every hour or two, in diphtheria.

See also *Mixtures of Tincture of Iron.*

POTASSII CITRAS (U. S., Br.)—CITRATE OF POTASSIUM.

Take of Citric acid	10 ounces.
Bicarbonate of potassium	14 ounces.
Water	sufficient.

Dissolve the acid in 1 pint of water, with the aid of a gentle heat, add the bicarbonate gradually, and when effervescence has ceased, filter, and evaporate to dryness, stirring constantly after a pellicle has begun to form, until the salt granulates.

It is a white, granular salt, of a slightly alkaline taste, and is very deliquescent. Used as a diaphoretic and refrigerant.

Dose: 20 to 40 grains.

PREPARATIONS.

Liquor Potassii Citratis (U. S.)—Solution of Citrate of Potassium.

Take of Citric acid ... ½ ounce.
Bicarbonate of potassium ... 330 grains.
Water ... ½ pint.

Dissolve the acid and bicarbonate in the water, and strain.

Dose: 2 to 4 drachms. Less agreeable than the following preparation.

Mistura Potassii Citratis (U. S.)—Mixture of Citrate of Potassium—Neutral Mixture.

Take of Lemon juice ... ½ pint.
Bicarbonate of potassium ... sufficient.

Add the bicarbonate gradually to the lemon juice until the acid is neutralized, then strain.

Dose: ½ to 1 ounce.

This and the preceding are often employed for diuretic, diaphoretic, and refrigerant effect, in febrile affections.

MIXTURES OF CITRATE OF POTASSIUM.

Take of Citrate of potassium ... 20 parts.
Spirit of lemon ... 2 parts.
Simple elixir ... 78 parts.

Mix. Dose: 1 drachm before meals, in the acute stage of gonorrhœa.

G. H. Fox.

Take of Citrate of potassium ... 1 ounce.
Water ... 4 ounces.

Mix. Dose: 1 drachm in a little water four times a day, in mercurial salivation. *E. L. Keyes.*

POTASSII IODIDUM (U. S. et al. Ph.)—IODIDE OF POTASSIUM.

Take of Potassa ... 6 ounces.
Iodine ... 16 ounces, or sufficient.
Charcoal ... 2 ounces.
Distilled water ... sufficient.

Dissolve the potassa in 3 pints of boiling distilled water, gradually add the iodine, stirring after each addition until the solution becomes colorless, and continue the additions until the liquid remains slightly colored from excess of iodine. Evaporate to dryness, stirring in the charcoal toward the close of the operation, powder, heat to dull redness for fifteen minutes in an iron crucible, then cool, dissolve the saline matter with distilled water, filter, evaporate, and crystallize.

Iodide of potassium occurs in white, transparent crystals, of an acrid, saline taste, and freely soluble in water and in alcohol. It is a powerful alterative and resolvent. Employed in syphilis, scrofula, chronic bronchitis, chronic rheumatism, etc.

Dose: 2 to 20 grains.

PREPARATIONS.

Unguentum Potassii Iodidi (U. S. et al. Ph.)—Ointment of Iodide of Potassium.

Take of	Iodide of potassium	60 grains.
	Boiling water	½ drachm.
	Lard	420 grains.

Dissolve the iodide in the water, in a warm mortar, then add the lard and mix.

As this preparation is liable to become rancid, Charles Rice proposes the use of petroleum ointment (*vaseline*, *cosmoline*, etc.) instead of lard.

Liniment of Iodide of Potassium and Soap (Br.).

Take of	Hard soap	1½ ounce.
	Iodide of potassium	1½ ounce.
	Glycerin	1 ounce.
	Oil of lemon	1 drachm.
	Distilled water	10 ounces.

Dissolve the soap in 7 ounces of the water by the heat of a water-bath. Dissolve the iodine and glycerin in the remainder of the water, mix the two solutions, and when cold, add the oil of lemon.

Mixtures of Iodide of Potassium.

Take of	Iodide of potassium	1 ounce.
	Solution of arsenite of potassium	1 drachm.
	Water	4 ounces.

Mix. Dose: 1 drachm every four or six hours. In hay asthma.

Bartholow.

Take of	Iodide of potassium	1 ounce.
	Compound tincture of cinchona	1 ounce.
	Compound tincture of cardamom	1½ ounce.
	Compound tincture of gentian	½ ounce.
	Simple syrup	4 ounces.
	Alcohol	4 ounces.
	Water	5 ounces.

Mix. Dose: 1 to 4 drachms.

Tonic, carminative, and dévoid of the sweet, and often nauseous character of the syrups and extracts of sarsaparilla which are used as vehicles. Used with great advantage in the later stages of syphilis. A favorite prescription with the late Dr. Bumstead. *F. R. Sturgis.*

Take of	Iodide of potassium	3 drachms.
	Tincture of tolu	1 drachm.
	Fluid extract of wild cherry	1 drachm.
	Syrup	1 ounce.
	Compound spirit of ether	2 ounces.
	Water	1 ounce.

Mix. Dose: 1 drachm. *E. G. Janeway.*

Take of	Iodide of potassium	3 drachms.
	Carbonate of ammonium	50 grains.
	Syrup of wild cherry	1½ ounce.
	Compound syrup of ether	1½ ounce.

Mix. Dose: 1 drachm. *W. H. Katzenbach.*

POTASSII NITRAS (U. S. et al. Ph.)—NITRATE OF POTASSIUM—NITRE—SALTPETRE.

Crude nitre occurs in India, and in some of the caves of this country. It is produced artificially by bringing together decaying animal and vegetable matter. Purified and prepared for medicinal use, it is in colorless crystals, of a sharp, cooling, saline taste, and freely soluble in water.

Nitre is diaphoretic, diuretic, refrigerant, and sedative, and is often administered in the early stages of acute inflammatory affections. The fumes of burning nitre, or of paper impregnated with it, are often inhaled with benefit in spasmodic asthma.

Dose: 10 to 15 grains.

PREPARATIONS.

Refrigerant Powder (Ger.).

Take of	Nitrate of potassium	1 part.
	Bitartrate of potassium	3 parts.
	White sugar	6 parts.

Mix. Dose: ½ to 1 drachm.

NITRATED PAPER (Ger., Fr.).

Take of Nitrate of potassium........................ 1 part.
Distilled water........................ 4 parts.

Soak bibulous paper in the solution, and then dry it.

The fumes of the burning paper are inhaled in spasmodic asthma.

NITROUS POWDERS.

Take of Nitrate of potassium........................ 15 grains.
Calomel .... ½ grain.
Tartar emetic........................ $\frac{1}{12}$ grain.
Sugar .... 2 grains.

Mix. To be placed dry on the tongue. Was used by Dr. Rush in what was probably fibroid phthisis, and was called Rush's Fever Powder.

J. R. Leaming.

POTASSII PERMANGANAS (U. S. et al. Ph.)—PERMANGANATE OF POTASSIUM.

Prepared by adding black oxide of manganese and chlorate of potassium to a solution of potassa, evaporating to dryness, and heating to redness. The residue is treated with water, the solution neutralized with sulphuric acid, evaporated, and crystallized.

Chiefly employed externally as a disinfectant and deodorizing dressing for foul and gangrenous ulcers and wounds. A weak solution has been employed as an injection in gonorrhœa, leucorrhœa, etc.

Dose: ¼ to 1 grain.

PREPARATIONS.

Liquor Potassii Permanganatis (U. S., Br.)—Solution of Permanganate of Potassium.

Take of Permanganate of potassium........................ 64 grains.
Distilled water........................ 1 pint.

Dissolve.

GARGLE OF PERMANGANATE OF POTASSIUM.

Take of Solution of permanganate of potassium....... 1 drachm.
Water........................ 10 ounces.

Mix. Stimulant and antiseptic. *G. M. Lefferts.*

Take of Permanganate of potassium........................ 1 grain.
Water .... 1 ounce.

Dissolve. In fetor of the breath, and ulcerous disease of the mouth.

Bartholow.

POTASSII SULPHAS (U. S. et al. Ph.)—SULPHATE OF POTASSIUM.

Obtained as a secondary product in the preparation of nitric acid, and in other chemical processes.

In medium doses it is laxative, in large doses purgative, but is seldom employed, as its action is harsh and painful. On account of its hardness it is used as a triturant in the preparation of Dover's powder.

Dose: 20 to 60 grains, laxative; 2 to 4 drachms, purgative.

POTASSII SULPHURETUM (U. S. et al. Ph.)—SULPHURET OF POTASSIUM.

Take of Sublimed sulphur.......................... 1 ounce.
Carbonate of potassium .... 2 ounces.

Rub the carbonate, previously dried, with the sulphur, and heat the mixture in a covered crucible until it ceases to swell, and is completely melted. Then pour upon a marble slab, and when cold, break into pieces, and preserve in a well-stopped bottle of green glass.

When recently prepared, sulphuret of potassium is of a liver-brown color. It dissolves in water, the solution exhaling the odor of hydrosulphuric acid. Its taste is nauseous and disagreeable. Seldom employed internally, its chief use being as a topical application in skin diseases.

PREPARATION.

Ointment of Sulphurated Potassium (Br.).

Take of Sulphurated potassium...................... 30 grains.
Prepared lard............................. 1 ounce.

Mix.

Lotions of Sulphuret of Potassium.

Take of Sulphuret of potassium..................... ½ ounce.
Lime water............................... 16 ounces.

Mix. Used in pityriasis, pustular and parasitic diseases.

Tilbury Fox.

Take of Sulphuret of potassium .... 1 drachm.
Sulphate of zinc........................ 1 drachm.
Rose water............................. 4 ounces.

Mix. Used in acne indurata. *Bulkley.*

POTASSII TARTRAS (U. S. et al. Ph.)—TARTRATE OF POTASSIUM.

Take of Pure carbonate of potassium................ 16 ounces.
Bitartrate of potassium...............36 ounces or sufficient.
Boiling water............................ 8 pints.

Dissolve the carbonate in the water, add the bitartrate gradually until the solution is neutralized; filter, evaporate, and crystallize. Lastly, pour off the mother-water, dry the crystals on bibulous paper, and preserve them in a well-stopped bottle.

It is diuretic and laxative, but is seldom employed.

Dose: 1 to 2 drachms, laxative; 2 to 8 drachms, purgative.

POTASSII ET SODII TARTRAS (U. S. et al. Ph.)—TARTRATE OF POTASSIUM AND SODIUM—ROCHELLE SALT.

Take of Carbonate of sodium...................... 12 ounces.
Bitartrate of potassium.................... 16 ounces.
Boiling water........................... 5 pints.

Dissolve the carbonate in the water, add the bitartrate gradually, filter, evaporate until a pellicle forms, then set aside to crystallize. Pour off the mother-water, and dry the crystals on bibulous paper.

Rochelle salt commonly occurs as a white powder, of a bitterish, cooling, saline taste, and is soluble in twice its weight of water.

It is a mild and pleasant laxative, especially applicable to febrile conditions.

Dose: ½ to 1 ounce.

PREPARATIONS.

Pulveres Effervescentes Aperientes (U. S., Ger.)—Aperient Effervescing Powders—Seidlitz Powders.

Take of Bicarbonate of sodium.................... 1 ounce.
Tartrate of potassium and sodium............ 3 ounces.
Tartaric acid ....420 grains.

Mix the bicarbonate with the tartrate, and divide the mixture into 12 equal parts. Divide the acid into the same number of parts. Lastly, keep the parts separately in papers of different colors.

A powder of each kind is dissolved separately, in three or four ounces of water, the solutions mixed, and administered while in a state of effervescence.

PRUNUS VIRGINIANA (U. S.)—WILD CHERRY.

The bark of Prunus serotina, Ehrhart (*Cerasus serotina, DC., Nat. ord., Rosaceæ*), the common wild black cherry of North America.

Wild cherry bark is used as a tonic and sedative. This latter property depends upon the hydrocyanic acid which it generates when infused with water.

It is employed in pulmonary diseases, especially in consumption. Its tonic properties improve the appetite and increase the strength, while its sedative influence is beneficial in moderating the cough and allaying nervous irritation.

Dose: 20 to 60 grains.

PREPARATIONS.

Extractum Pruni Virginianæ Fluidum (U. S.)—Fluid Extract of Wild Cherry.

Take of		
	Wild cherry	16 ounces.
	Glycerin	4 ounces.
	Water	½ pint.
	Stronger alcohol	sufficient.

Mix the glycerin and water, and macerate the wild cherry in the mixture for four days. Transfer to a percolator, and pour on stronger alcohol until 12 fluid ounces are obtained; reserve this portion. Continue the percolation until 20 ounces more are obtained, evaporate to 4 ounces, filter, and add to the reserved portion.

Dose: 20 to 60 minims.

Contains the tonic properties of the bark, but has little sedative action.

Infusum Pruni Virginianæ (U. S.)—Infusion of Wild Cherry.

Take of		
	Wild cherry	½ ounce.
	Water	sufficient.

Moisten, pack, and percolate to 1 pint.

Dose: 2 to 3 ounces. The best, and indeed, only preparation of wild cherry to use for sedative effect.

Syrupus Pruni Virginianæ (U. S.)—Syrup of Wild Cherry.

Take of		
	Wild cherry	5 ounces.
	Sugar	28 ounces.
	Water	sufficient.

Macerate the bark in water for twenty-four hours, then percolate until 1 pint is obtained. In this dissolve the sugar by agitation, without heat.

Dose: 1 to 4 drachms. Used as a vehicle in cough mixtures.

PULSATILLA (Fr., Ger.)—PULSATILLA.

The herb of Anemone pulsatilla, L. (*Nat. ord., Ranunculaceæ*), a small plant indigenous to Central and Northern Europe.

It is employed in a variety of cases, but mainly in acute and subacute inflammations of the mucous membranes, as those of the eyes, ears, uterus, etc. It has also been used with happiest effect in dysmenorrhœa, and in gonorrhœal orchitis.

Dose: Of the fresh plant, 1 to 5 grains.

PREPARATION.

Tincture of Pulsatilla.

Take of Pulsatilla .... 1 part.
Alcohol .... 2 parts.

Prepare the tincture by maceration.

Dose: 5 to 15 minims. *Phillips.*

Preparations of pulsatilla made in this country, from the plant imported in the dry state, are not to be relied upon.

QUASSIA (U. S. et al. Ph.)—QUASSIA.

The wood of Picræna excelsa, Lindley (*Quassia excelsa, Swartz. Simaruba excelsa, DC., Nat ord., Simarubeæ*), a tree fifty or sixty feet in height, indigenous to Jamaica and other islands of the West Indies.

Quassia is a pure and simple bitter, without irritant or astringent effects, and is used as a tonic in dyspepsia, loss of appetite, etc.

Dose: 30 to 60 grains.

PREPARATIONS.

Extractum Quassiæ (U. S. et al. Ph.)—Extract of Quassia.

Take of Quassia .... 12 ounces.
Water .... sufficient.

Exhaust the quassia by percolation with water, boil to three-fourths of its bulk, strain, and evaporate to a proper consistence.

Dose: 3 to 5 grains.

Infusum Quassiæ (U. S., Br.)—Infusion of Quassia.

Take of Quassia .... 120 grains.
Water .... 1 pint.

Macerate for twelve hours, and strain.

Dose: 1 to 2 ounces. Used as a vehicle for other tonics.

Tinctura Quassiæ (U. S., Br., Fr.)—Tincture of Quassia.

Take of Quassia	2 ounces.
Diluted alcohol	sufficient.

Moisten, pack, and percolate to 2 pints.
Dose: ½ to 2 drachms.

QUERCUS (U. S. et al. Ph.)—OAK BARK.

The bark of Quercus alba, L., and of Quercus coccinea, Wang., var. tinctoria, Bartram (*Nat. ord., Cupuliferæ*), both of which are indigenous to the United States.

Oak bark contains a large percentage of tannic acid, and to this it owes its medicinal properties.

Dose: 30 to 60 grains, in decoction.

PREPARATIONS.

Decoctum Quercus Albæ (U. S., Br.)—Decoction of White Oak.

Take of White oak	1 ounce.
Water	sufficient.

Boil half an hour, strain, and add sufficient water through the strainer to make 1 pint.

Dose: 1 to 2 ounces.

Used topically as a gargle in sore throat, as an injection in leucorrhœa, as a wash for bed-sores, etc.

QUILLAIA—SOAP-BARK.

The inner bark of Quillaia saponaria, Molina (*Nat. ord., Rosaceæ*), a tree growing in Peru and Chili. It contains a vegetable soap principle, termed *saponin*, which is also found in some other plants. It makes a lather or froth with water, and has been used to stimulate the growth of the hair. In pharmacy it is employed as an emulsifying agent.

Tincture of Quillaia.

Take of Quillaia in fine powder	4 ounces.
Alcohol (sp. gr. 0.820)	sufficient.

Moisten, pack, and percolate to 16 ounces.

Probably the most useful preparation. Much used in England as an emulsifying agent for fixed oils, oleoresins, resins, etc. *Charles Rice.*

QUINIA (Fr., Ger.)—QUININE.

This alkaloid is not recognized by the United States and British Pharmacopœias. Owing to its difficult solubility, it acts less speedily than many of its salts. Of these, a large number are prepared, but the Pharmacopœias wisely reject most of them, since, as Bartholow justly remarks, "the curative value of the preparations of quinine depends on the base, and not on the acid combined with it."

Quiniæ Bisulphas (Fr., Ger.)—Bisulphate of Quinia.

Take of Sulphate of quinia........................100 parts.
Sulphuric acid........................ 12 parts.
Distilled water........................ sufficient.

Dissolve the sulphate in the acid, previously diluted with a sufficient quantity of water, evaporate, and crystallize.

It is in prismatic, white, shining crystals, of a very bitter taste, soluble in about 10 parts of water and in 2 parts of alcohol. It contains about one-sixth less quinia than the sulphate, and should be administered in proportionally larger doses.

Dose: 1 to 20 grains.

Hypodermic Injection of Bisulphate of Quinia.

Take of Bisulphate of quinia........................ 50 grains.
Diluted sulphuric acid....................100 minims.
Water........................ 1 ounce.
Carbolic acid (liq.)........................ 5 minims.

Dissolve the bisulphate in the sulphuric acid and water, by the aid of heat, filter, and add the carbolic acid.

Twelve minims contain 1 grain of the bisulphate.

Lente's Solution. Bartholow.

Quiniæ Hydrobromas—Hydrobromate of Quinia.

Take of Sulphate of quinia........................100 parts.
Bromide of potassium........................27½ parts.
Distilled water........................100 parts.
Alcohol........................400 parts.

Triturate the sulphate and bromide with the water, in a mortar, then heat gently in a flask placed on a water-bath, and, after awhile, add the alcohol, and digest for an hour. Filter the hot solution, and set aside to crystallize.

Dose: 1 to 20 grains. *Charles Rice.*

Quiniæ Hydrochloras (Ger.)—Hydrochlorate of Quinia.

Take of	Sulphate of quinia	300 parts.
	Carbonate of sodium	sufficient.
	Hydrochloric acid	sufficient.
	Sulphuric acid	sufficient.
	Water	sufficient.

Dissolve the sulphate of quinia in water with the aid of sulphuric acid, and precipitate the solution with the carbonate of sodium. Wash the precipitate, and while still moist, add it in one lot to 78 parts of hydrochloric acid previously diluted with 1,000 parts of water, and heated to a temperature not exceeding 86°. Let it stand awhile, then heat to 140°, neutralize exactly by the addition of acid or quinia, as may be required, and set aside to crystallize. *Charles Rice.*

It is in white crystals of a silky lustre, of a very bitter taste, soluble in 20 parts of cold water and in 2 or 3 parts of alcohol.

Dose: 1 to 20 grains.

Quiniæ Sulphas (U. S. et al. Ph.)—Sulphate of Quinia.

Take of	Yellow cinchona	48 ounces.
	Hydrochloric acid (by weight)	3½ ounces.
	Lime	5 ounces.
	Animal charcoal	sufficient.
	Sulphuric acid	sufficient.
	Alcohol	sufficient.
	Water	sufficient.
	Distilled water	sufficient.

Boil the cinchona in 13 pints of water mixed with one-third of the hydrochloric acid, and strain. Repeat the process twice, mix the decoctions, and while hot, gradually add the lime previously mixed with 2 pints of water, stirring constantly till the quinia is precipitated. Wash, dry, powder, and digest the precipitate in boiling alcohol. Decant, and repeat the digestion until the alcohol is no longer rendered bitter. Mix the liquids and distil off the alcohol until a viscid mass remains. Boil this in 4 pints of distilled water, and add enough sulphuric acid to dissolve the quinia. Then add 1½ ounce of animal charcoal, boil two minutes, filter while hot, and set aside to crystallize. The crystals may be further purified by dissolving them in boiling water acidulated with sulphuric acid, adding a little animal charcoal, and recrystallizing.

Sulphate of quinia is a colorless salt, in light, silky crystals, of an intensely bitter taste, soluble in 700 parts of cold or 30 parts of boiling water, much more soluble in alcohol and in water acidulated with sulphuric acid. Its therapeutic effects are those of cinchona, which see.

Dose: 1 to 20 grains.

PREPARATIONS.

Pilulæ Quiniæ Sulphatis (U. S., Br., Fr.)—Pills of Sulphate of Quinia.

Take of Sulphate of quinia........................ 24 grains.
Clarified honey, sufficiently inspissated....... 14 grains.

Mix. Form a pilular mass, and divide into 24 pills.

Dose: 1 to 20 pills.

Tincture of Quinia (Br.).

Take of Sulphate of quinia........................160 grains.
Tincture of orange peel.................... 20 ounces.

Dissolve with aid of a gentle heat, allow the solution to stand for three days, shaking it occasionally, then filter.

Dose: ½ to 2 drachms.

Ammoniated Tincture of Quinia (Br.).

Take of Sulphate of quinia........................160 grains.
Solution of ammonia....................... 2½ ounces.
Proof spirit..............................17½ ounces.

Dissolve the sulphate in the spirit with the aid of a gentle heat, and add the solution of ammonia.

Dose: ½ to 2 drachms.

Wine of Quinia (Br.).

Take of Sulphate of quinia........................ 20 grains.
Citric acid................................ 30 grains.
Orange wine................................ 20 ounces.

Dissolve, first the citric acid, and then the sulphate in the wine; allow the solution to stand for three days, shaking it occasionally, then filter.

Dose: ½ to 1 ounce.

Mixtures of Sulphate of Quinia.

Take of Sulphate of quinia........................ 30 grains.
Diluted sulphuric acid...................... sufficient.
Water...................................... 2 ounces.
Tincture of chloride of iron................ ½ ounce.
Spirit of chloroform....................... 6 drachms.
Glycerin...................................to 4 ounces.

Mix. Dose: 1 drachm. *A. L. Loomis.*

Take of Sulphate of quinia........................ 1 drachm.
Pyrophosphate of iron...................... 1 drachm.
Strychnia.................................. 1 grain.
Diluted phosphoric acid.................... 2 drachms.
Syrup of ginger............................ 2 ounces.
Water......................................to 4 ounces.

Mix. Dose: 1 drachm. *W. A. Hammond.*

Take of Sulphate of quinia........................... 30 grains.
Diluted sulphuric acid........................ sufficient.
Water.. 2 ounces.
Tincture of chloride of iron.................. 2 drachms.

Mix. Dose: 1 drachm. *Hospital Formulary.*

Take of Sulphate of quinia........................... ½ drachm.
Elixir of taraxacum.......................... 2 ounces.

Mix. Dose: 1 drachm every two to four hours, and hourly 1 drachm of the following:

Take of Tincture of chloride of iron................. 2 drachms.
Chlorate of potassium........................ 2 drachms.
Syrup.. 4 ounces.

Mix. The dose prescribed is for a child of five years. In diphtheria.
J. Lewis Smith.

Take of Sulphate of quinia........................... ½ drachm.
Sulphate of strychnia........................ 1 grain.
Diluted sulphuric acid........................ sufficient.
Tincture of chloride of iron.................. 3 drachms.
Glycerin....................................... 5 drachms.
Water..to 4 ounces.

Mix. Dose: 1 drachm.

Take of Sulphate of quinia........................... 20 grains.
Diluted sulphuric acid........................ sufficient.
Carbolic acid (liq.).......................... 10 to 40 minims.
Water.. 5 ounces.

Mix. Dose: ½ ounce three times daily before meals, in a wineglass of water. To relieve the pains of mammary cancer. *F. A. Burrall.*

PILLS OF SULPHATE OF QUINIA.

Take of Sulphate of quinia........................... 40 grains.
Sulphate of iron.............................. 20 grains.
Extract of nux vomica......................... 5 grains.

Mix, and divide into 20 pills.
Dose: 1 pill. *Hospital Formulary.*

Take of Sulphate of quinia........................... 2 drachms.
Sulphate of morphia........................... 3 grains.
Strychnia...................................... 2 grains.
Arsenious acid................................ 3 grains.
Extract of aconite............................ 30 grains.

Mix, and divide into 60 pills.
Dose: 1 pill. *Gross's Neuralgic Pills.*

QUINIA FOR INUNCTION.

Take of Sulphate of quinia	1 drachm.
Oleic acid (pure)	1 ounce.
Olive oil	2 ounces.

Dissolve the quinia in the acid with the aid of a gentle heat, and add the oil. If properly prepared, the solution will remain clear.

For inunction in cases of debility, especially in children. It may be applied once or twice daily to the entire surface, and should be well rubbed in. *Andrew H. Smith.*

Quiniæ Valerianas (U. S., Fr., Ger.)—Valerianate of Quinia.

Valerianate of quinia is supposed to exert the combined influence of quinia and valerianic acid, but is in no way better than extemporaneous mixtures of the two. The mode of its preparation is therefore omitted.

Dose: 1 to 3 grains.

Quinidæ Sulphas—Sulphate of Quinidia.

Obtained by evaporating the mother-liquor left from the crystallization of sulphate of quinia.

Sulphate of quinidia is in crystals resembling those of the sulphate of quinia, but is more soluble than that salt, and possesses nearly the same virtues as a tonic and antiperiodic.

Dose: 1 to 20 grains.

RESINA (U. S. et al. Ph.)—RESIN—ROSIN.

The residue after the distillation of the volatile oil from the turpentine of Pinus palustris, Mill., and of other species of Pinus.

Rosin is a brittle, pulverizable, translucent resin, tasteless, and of a feeble terebinthinate odor. It enters into the composition of cerates and plasters.

PREPARATIONS.

Ceratum Resinæ (U. S., Br.)—Resin Cerate—Basilicon Ointment.

Take of Resin	10 ounces.
Yellow wax	4 ounces.
Lard	16 ounces.

Melt together, strain, and stir till cool.

Ointment of Resin, Br.

Ceratum Resinæ Compositum (U. S., Ger.)—Compound Resin Cerate.

Take of Resin .... 12 ounces.
Suet .... 12 ounces.
Yellow wax .... 12 ounces.
Turpentine .... 6 ounces.
Flaxseed oil (by weight) .... 7 ounces.

Melt together, strain, and stir till cool. *Basilicon Ointment, Ger.*

Emplastrum Resinæ (U. S., Br.)—Resin plaster—Adhesive Plaster.

Take of Resin .... 6 ounces.
Lead plaster .... 36 ounces.

To the lead plaster, melted with a gentle heat, add the resin, and mix.

RHAMNUS FRANGULA—ALDER BUCKTHORN.

The bark of Rhamnus frangula, L. (*Nat. ord., Rhamnaceæ*), a shrub indigenous to the Old World.

Employed as a purgative, and is recommended in habitual constipation.

The bark of Rhamnus Purshiana, DC., a shrub indigenous to the Pacific Coast of North America, has similar properties.

Dose: ½ to 2 drachms.

Fluid Extract of Rhamnus Frangula.

Take of Rhamnus frangula .... 16 parts.
Stronger alcohol, 4 parts—water, 15 parts .... sufficient.

Moisten, pack, and percolate 14 ounces, which set aside. Then percolate 24 ounces more, evaporate to 2 ounces, mix with the reserved portion, and filter.

Dose: ½ to 2 drachms. *Charles Rice.*

Mixture of Fluid Extract of Rhamnus Frangula.

Take of Fluid extract of rhamnus frangula .... 1 ounce.
Peppermint water .... 1 ounce.

Mix. Dose: 1 to 2 drachms for a child from two to eight years old. A pleasant and efficacious laxative. *A. A. Smith.*

RHEUM (U. S. et al. Ph.)—RHUBARB.

The root of Rheum officinale, Baillon, R. palmatum, L., and probably of other species (*Nat. ord., Polygonaceæ*), plants indigenous to Asia, and much resembling the common garden rhubarb in general appearance.

Rhubarb is a purgative which acts without violence, and is even said to have a tonic effect. Free purgation by rhubarb is followed by constipation, but laxative doses have a tendency to overcome habitual constipation. It is an excellent cathartic for pregnant women, for children when suffering from indigestion, and for persons suffering from piles attended with constipation.

Dose: 5 to 10 grains, laxative; 10 to 40 grains, purgative.

PREPARATIONS.

Extractum Rhei (U. S. et al. Ph.)—Extract of Rhubarb.

Take of Rhubarb	12 ounces.
Alcohol	1 pint.
Diluted alcohol	sufficient.

Percolate the rhubarb with the alcohol, continuing with diluted alcohol until 12 ounces of tincture are obtained. Set this in a warm place, and allow it to evaporate to 6 ounces. Continue the percolation with diluted alcohol until the rhubarb is exhausted, evaporate, at or below 160°, to the consistence of syrup; mix with the tincture first obtained, and evaporate to a proper consistence.

Dose: 5 to 15 grains.

Extractum Rhei Fluidum (U. S.)—Fluid Extract of Rhubarb.

Take of Rhubarb	16 ounces.
Glycerin	2 ounces.
Alcohol	sufficient.
Water	sufficient.

Mix the glycerin with 14 ounces of alcohol, moisten the rhubarb with 4 ounces of the mixture, and proceed according to the general formula, page 161. Finish the percolation with a menstruum of 2 parts of alcohol and 1 part of water.

Dose: 5 to 10 minims, laxative; 10 to 40 minims, purgative.

Infusum Rhei (U. S., Br.)—Infusion of Rhubarb.

Take of Rhubarb	2 drachms.
Boiling water	½ pint.

Digest for an hour, in a covered vessel, and strain.

Dose: 1 to 2 ounces.

Pilulæ Rhei (U. S.)—Pills of Rhubarb.

Take of Rhubarb	72 grains.
Soap	24 grains.

Beat into a mass with water, and divide into 24 pills.

Dose: 1 to 5, as a laxative.

Pilulæ Rhei Compositæ (U. S., Br.)—Compound Pills of Rhubarb.

Take of Rhubarb............................ 48 grains.
Socotrine aloes............................ 36 grains.
Myrrh............................ 24 grains.
Oil of peppermint............................ 3 minims.

Beat into a mass with water, and divide into 24 pills.

Dose: 2 to 4 pills.

Pulvis Rhei Compositus (U. S., Br.)—Compound Powder of Rhubarb.

Take of Rhubarb............................ 4 ounces.
Magnesia............................ 12 ounces.
Ginger............................ 2 ounces.

Rub together until thoroughly mixed.

Dose: ½ to 1 drachm.

Syrupus Rhei (U. S.)—Syrup of Rhubarb.

Take of Fluid extract of rhubarb............................ 3 ounces.
Syrup............................ 29 ounces.

Mix. Dose: 1 to 2 drachms.

Syrupus Rhei Aromaticus (U. S. et al. Ph.)—Aromatic Syrup of Rhubarb.

Take of Rhubarb............................ 2½ ounces.
Cloves............................ ½ ounce.
Cinnamon............................ ½ ounce.
Nutmeg............................ 2 drachms.
Syrup............................ 6 pints.
Diluted alcohol............................ sufficient.

Mix the powders, and percolate with diluted alcohol until 1 pint of tincture is obtained. Add this to the syrup, previously heated, and mix.

Dose: 1 to 2 drachms.

Tinctura Rhei (U. S. et al. Ph.)—Tincture of Rhubarb.

Take of Rhubarb............................ 3 ounces.
Cardamom............................ ½ ounce.
Diluted alcohol............................ sufficient.

Mix the powders, and percolate with diluted alcohol until 2 pints of tincture are obtained.

Dose: 1 to 2 drachms, laxative; 2 to 8 drachms, purgative.

Tinctura Rhei et Sennæ (U. S.)—Tincture of Rhubarb and Senna.

Take of Rhubarb	1 ounce.
Senna	2 drachms.
Coriander	1 drachm.
Fennel	1 drachm.
Liquorice	½ drachm.
Raisins, deprived of seeds	6 ounces.
Diluted alcohol	3 pints.

Macerate seven days, express and filter.

Dose: 15 to 60 minims.

Vinum Rhei (U. S. et al. Ph.)—Wine of Rhubarb.

Take of Rhubarb	2 ounces.
Canella	1 drachm.
Sherry wine	14 ounces.
Diluted alcohol	sufficient.

Mix 2 ounces of diluted alcohol with the wine, and percolate the powders with the mixture, continuing the process with diluted alcohol until 1 pint of filtered liquid is obtained.

Dose: 1 to 4 drachms.

COMPOUND EXTRACT OF RHUBARB (Ger.).

Take of Extract of rhubarb	3 parts.
Extract of aloes	1 part.
Distilled water	4 parts.
Jalap soap	1 part.
Diluted alcohol	4 parts.

Soften the extracts with the water, then add the soap, previously dissolved in the alcohol, and evaporate to a dry extract.

AQUEOUS TINCTURE OF RHUBARB (Ger.).

Take of Rhubarb	10 parts.
Borax	1 part.
Pure carbonate of potassium	1 part.
Boiling distilled water	85 parts.
Alcohol	10 parts.
Cinnamon water	15 parts.

Mix the rhubarb, borax, and carbonate, add the water, macerate fifteen minutes, then add the alcohol. After one hour and a quarter, express, filter, and add the cinnamon water.

MIXTURES OF RHUBARB AND SODA.

Take of Powdered rhubarb	½ ounce.
Bicarbonate of sodium	1 drachm.
Spirit of peppermint	2 drachms.
Water	to 4 ounces.

Mix. Dose: ½ ounce. *Hospital Formulary.*

Take of Fluid extract of rhubarb .... 256 minims.
Fluid extract of ipecac .... 51 minims.
Bicarbonate of sodium .... 512 grains.
Glycerin .... 12 ounces.
Peppermint water .... 2 pints.

Mix. Dose : ½ to 1 drachm two or three times daily, for children.

E. R. Squibb.

ROSA (U. S. et al. Ph.)—ROSE.

The petals of Rosa centifolia, L., pale rose, and of Rosa Gallica, L., red rose (*Nat. ord., Rosaceæ*), small shrubs indigenous to the warmer portions of Europe and Western Asia, but cultivated in many varieties all over the world.

Rose petals are used chiefly as a perfume, though those of the red rose are slightly astringent, and are used in collyria, gargles, etc.

PREPARATIONS.

Aqua Rosæ (U. S. et al. Ph.)—Rose Water.

Take of Recent pale rose .... 48 ounces.
Water .... 16 pints.

Mix, and distil 8 pints. Used as a vehicle.

When it is desirable to keep the rose for some time before distilling, mix it with half its weight of chloride of sodium.

Confectio Rosæ (U. S., Br., Fr.)—Confection of Rose.

Take of Red rose .... 4 ounces.
Sugar .... 30 ounces.
Clarified honey (by weight) .... 6 ounces.
Rose water .... 8 ounces.

Rub the rose with the water heated to 150°, then gradually add the sugar and honey, and beat well together.

Dose : ½ to 1 drachm. Used chiefly as an excipient.

Infusum Rosæ Compositum (U. S., Br.)—Compound Infusion of Rose.

Take of Red rose .... ½ ounce.
Diluted sulphuric acid .... 3 drachms.
Sugar .... 1½ ounce.
Boiling water .... 2½ pints.

Macerate the rose in the water and acid for half an hour, then add the sugar, and strain.

Dose : 1 to 3 ounces. *Acid Infusion of Roses, Br.*

Mel Rosæ (U. S., Fr., Ger.)—Honey of Rose.

Take of Red rose	2 ounces.
Clarified honey (by weight)	25 ounces.
Diluted alcohol	sufficient.

Percolate the rose with diluted alcohol until 6 drachms of filtered liquid have passed; set this aside. Continue the percolation until ½ pint more of liquid is obtained; evaporate this to 10 drachms, add the reserved liquid, and mix with the honey.

Used as a vehicle.

Oleum Rosæ (U. S., Fr., Ger.)—Oil of Rose.

The oil obtained from the petals of Rosa centifolia by distillation. Used as a perfume.

Syrupus Rosæ Gallicæ (U. S., Br., Fr.)—Syrup of Red Rose.

Take of Red rose	2 ounces.
Sugar	18 ounces.
Diluted alcohol	sufficient.
Water	sufficient.

Percolate the rose with the alcohol until 1 ounce has passed; set this aside. Continue the percolation until 5 ounces more are obtained; evaporate this to 1½ ounce, and mix it with 7 ounces of water. Then add the sugar, dissolve with a gentle heat, strain while hot, cool, and add the reserved tincture.

Dose: 1 to 2 drachms.

Unguentum Aquæ Rosæ (U. S., Ger.)—Ointment of Rose Water.

Take of Expressed oil of almonds	3½ ounces.
Spermaceti	1 ounce.
White wax	2 drachms.
Rose water	2 ounces.

Melt together the oil, spermaceti, and wax, then gradually add the water, and stir until cool.

RUBUS (U. S.)—BLACKBERRY.

The bark of the root of Rubus Canadensis, L., and of Rubus villosus, Ait. (*Nat. ord., Roseaceæ*), the former a trailing, the latter an erect shrub, covered with prickles, indigenous to North America, and very common.

Blackberry root is a mild, but efficient astringent, and is very useful in diarrhœa and dysentery, especially when occurring in children.

Dose: ½ to 1 drachm.

PREPARATIONS.

Extractum Rubi Fluidum (U. S.)—Fluid Extract of Blackberry.

Take of Blackberry .... 16 ounces.
Glycerin .... 4 ounces.
Alcohol .... sufficient.
Water .... sufficient.

Mix 8 ounces of alcohol, 3 of glycerin, and 5 of water, moisten the powder with 4 ounces of the mixture, and proceed according to the general formula, page 161, finishing the percolation with diluted alcohol.

Dose: ½ to 1 drachm.

Syrupus Rubi (U. S.)—Syrup of Blackberry.

Take of Fluid extract of blackberry .... ½ pint.
Syrup .... 1½ pint.

Mix.

Dose: 1 to 2 drachms.

SABINA (U. S. et al. Ph.)—SAVIN.

The tops of Juniperus Sabina, L. (*Nat. ord.*, *Coniferæ*), a small evergreen shrub indigenous to Europe and Asia.

Applied locally, savin is an irritant; taken internally, it is a general stimulant, and in overdoses may cause vomiting, purging, suppression of urine, hæmaturia, etc., and even convulsions and death.

It is used internally, almost exclusively as an emmenagogue, and for this purpose the oil is generally employed. A cerate or ointment of savin is used to maintain discharge from blistered surfaces.

Dose: 5 to 15 grains.

PREPARATIONS.

Ceratum Sabinæ (U. S., Br., Ger.)—Savin Cerate.

Take of Fluid extract of savin .... 3 ounces.
Resin cerate .... 12 ounces.

Melt the cerate, add the fluid extract, and stir until the alcohol has evaporated, and the cerate has become cool.

Extractum Sabinæ Fluidum (U. S.)—Fluid Extract of Savin.

Take of Savin .... 16 ounces.
Stronger alcohol .... sufficient.

Moisten the savin with ½ pint of stronger alcohol, and proceed according to the general formula, page 161.

Dose: 5 to 15 minims.

Oleum Sabinæ (U. S., Br., Ger.)—Oil of Savin.

The oil distilled from the fresh tops of savin.

Dose: 1 to 5 minims.

TINCTURE OF SAVIN (Br.).

Take of	Savin tops, dried	2½ ounces.
	Proof spirit	20 ounces.

Macerate the savin in 15 ounces of the spirit, then transfer to a percolator and percolate with the remainder. Afterward express, filter, and add sufficient proof spirit to make 20 ounces.

Dose: 20 to 60 minims.

SACCHARUM (U. S. et al. Ph.)—SUGAR.

A sweet, crystalline principle obtained from the sugar cane, Saccharum officinarum, L. (*Nat. ord.*, *Gramineæ*), a perennial plant which has been cultivated in warm countries from time immemorial.

Sugar possesses some virtue as a demulcent, but is chiefly used in medicine to give an agreeable taste to mixtures, etc., to protect mineral preparations from oxidation, and to preserve vegetable substances.

PREPARATIONS.

Syrupus (U. S. et al. Ph.)—Syrup—Simple Syrup.

Take of	Sugar	36 ounces.
	Distilled water	sufficient.

Dissolve the sugar, with the aid of heat, in 20 ounces of distilled water, heat to the boiling point, and strain while hot. Then incorporate with it sufficient distilled water, added through the strainer, to make the syrup measure 2 pints and 12 ounces, or weigh 55 ounces. Thus prepared, it has the sp. gr. 1.317.

Syrupus Fuscus (U. S., Br.)—Molasses—Treacle.

The uncrystallized residue of the refining of sugar.

Used as an excipient in pills, etc.

SACCHARUM LACTIS (U. S. et al. Ph.)—SUGAR OF MILK.

A sweet, crystalline principle existing in milk, and obtained by concentrating whey. Sugar of milk occurs in hard crystals, less sweet and less soluble than cane sugar. On account of its hardness, it is often used as a triturant in reducing drugs to a state of minute subdivision.

SALIX (U. S., Fr.)—WILLOW.

The bark of Salix alba, L. (*Nat. ord., Salicaceæ*), a tree indigenous to Europe, but cultivated, and sparingly naturalized in this country.

All the willows contain, in addition to tannin, a crystalline principle termed *Salicin*, to which most of their medicinal effects are due. Willow bark is tonic, antiperiodic, and somewhat astringent. Salicin has been used with considerable success as a substitute for the alkaloids of cinchona bark in the treatment of intermittent fever.

Dose: ½ to 2 drachms.

Salicin.

Prepared by treating a boiling, concentrated decoction of willow bark with litharge until colorless, filtering, evaporating, and crystallizing.

It is in white, shining scales or needles, of a persistently bitter taste, and soluble in 20 parts of water.

Dose: 2 to 20 grains.

SANGUINARIA (U. S.)—BLOODROOT.

The rhizome of Sanguinaria Canadensis, L. (*Nat. ord., Papaveraceæ*), a small herb with a perennial root, indigenous to North America, blooming early in spring.

All parts of the plant, when wounded, exude an orange-red juice, of a peculiar, acrid taste. Bloodroot is alterative, expectorant, and, in full doses, emetic, this latter action being violent and depressing. It is used chiefly in diseases of the respiratory organs. The powdered root has been used as a stimulant to unhealthy ulcers.

Dose: 1 to 5 grains, alterative and expectorant; 10 to 20 grains, emetic.

PREPARATIONS.

Acetum Sanguinariæ (U. S.)—Vinegar of Bloodroot.

Take of Bloodroot	4 ounces.
Diluted acetic acid	sufficient.

Percolate the bloodroot with the acid until 2 pints are obtained.

It may also be prepared by macerating the bloodroot in 2 pints of diluted acetic acid for seven days, expressing, and filtering the liquid.

Dose: 15 to 60 minims, alterative and expectorant; 3 to 4 drachms, emetic.

Tinctura Sanguinariæ (U. S.)—Tincture of Bloodroot.

Take of Bloodroot .... 4 ounces.
Alcohol .... sufficient.
Water .... sufficient.

Mix 3 measures of alcohol with 1 of water, and percolate the bloodroot with the mixture until 2 pints of tincture are obtained.

Dose: 15 to 60 minims, alterative and expectorant; 3 to 4 drachms, emetic.

SANGUINARINE.

Take of Bloodroot .... 16 parts.
Alcohol (sp. gr. 0.835) .... sufficient.

Moisten, pack, and percolate until 16 parts of tincture are obtained, or until the bloodroot is exhausted. Mix 4 parts of distilled water with the percolate, and evaporate the alcohol. To the residue add 12 parts of cold distilled water, and allow the mixture to stand in a cool place for twenty-four hours. Then filter, add an excess of water of ammonia, wash the precipitate with cold water, and dry it.

A bluish powder, exceedingly irritating when the dust is inhaled, exciting violent sneezing, and inflammation of the mucous surfaces. All its salts are red.

Though not absolutely pure sanguinarine, it is the article sold and used under that name.

Dose: 1 to 2 grains. *J. U. Lloyd.*

MIXTURES OF BLOODROOT.

Take of Tincture of bloodroot .... 1 drachm.
Camphorated tincture of opium .... 2 drachms.
Syrup of squill .... 2 drachms.
Syrup of tolu .... 2 drachms.
Water .... to 2 ounces.

Mix. Dose: 1 drachm. Expectorant. *Hospital Formulary.*

Take of Tincture of bloodroot .... 1 drachm.
Tincture of lobelia .... 1 drachm.
Wine of ipecac .... 2 drachms.
Syrup of tolu .... ½ ounce.

Mix. Dose: 1 drachm every three hours, as an expectorant.

Bartholow.

SANTALUM (U. S., Br., Fr.)—RED SAUNDERS—RED SANDAL-WOOD.

The wood of Pterocarpus santalinus, L. (*Nat. ord., Leguminosæ*), a small tree indigenous to India.

Used as a coloring agent in tinctures, etc.

SANTALUM ALBUM (Fr.)—SANDAL-WOOD.

The wood of Santalum album, L. (*Nat. ord.*, *Santalaceæ*) a small tree indigenous to Southern India and the islands of the Eastern Archipelago.

Its most important constituent is a volatile oil, which has the odor of the wood, and possesses its medicinal virtues.

Dose: 1 to 2 drachms.

Oleum Santali—Oil of Sandal-wood.

Obtained from sandal-wood by distillation. Used as a substitute for copaiba in the treatment of gonorrhœa.

Dose: 20 to 40 minims. Generally administered in capsules.

SANTONICA (U. S. et al. Ph.)—LEVANT WORMSEED.

The unexpanded flowers of a variety of Artemisia maritima, L. (*Nat. ord.*, *Compositæ*), a low, shrubby plant indigenous to Asia.

Levant wormseed owe their medicinal efficacy to a peculiar, crystalline principle termed *santonin*, which is a very efficient anthelmintic for lumbrici.

Dose: 10 to 30 grains.

PREPARATIONS.

Santoninum (U. S. et al. Ph.)—Santonin.

Take of		
Take of	Santonica	48 ounces.
	Lime, recently slaked	18 ounces.
	Animal charcoal	sufficient.
	Diluted alcohol	sufficient.
	Acetic acid	sufficient.
	Alcohol	sufficient.

Digest the santonica and lime with 12 pints of diluted alcohol for twenty-four hours, and express. Repeat the process twice with the residue, mix the tinctures, reduce to 8 pints by distilling off the alcohol, filter, evaporate to one-half, gradually add acetic acid to slight excess, and set aside for forty-eight hours. Wash and dry the precipitate, then boil it with ten times its weight of alcohol, digest several times with animal charcoal, filter while hot, washing the charcoal with hot alcohol, then set aside in a dark place to crystallize. Lastly, dry the crystals on bibulous paper in the dark.

It is in colorless crystals, nearly tasteless, and but very slightly soluble in cold water.

Dose: 3 to 6 grains.

Trochisci Santonini (U. S., Fr., Ger.)—Troches of Santonin.

Take of Santonin	¼ ounce.
Sugar	18 ounces.
Tragacanth	½ ounce.
Orange flower water	sufficient.

Rub the powders together, then with the orange flower water form a mass, to be divided into 480 troches.

SAPO (U. S. et al. Ph.)—SOAP.

Soaps are formed by combining oils or fats with alkalies. The only one recognized by the United States Pharmacopœia is that made with soda and olive oil, the white castile soap of commerce, while not only this, but several other varieties are officinal in the European Pharmacopœias.

Soap is mildly laxative and antacid, but is seldom used alone for these purposes. It is, however, often combined with resinous substances in pills, increasing their action by hastening their solution. Externally it is employed in plasters and liniments. Here it acts by softening the epidermis, and thus favoring the absorption of the substances with which it is combined.

Dose: 5 to 30 grains.

PREPARATIONS.

Ceratum Saponis (U. S.)—Soap Cerate.

Take of Soap plaster	2 ounces.
Yellow wax	2½ ounces.
Olive oil (by weight)	4 ounces.

Melt together the plaster and wax, add the oil, and, after continuing the heat a few minutes, stir until cool.

Emplastrum Saponis (U. S. et al. Ph.)—Soap Plaster.

Take of Soap, sliced	4 ounces.
Lead plaster	36 ounces.
Water	sufficient.

Rub the soap with water until brought to a semi-liquid state, then add the plaster, previously melted, and boil to a proper consistence.

Linimentum Saponis (U. S. et al. Ph.)—Soap Liniment.

Take of Soap, in shavings	4 ounces.
Camphor	2 ounces.
Oil of rosemary	½ ounce.
Water	6 ounces.
Alcohol	2 pints.

Digest the soap in the water until dissolved; dissolve the camphor and oil in the alcohol, mix the two solutions, and filter.

Pilula Saponis Composita (U. S., Br.)—Compound Pill of Soap.

Take of Opium, in fine powder .... 60 grains.
Soap, in fine powder .... ½ ounce.

Beat together with water so as to form a pilular mass.

Dose: 3 to 5 grains. See also *Pilulæ Opii.*

Curd Soap (Br.).

A soap made with soda and a purified animal fat, consisting principally of stearin.

Medicinal Soap (Ger.).

Take of Solution of caustic soda .... 60 parts.
Olive oil ....100 parts.

Digest until a soap is formed, then dissolve it in 300 parts of distilled water, and add a solution of 25 parts of common salt in 75 parts of distilled water. Boil until the soap has separated from the liquid portion, cool, wash with distilled water, dissolve again in 60 parts of hot distilled water, and pour into moulds.

Soft Soap (Br., Ger.)—Green Soap.

Soap made with olive oil and potash. It is yellowish green, inodorous, of a gelatinous consistence (British); a lubricious, yellowish green mass, of a nauseous smell (German).

Ammoniated Soap Liniment (Ger.).

Take of Common hard soap .... 1 part.
Water .... 30 parts.
Alcohol .... 10 parts.
Water of ammonia .... 15 parts.

Digest the soap in the water and alcohol until dissolved, then add the water of ammonia, and mix.

Soap Cerate Plaster (Br.)

Take of Hard soap .... 10 ounces.
Yellow wax ....12½ ounces.
Olive oil .... 20 ounces.
Oxide of lead .... 15 ounces.
Vinegar .... 1 gallon.

Boil the oxide and vinegar together on a water-bath, until the oxide has combined with the acid, then add the soap, and boil until the moisture has evaporated. Finally, add the wax and oil melted together, and heat, stirring constantly, until reduced to the proper consistence for a plaster.

Spirit of Soap (Ger.).

Take of Castile soap	1 part.
Alcohol	3 parts.
Rose water	2 parts.

Dissolve by digestion with a gentle heat, and filter.

Lotions of Soap.

Take of Soft soap	1 ounce.
Boiling water	16 ounces.

Mix, and perfume to taste.

Used in second stage of eczema, to counteract the infiltration.

Tilbury Fox.

Take of Soft soap	1 ounce.
Alcohol	1 ounce.
Oil of cade	1 ounce.
Oil of lavender	$1\frac{1}{2}$ drachm.

Mix. Used like the preceding, in eczema. *McCall Anderson.*

SARSAPARILLA (U. S. et al. Ph.)—SARSAPARILLA.

The root of Smilax officinalis, H. B. K. and of other species (*Nat. ord., Smilaceæ*), climbing plants of the West Indies and Central America.

Sarsaparilla was formerly held in high repute as a tonic and alterative, and was largely used in scrofula, syphilis, etc., but it is not much esteemed at present.

Dose: 30 to 60 grains.

PREPARATIONS.

Decoctum Sarsaparillæ Compositum (U. S., Br.)—Compound Decoction of Sarsaparilla.

Take of Sarsaparilla	6 ounces.
Bark of sassafras root	1 ounce.
Guaiacum wood	1 ounce.
Liquorice root	1 ounce.
Mezereon	3 drachms.
Water	sufficient.

Boil in 4 pints of water for fifteen minutes, then digest for two hours in a covered vessel at about 200°, strain, and add sufficient water through the strainer to make 4 pints.

Dose: 2 to 4 ounces.

The German Pharmacopœia has two compound decoctions: *Zittmann's Stronger* and *Milder;* but as their formulæ are long, while the products are not much used in this country, they are omitted.

Extractum Sarsaparillæ Compositum Fluidum (U. S.)—Compound Fluid Extract of Sarsaparilla.

Take of	Sarsaparilla	16 ounces.
	Liquorice root	2 ounces.
	Sassafras	2 ounces.
	Mezereon	6 drachms.
	Glycerin	½ pint.
	Alcohol	sufficient.
	Water	sufficient.

Mix ½ pint of alcohol with 4 ounces each of glycerin and water, moisten the powders with 6 ounces of the mixture, and proceed according to the general formula, page 161. Continue the percolation with diluted alcohol until 2 pints are obtained, reserve the first 12 ounces, add 4 ounces of glycerin to the remainder, evaporate to 6 ounces, and mix with the reserved portion.

Dose: 30 to 60 minims.

Extractum Sarsaparillæ Fluidum (U. S., Br.)—Fluid Extract of Sarsaparilla.

Take of	Sarsaparilla	16 ounces.
	Glycerin	½ pint.
	Alcohol	sufficient.
	Water	sufficient.

Proceed as in the preceding preparation until 26 ounces are obtained; reserve the first 10 ounces, add 4 ounces of glycerin to the remainder, evaporate to 6 ounces, and mix with the reserved portion.

Dose: 30 to 60 minims.

Syrupus Sarsaparillæ Compositus (U. S., Fr., Ger.)—Compound Syrup of Sarsaparilla.

Take of	Sarsaparilla	24 ounces.
	Guaiacum wood	3 ounces.
	Pale rose	2 ounces.
	Senna	2 ounces.
	Liquorice root	2 ounces.
	Oil of sassafras	5 minims.
	Oil of anise	5 minims.
	Oil of gaultheria	3 minims.
	Sugar	96 ounces.
	Water	1 pint.
	Diluted alcohol	sufficient.

Macerate the solid ingredients, except the sugar, in 3 parts of diluted alcohol for four days, then percolate until 6 pints of tincture are obtained. Evaporate to 3 pints, add the water, filter, then add the sugar; dissolve with the aid of heat, and strain while hot. Lastly, rub the oils with a small portion of the syrup, then mix with the remainder.

Dose: 1 to 4 drachms.

SASSAFRAS (U. S. et al. Ph.)—SASSAFRAS.

The bark of the root of Sassafras officinale, Nees (*Nat. ord., Lauraceæ*), a tree indigenous to North America.

It is an agreeable aromatic, and is used for flavoring. Its volatile oil (*Oleum Sassafras*) is commonly employed instead of the bark.

The pith of the young branches (*Sassafras Medulla, U. S.*), abounds in mucilage, which is highly esteemed as a demulcent.

PREPARATION.

Mucilago Sassafras Medullæ (U. S.)—Mucilage of Sassafras Pith.

Take of Sassafras pith	120 grains.
Water	1 pint.

Macerate three hours, and strain.

SCAMMONIUM (U. S. et al. Ph.)—SCAMMONY.

A resinous exudation from the root of Convolvulus scammonium, L. (*Nat. ord., Convolvulaceæ*), a twining plant indigenous to Syria, Asia Minor, and Greece.

Scammony is a very active hydragogue cathartic, but as it acts with great violence, it is seldom used except in combination with other drugs which modify its action. It owes its virtues to a resin (*Resina Scammonii*), which is generally used instead of the crude drug. It is one of the constituents of compound extract of colocynth.

Dose: 5 to 15 grains.

PREPARATIONS.

Resina Scammonii (U. S. et al. Ph.)—Resin of Scammony.

Take of Scammony	6 ounces.
Alcohol	sufficient.
Water	sufficient.

Digest the scammony with successive portions of boiling alcohol until exhausted. Mix the tinctures, reduce to a syrupy consistence by distilling off the alcohol, add the residue to 1 pint of water, collect, wash and dry the precipitate.

Dose: 4 to 8 grains.

The British and German Pharmacopœias prepare this resin from scammony root, though the former permits its preparation from scammony also.

22

Confection of Scammony (Br.).

Take of Scammony	3 ounces.
Ginger	1½ ounce.
Oil of caraway	1 drachm.
Oil of cloves	½ drachm.
Syrup	3 ounces.
Clarified honey (by weight)	1½ ounce.

Rub the powders with the syrup and the honey, then add the oil, and mix.

Dose : 10 to 30 grains.

Scammony Mixture (Br.).

Take of Resin of scammony	4 grains.
Milk	2 ounces.

Triturate the scammony with the milk, added gradually.

Dose : ½ to 2 ounces, for a child.

Compound Pill of Scammony (Br.).

Take of Resin of scammony	1 ounce.
Resin of jalap	1 ounce.
Curd soap	1 ounce.
Strong tincture of ginger	1 ounce.
Rectified spirit	2 ounces.

Mix, and dissolve with a gentle heat; then evaporate on a water-bath to a proper consistence.

Dose: 2 to 15 grains.

Compound Powder of Scammony (Br.).

Take of Scammony	4 ounces.
Jalap	3 ounces.
Ginger	1 ounce.

Mix thoroughly, pass through a fine sieve, and finally rub lightly in a mortar.

Dose : 10 to 20 grains.

Tincture of Scammony (Fr.).

Take of Scammony	1 part.
Alcohol (80%)	5 parts.

Macerate ten days, agitating occasionally, and filter.

Dose : ½ to 1 drachm.

SCILLA (U. S. et al. Ph.)—SQUILL.

The bulb of Scilla maratima, L. (*Nat. ord.*, *Liliaceæ*), a small plant indigenous to Southern Europe and the adjacent parts of Asia. It is usually sliced and dried, and occurs in pieces which are brittle when dry, but flexible when damp. It has a bitter, nauseous taste, with some acridity, and but little odor.

Squill is diuretic and expectorant, and in overdoses acts as a violent emetic. It is used in chronic and subacute bronchial affections, cardiac dropsy, etc.

Dose: 1 to 12 grains.

PREPARATIONS.

Acetum Scillæ (U. S. et al. Ph.)—Vinegar of Squill.

Take of Squill	4 ounces.
Diluted acetic acid	sufficient.

Moisten the squill with 8 ounces of the acid, and, after it has ceased to swell, pack, and percolate until the filtered liquid measures 2 pints.

It may also be prepared by macerating the squill in 2 pints of diluted acetic acid for seven days, expressing, and filtering.

Dose: 15 to 30 minims.

Extractum Scillæ Fluidum (U. S.)—Fluid Extract of Squill.

Take of Squill	16 ounces.
Glycerin	2 ounces.
Alcohol	sufficient.
Water	sufficient.

Mix the glycerin with 14 ounces of alcohol, moisten the squill with 4 ounces of the mixture, and proceed according to the general formula, page 161. Finish the percolation with a menstruum of 2 parts of alcohol and 1 part of water.

Dose: 1 to 12 minims.

Pilulæ Scillæ Compositæ (U. S., Br.)—Compound Pills of Squill.

Take of Squill	12 grains.
Ginger	24 grains.
Ammoniac	24 grains.
Soap	36 grains.
Syrup	sufficient.

Mix the powders, beat into a mass with syrup, and divide into 24 pills.

Dose: 1 to 3 pills.

Syrupus Scillæ (U. S., Br.)—Syrup of Squill.

Take of	Vinegar of squill	1 pint.
	Sugar	24 ounces.

Dissolve with the aid of heat, and strain while hot.
Dose: ½ to 1 drachm.

Syrupus Scillæ Compositus (U. S.)—Compound Syrup of Squill.

Take of	Squill	4 ounces.
	Seneka	4 ounces.
	Tartrate of antimony and potassium	48 grains.
	Sugar	42 ounces.
	Diluted alcohol	sufficient.
	Water	sufficient.

Macerate the squill and seneka in ½ pint of diluted alcohol for four days, then percolate until 1 pint is obtained. Boil a few minutes, evaporate on a water-bath to 2 pints, add 14 ounces of boiling water, and filter. Dissolve the sugar in the solution, heat to the boiling-point, strain, add the tartrate, and pour sufficient water through the strainer to make the syrup measure 3 pints.
Dose: 10 to 60 minims.

Tinctura Scillæ (U. S. et al. Ph.)—Tincture of Squill.

Take of	Squill	4 ounces.
	Diluted alcohol	sufficient.

Moisten, pack, and percolate to 2 pints.
Dose: 10 to 20 minims.

Oxymel of Squill (Br., Fr., Ger.).

Take of	Vinegar of squill	20 ounces.
	Clarified honey	2 pounds.

Mix, and evaporate on a water-bath until the product has the sp. gr. 1.32.
Dose: ½ to 1 drachm.

Mixtures of Squill.

Take of	Vinegar of squill	½ ounce.
	Fluid extract of ipecac	½ drachm.
	Deodorized tincture of opium	1 drachm.
	Syrup of tolu	10 drachms.

Mix. Dose: 1 drachm every two, three, or four hours. In bronchial catarrh. *Bartholow.*

Take of Vinegar of squill	2 drachms.
Tincture of digitalis	30 drops.
Water	4 ounces.

Mix. Dose: 1 drachm three or four times a day, for children two years old. In pleurisy with effusion. *Meigs and Pepper.*

PILLS OF SQUILL.

Take of Squill	6 grains.
Ipecac	6 grains.
Extract of hyoscyamus	3 grains.
Sulphate of morphia	½ to 1 grain.

Mix, and divide into 12 pills.

Dose: 1 pill every four hours. In bronchial catarrh. *Bartholow.*

SENEGA (U. S. et al. Ph.)—SENEKA—SENEGA SNAKEROOT.

The root of Polygala Senega, L. (*Nat. ord.*, *Polygalaceæ*), a small herb indigenous to the United States. When fresh it has a peculiar, disagreeable odor, and a taste which is at first sweetish, but afterward acrid and irritating.

Senega is a stimulating expectorant, and, in large doses, an emeto-cathartic. It is used chiefly in chronic bronchial affections, but occasionally as an emmenagogue, and, in large doses, in acute rheumatism.

Dose: 5 to 20 grains.

PREPARATIONS.

Decoctum Senegæ (U. S.)—Decoction of Seneka.

Take of Seneka, bruised	1 ounce.
Water	sufficient.

Boil fifteen minutes, strain, and add sufficient water through the strainer to make 1 pint.

Dose: 1 to 2 ounces.

Very similar to *Infusion of Seneka, Br.*

Extractum Senegæ (U. S., Fr., Ger.)—Extract of Seneka.

Take of Seneka	12 ounces.
Diluted alcohol	sufficient.

Moisten, pack, and percolate until 3 pints of tincture are obtained. Evaporate this on a water-bath to the proper consistence.

Dose: 1 to 3 grains.

Extractum Senegæ Fluidum (U. S.)—Fluid Extract of Seneka.

Take of Seneka	16 ounces.
Glycerin	4 ounces.
Alcohol	sufficient.
Water	sufficient.

Mix 8 ounces of alcohol, 3 of glycerin, and 5 of water, moisten the seneka with 4 ounces of the mixture, and proceed according to the general formula, page 161. Finish the percolation with diluted alcohol, reserve 14 ounces, and add 1 ounce of glycerin to the remainder, before evaporation.

Dose: 5 to 20 minims.

Syrupus Senegæ (U. S., Ger.)—Syrup of Seneka.

Take of Seneka	4 ounces.
Sugar	15 ounces.
Diluted alcohol	2 pints.

Percolate the seneka with the diluted alcohol, evaporate the tincture on a water-bath, at or below 160°, to ½ pint; then filter, add the sugar, dissolve with a gentle heat, and strain while hot.

Dose: 1 to 2 drachms.

Tincture of Seneka (Br.).

Take of Seneka	2½ ounces.
Proof spirit	20 ounces.

Macerate the seneka forty-eight hours in 15 ounces of the spirit, then percolate with the remainder, express, filter, and add sufficient proof spirit to make 20 ounces.

Dose: ½ to 2 drachms.

Mixture of Seneka.

Take of Syrup of seneka	2 drachms.
Solution of acetate of ammonia	4 drachms.
Syrup of wild cherry	1 ounce.

Mix. Dose: 1 drachm every 3 hours. In pneumonia of children.

H. S. Dessau.

SENNA (U. S. et al. Ph.)—SENNA.

The senna of commerce is furnished by two species of Cassia, C. acutifolia, Delile, and C. angustifolia, Vahl (*Nat. ord., Leguminosæ*), both of which are shrubs indigenous to Northern Africa.

Senna is a prompt and efficient cathartic, operating, however, with some griping, which is generally counteracted by the use of aromatics, or neutral salts in combination with it.

Dose: ½ to 2 drachms.

PREPARATIONS.

Confectio Sennæ (U. S. et al. Ph.)—Confection of Senna.

Take of		
Take of	Senna	8 ounces.
	Coriander	4 ounces.
	Purging cassia	16 ounces.
	Tamarind	10 ounces.
	Prune	7 ounces.
	Fig	12 ounces.
	Sugar	30 ounces.
	Water	sufficient.

Reduce the cassia, tamarind, prune, and fig to a pulpy liquid by digestion with water on a water-bath, rub through a seive, dissolve the sugar in it by a gentle heat, and evaporate until it weighs 84 ounces. Lastly, add the senna and coriander, and mix.

Dose: 1 to 3 drachms.

Extractum Sennæ Fluidum (U. S.)—Fluid Extract of Senna.

Take of		
Take of	Senna	16 ounces.
	Glycerin	½ pint.
	Alcohol	sufficient.
	Water	sufficient.

Mix 8 ounces of alcohol with 4 each of glycerin and water, moisten the senna with 8 ounces of the mixture, and proceed according to the general formula, page 161. Continue the percolation with diluted alcohol until 26 ounces are obtained, reserve the first 10 ounces, mix 4 ounces of glycerin with the remainder, evaporate to 6 ounces, and mix with the reserved portion.

Dose: ½ to 2 drachms.

See also *Extractum Spigeliæ et Sennæ Fluidum*.

Infusum Sennæ (U. S., Br.)—Infusion of Senna.

Take of		
Take of	Senna	1 ounce.
	Coriander	1 drachm.
	Boiling water	1 pint.

Macerate an hour in a covered vessel, and strain.

Dose: 1 to 4 ounces.

The British Pharmacopœia directs: senna, 1 ounce; ginger, 30 grains; boiling distilled water, 10 ounces.

Compound Infusion of Senna (Ger.).

Take of		
Take of	Senna	2 parts.
	Tartrate of potassium and sodium	2 parts.
	Manna	3 parts.
	Boiling water	12 parts.

Infuse the senna in the water for five minutes, express, dissolve the salt and manna in the infusion, and strain.

Extract of Senna (Fr.).

Take of Senna	1 part.
Boiling distilled water	8 parts.

Infuse the senna in 6 parts of the water for 12 hours, and express. Repeat the operation with the remainder of the water, evaporate the infusions separately to a syrupy consistence, then mix, and evaporate to a soft extract.

Dose: 10 to 30 grains.

Compound Mixture of Senna (Br.).

Take of Sulphate of magnesium	4 ounces.
Extract of liquorice	½ ounce.
Tincture of senna	2½ ounces.
Compound tincture of cardamoms	10 drachms.
Infusion of senna	sufficient.

Dissolve the sulphate and the extract in 14 ounces of infusion of senna, with a gentle heat, add the tinctures, and sufficient infusion of senna to make 20 ounces.

Dose: 1 to 1½ ounce.

Syrup of Senna (Br.).

Take of Senna	16 ounces.
Oil of coriander	3 minims.
Refined sugar	24 ounces.
Distilled water	100 ounces, or sufficient.
Rectified spirits	2 ounces.

Digest the senna in 70 ounces of the water for twenty-four hours at 120°, express and strain. Digest the marc in 30 ounces of water for six hours, express and strain. Evaporate the mixed liquids on a water-bath to 10 ounces, and, when cold, add the spirit, previously mixed with the oil. Filter, adding sufficient water through the filter to make 16 ounces, then add the sugar, and dissolve with a gentle heat.

Dose: 1 to 4 drachms.

Tincture of Senna (Br., Fr.).

Take of Senna	2½ ounces.
Raisins, freed from seeds	2 ounces.
Caraway	½ ounce.
Coriander	½ ounce.
Proof spirit	20 ounces.

Obtain, by maceration and percolation, 20 ounces.

Dose: 1 to 4 drachms. See also *Tinctura Rhei et Sennæ*.

The French Codex obtains, by percolation, 5 parts of tincture from 1 part of senna.

MIXTURES OF SENNA.

Take of Senna	2 drachms.
Coffee	1 drachm.
Hot milk	3 ounces.
Boiling water	3 ounces.

Infuse. Dose: The whole may be drunk after twelve hours.

Bartholow.

Take of Senna	½ ounce.
Sulphate of magnesium	1 ounce.
Manna	1 ounce.
Fennel	1 drachm.
Boiling water	½ pint.

Macerate in a covered vessel till cool.

Dose: ⅓ of the mixture every four or five hours till it operates. Known as *Black Draught.* *Wood.*

SERPENTARIA (U. S. et al. Ph.)—VIRGINIA SNAKEROOT.

The root of Aristolochia Serpentaria, L. (*Nat. ord., Aristolochiaceæ*), an herbaceous perennial indigenous to the United States, growing most abundantly near the Alleghany Mountains.

It is diaphoretic, diuretic, tonic, and stimulant. The infusion, taken warm, is used as a diaphoretic in acute febrile diseases.

Dose: 15 to 30 grains.

PREPARATIONS.

Extractum Serpentariæ Fluidum (U. S.)—Fluid Extract of Serpentaria.

Take of Serpentaria	16 ounces.
Alcohol	sufficient.

Moisten the serpentaria with 4 ounces of the alcohol, and proceed according to general formula, page 161.

Dose: 15 to 30 minims.

Infusum Serpentariæ (U. S., Br.)—Infusion of Serpentaria.

Take of Serpentaria	½ ounce.
Water	sufficient.

Moisten, pack, and percolate to 1 pint.

Dose: 1 to 2 ounces.

Tinctura Serpentariæ (U. S., Br.)—Tincture of Serpentaria.

Take of Serpentaria	4 ounces.
Diluted alcohol	sufficient.

Moisten, pack, and percolate to 2 pints.

Dose: $\frac{1}{2}$ to 2 ounces.

SEVUM (U. S. et al. Ph.)—SUET.

The prepared fat of the sheep (*Ovis Aries*, L.) Mutton suet is a white, solid, nearly odorless fat. It is used as a dressing for ulcers and excoriated surfaces, and as an ingredient of ointments, cerates, etc.

SINAPIS (U. S. et al Ph.)—MUSTARD.

The seed of Sinapis alba, L., and of Sinapis nigra, L. (*Nat. ord., Cruciferæ*), white, and black mustard, respectively, both of which are indigenous to the Old World, but are cultivated and have become naturalized here.

Both kinds of mustard seed yield, upon pressure, a fixed oil. The black seeds, by distillation, yield a volatile oil, of an acrid, irritating character; the white seeds yield no volatile oil, but, treated with water, a white crystalline principle is obtained, which corresponds to the volatile oil obtained from the black seeds.

Mustard is an acrid stimulant and irritant. Internally it is used as an emetic, and, in small doses, as a stomachic. Externally it is used as a rubefacient.

Dose: $\frac{1}{2}$ to 2 drachms.

PREPARATIONS.

Charta Sinapis (U. S., Br.)—Mustard Paper.

Take of Black mustard	90 grains.
Solution of gutta-percha	sufficient.

Mix the mustard with sufficient of the solution to give it a semi-liquid consistence, spread upon pieces of paper four inches square, and allow the surface to dry. Before applying to the skin, dip for fifteen seconds in warm water.

MUSTARD POULTICE (Br.).

Take of Mustard	$2\frac{1}{2}$ ounces.
Linseed meal	$2\frac{1}{2}$ ounces.
Boiling water	10 ounces.

Mix the linseed meal with the water, then add the mustard, with constant stirring.

COMPOUND LINIMENT OF MUSTARD (Br.).

Take of Oil of mustard	1 drachm.
Ethereal extract of mezereon	40 grains.
Camphor	120 grains.
Castor oil	5 drachms.
Rectified spirit	4 drachms.

Dissolve the extract and camphor in the spirit, and add the oils.

OIL OF MUSTARD (Br., Ger.).

The oil distilled from the seeds of black mustard, after expression of the fixed oil.

Used externally as a rubefacient.

EXPRESSED OIL OF MUSTARD.

The oil expressed from the seeds of both the officinal species of mustard.

Used as a vehicle, etc.

SODIUM—SODIUM.

SODA (U. S., Br.)—SODA—CAUSTIC SODA.

Take of solution of soda a convenient quantity. Evaporate rapidly in an iron vessel until ebullition ceases and the soda melts. Pour this on a flat stone, and when it has congealed, break in pieces, and preserve in a well-stopped bottle.

Soda is very soluble in water and in alcohol. Exposed to the air it first becomes moist, and afterward effloresces. It is a powerful alkali, and is occasionally used as an escharotic.

For internal use the officinal solution is generally employed.

PREPARATIONS.

Liquor Sodæ (U. S. et al. Ph.)—Solution of Soda.

Take of Carbonate of sodium	26 ounces.
Lime	8 ounces.
Distilled water	sufficient.

Dissolve the carbonate in 3½ pints of distilled water, and heat to the boiling point. Mix the lime with 3 pints of distilled water, heat to the boiling point, add the solution of the carbonate, strain, and add sufficient distilled water through the strainer to make 6 pints.

It is a colorless liquid, of an extremely acrid taste, and a strong, alkaline reaction. Used in preparations, and occasionally as an antacid.

Dose: 5 to 20 minims.

Liquor Sodæ Chlorinatæ (U. S. et al. Ph.)—Solution of Chlorinated Soda.

Take of Chlorinated lime........................... 12 ounces.
Carbonate of sodium....................... 24 ounces.
Water.................................. 12 pints.

Dissolve the carbonate in 3 pints of the water, and mix the chlorinated lime thoroughly with the remainder. Let the latter mixture stand twenty-four hours, then decant the clear liquid, place the residue on a strainer, and allow it to drain until enough liquid has passed to make, with that decanted, 8 pints. Mix this with the solution of the carbonate, and strain, adding water, if necessary, to make 11½ pints.

It is a greenish yellow, transparent liquid, having a slight odor of chlorine, and a sharp, saline taste. Used internally as an antiseptic in low fevers, etc., and topically as a disinfectant application to foul ulcers and wounds, in diphtheria, scarlatina, etc.

The British preparation is made by passing chlorine through a solution of carbonate of sodium.

Dose: 30 to 60 minims.

PREPARATIONS.

Chlorine Poultice (Br.).

Take of Solution of chlorinated soda................ 2 ounces.
Linseed meal.............................. 4 ounces.
Boiling water.............................. 8 ounces.

Mix the linseed meal with the water, then add the solution of chlorinated soda, with constant stirring.

Gargle of Chlorinated Soda.

Take of Solution of chlorinated soda................. 24 minims.
Water.................................to 1 ounce.

Mix. Disinfectant. Very useful in sloughing phagedæna, and putrid conditions of the throat. *London Throat Hospital.*

Lotion of Chlorinated Soda.

Take of Solution of chlorinated soda .... ½ ounce.
Water .... 3½ ounces.

Mix. *N. Y. Dispensary.*

SODII ACETAS (U. S. et al. Ph.)—ACETATE OF SODIUM.

Prepared by neutralizing acetic acid with carbonate or bicarbonate of sodium.

Used as an alkaline diuretic for the same purposes as acetate of potassium, but is milder in its action, and less apt to derange the stomach.

Dose: 20 grains to 2 drachms.

SODII ARSENIAS (U. S., Br., Fr.)—ARSENIATE OF SODIUM.

Take of Arsenious acid............................ 2 ounces.
Nitrate of sodium ....816 grains.
Dried carbonate of sodium ....528 grains.
Distilled water, boiling hot................. ½ pint.

Mix the powders, put the mixture into a large clay crucible, cover with a lid, and expose to a red heat until fusion has taken place. Then pour on a porcelain slab, and when solidified, and while still warm, dissolve in the water, filter, and set aside to crystallize. Drain the crystals, dry them rapidly on filtering paper, and keep in a well-stopped bottle.

It is in colorless, transparent crystals, slightly efflorescent, and soluble in water. Produces the therapeutic effects of arsenic, but is not much used.

Dose: $\frac{1}{16}$ to $\frac{1}{8}$ grain.

PREPARATIONS.

Liquor Sodii Arseniatis (U. S., Br., Fr.)—Solution of Arseniate of Sodium.

Take of Arseniate of sodium, dried at 300°........... 64 grains.
Distilled water............................ 1 pint.

Dissolve.

Dose: 3 to 10 minims.

Pills of Arseniate of Sodium.

Take of Arseniate of sodium .... 2 grains.
Extract of hops........................... 20 grains.
Sulphate of iron.......................... 20 grains.
Extract of nux vomica .... 3 grains.

Mix, and divide into 24 pills.

Dose: 1 pill three times a day. In chronic eczema, and psoriasis.

Tilbury Fox.

Take of Arseniate of sodium....................... $\frac{1}{16}$ grain.
Sugar of milk............................ 1 grain.
Extract of hops.......................... 3 grains.

Make 1 pill. *London Ophthalmic Hospital.*

SODII BENZOAS—BENZOATE OF SODIUM.

Prepared by neutralizing solution of soda with benzoic acid, evaporating, and crystallizing.

It is in efflorescent, crystalline masses, soluble in water. Its effects are similar to those of benzoic acid.

Dose: 10 to 30 grains.

Inhalation of Benzoate of Sodium.

Take of Benzoate of sodium	20 grains.
Water	1 ounce.

Dissolve. Antiseptic. Used by means of an atomizer.

London Throat Hospital.

SODII BICARBONAS (U. S. et al. Ph.)—BICARBONATE OF SODIUM.

Take of Commercial bicarbonate of sodium	64 ounces.
Distilled water	6 pints.

Introduce the bicarbonate into a percolator, cover with wet muslin, and pour the water gradually upon it. When the water has ceased to drop, remove the salt from the percolator, and dry it on bibulous paper.

Commercial bicarbonate of sodium is prepared by passing carbonic acid through a solution of carbonate of sodium.

Bicarbonate of sodium is a white, opaque powder, wholly soluble in water. It is the least irritating of the alkaline carbonates, and is frequently employed as an antacid in dyspepsia, in gout, rheumatism, etc.

Dose: 10 to 60 grains.

PREPARATIONS.

Pulveres Effervescentes (U. S., Fr., Ger.)—Effervescing Powders.

Take of Bicarbonate of sodium	360 grains.
Tartaric acid	300 grains.

Divide each into 12 equal parts, and keep the parts of the bicarbonate, and those of the acid, in papers of different colors.

A powder of each kind is dissolved in 3 or 4 ounces of water, the solutions mixed, and administered while in a state of effervescence.

Trochisci Sodii Bicarbonatis (U. S., Br., Fr.)—Troches of Bicarbonate of Sodium.

Take of Bicarbonate of sodium	3 ounces.
Sugar	9 ounces.
Nutmeg	60 grains.
Mucilage of tragacanth	sufficient.

Rub together the bicarbonate, sugar, and nutmeg, then with the mucilage form a mass, to be divided into 480 troches.

Dose: 1 to 6 troches.

Effervescent Citro-Tartrate of Soda (Br.).

Take of Bicarbonate of sodium.................... 17 ounces.
Tartaric acid.............................. 8 ounces.
Citric acid................................ 6 ounces.

Mix thoroughly, place in a dish heated to between 200° and 220°, and when particles begin to aggregate, stir until they assume a granular form.

Dose: 60 grains to ¼ ounce.

Effervescing Solution of Soda (Br.).

Take of Bicarbonate of sodium.................... 30 grains.
Water................................... 20 ounces.

Dissolve and filter, then pass into the solution as much carbonic acid as can be introduced with a pressure of seven atmospheres. Keep in bottles securely closed.

Lotion of Bicarbonate of Sodium.

Take of Bicarbonate of sodium.................... 1 drachm.
Glycerin................................. 1½ drachm.
Elder-flower water........................ 6½ ounces.

Mix. Used in eczema, lichen, and urticaria, to allay itching.

Tilbury Fox.

Mixtures of Bicarbonate of Sodium.

Take of Bicarbonate of sodium.................... 1 ounce.
Tincture of ginger......................... 2 drachms.
Compound tincture of gentian............... 1 ounce.
Water.................................... 5 ounces.

Mix. Dose: 2 drachms. *Hospital Formulary.*

Take of Bicarbonate of sodium.................... 1 ounce.
Compound tincture of cardamom............ 2 ounces.
Compound tincture of gentian.............. 2 ounces.
Peppermint water......................... 3 pints.

Mix. *N. Y. Dispensary.*

Powders of Bicarbonate of Sodium.

Take of Bicarbonate of sodium.................... 2½ grains.
Calomel.................................. ½ grain.
Aromatic chalk powder..................... 5 grains.

Mix. One dose. *Guy's Hospital.*

Take of Bicarbonate of sodium.................... 1 drachm.
Opium.................................... 1 grain.
Ipecac................................... 1 grain.

Mix, and divide into 20 powders.

Dose: 1 powder every three or four hours.

Hospital for Ruptured and Crippled.

SODII BORAS (U. S. et al. Ph.)—BORATE OF SODIUM—BORAX.

Borax occurs in the waters of certain lakes in Thibet and Persia, and also of one in California. It is also prepared from certain minerals, as boracite, etc.

Borax occurs in colorless crystals, of a mildly alkaline taste, and soluble in water. It is mildly refrigerant and diuretic, and is occasionally used in rheumatic and gouty affections. Its chief use, however, is as a topical application in aphthæ, sore nipples, leucorrhœa, skin diseases, etc.

Dose: 5 to 40 grains.

PREPARATIONS.

Glyceritum Sodii Boratis (U. S., Br.)—Glycerite of Borate of Sodium.

Take of Borate of sodium	2 ounces.
Glycerin	½ pint.

Rub together until the borate of sodium is dissolved.

Mel Sodii Boratis (U. S., Br.)—Honey of Borate of Sodium.

Take of Borate of sodium	60 grains.
Clarified honey (by weight)	1 ounce.

Mix.

Gargles of Borax.

Take of Borax	2 drachms.
Yeast	½ ounce.
Honey	½ ounce.
Water	to 8 ounces.

Mix. *Bell's Gargle—Hospital Formulary.*

Lotions of Borax.

Take of Borax	1 drachm.
Hydrocyanic acid	2 drachms.
Rose water	8 ounces.

Mix. Used in pruritus of old people. *Neligan.*

Take of Borax	½ ounce.
Sulphate of morphia	6 grains.
Rose water	8 ounces.

Mix. Used in pruritus vulvæ. *C. D. Meigs.*

Take of Borax	40 grains.
Oxide of zinc	1 drachm.
Solution of subacetate of lead	2 drachms.
Lime water	6 to 8 ounces.

Mix. Used in eczema and herpes. *Tilbury Fox.*

OINTMENT OF BORAX.

Take of	Borax	1 to 2 drachms.
	Glycerin	1 drachm.
	Lard	1 ounce.

Mix. Used in eczema, erythema, intertrigo, lichen, and parasitic diseases. *Tilbury Fox.*

SODII BORO-BENZOAS—BORO-BENZOATE OF SODIUM.

Take of	Borate of sodium	3 ounces.
	Benzoate of sodium	4 ounces.
	Water	sufficient.

Dissolve the borate and the benzoate in the water, and evaporate slowly, stirring toward the end of the process so as to obtain a granular salt.

"This mixture was proposed by the writer two years ago, as a means of combining the stimulant and antizymotic powers of borax and benzoic acid." *W. M. Chamberlain.*

LOTION OF BORO-BENZOATE OF SODIUM.

Take of	Boro-benzoate of sodium	1 ounce.
	Fluid extract of hydrastis	1 ounce.
	Glycerin	1 drachm.
	Carbolic acid	20 minims.
	Camphor water	6 ounces.
	Water	6 ounces.

Mix. Used in naso-pharyngeal catarrh. After cleansing the parts, apply the lotion, either by means of the syringe or the nasal spray-bulb, three times daily. Under its use the secretion diminishes, and the swelling and œdema of the nasal membranes disappear.

"This formula has been employed by several competent observers, and the general report has been favorable." *W. M. Chamberlain.*

SODII BROMIDUM—BROMIDE OF SODIUM.

Prepared by subjecting iron filings, mixed with water, to the action of bromine, precipitating with solution of carbonate of sodium, filtering, and crystallizing.

Bromide of sodium occurs in colorless, transparent crystals, of a slightly alkaline taste, and freely soluble in water. Its effects are similar to those of bromide of potassium, and it is administered in the same manner.

Dose: 5 to 60 grains.

SODII CARBONAS (U. S. et al. Ph.)—CARBONATE OF SODIUM.

Carbonate of sodium exists in many mineral waters, in the ashes of many plants, and in numerous minerals. It occurs in commerce in large, colorless crystals, or crystalline masses, of a strong, alkaline taste, and freely soluble in water. It is less irritating than carbonate of potassium, and may be used for the same purposes, though the bicarbonate is generally preferred.

Dose: 10 to 60 grains.

PREPARATIONS.

Sodii Carbonas Exsiccata (U. S., Br., Ger.)—Dried Carbonate of Sodium.

Take of carbonate of sodium, a convenient quantity. Heat in an iron vessel, with constant stirring, until it is thoroughly dried.

Dose: 5 to 20 grains.

LOTION OF CARBONATE OF SODIUM.

Take of	Carbonate of sodium	½ ounce.
	Sulphate of morphia	6 grains.
	Elder-flower water	1 ounce.

Mix. Used in eczema, lichen, and urticaria, to allay itching.

Tilbury Fox.

SODII CHLORIDUM (U. S. et al. Ph.)—CHLORIDE OF SODIUM—COMMON SALT.

Salt occurs native as rock salt, and exists in large proportions in sea-water, and in the waters of most saline springs. Its chief use in medicine is as a topical application in catarrhal affections of the mucous membranes, etc.

INHALATION OF SALT.

Take of	Salt	40 grains.
	Water	8 ounces.

Dissolve. Stimulant. Used by means of a steam-atomizer.

G. M. Lefferts.

SODII HYPOPHOSPHIS (U. S., Br.)—HYPOPHOSPHITE OF SODIUM.

Prepared by precipitating a solution of hypophosphite of calcium with carbonate of sodium, filtering, and evaporating.

Hypophosphite of sodium occurs as a white, granular salt, of a bitter, saline taste, and freely soluble in water. It is used in chronic bronchitis, phthisis, nervous diseases, etc.

Dose: 5 to 10 grains. *See Compound Syrup of Hypophosphites.*

MIXTURE OF HYPOPHOSPHITE OF SODIUM.

Take of Hypophosphite of sodium	5 grains.
Spirit of chloroform	10 minims.
Syrup	½ drachm.
Camphor water	1 ounce.

Mix. One dose. *London Chest Hospital.*

SODII HYPOSULPHIS (U. S. et al. Ph.)—HYPOSULPHITE OF SODIUM.

Prepared by heating sulphur with dried carbonate of sodium, dissolving in water, adding sulphur, and boiling. The resulting solution is filtered, evaporated, and crystallized.

Hyposulphite of sodium is in large, colorless crystals, of a slightly alkaline, bitter, sulphurous taste, and freely soluble in water.

Owing to its power in arresting fermentation it is used in certain forms of dyspepsia, in zymotic diseases, and externally in cutaneous affections characterized by vegetable parasites, etc.

Dose: 5 to 20 grains.

LOTION OF HYPOSULPHITE OF SODIUM.

Take of Hyposulphite of sodium	1 ounce.
Glycerin	1 ounce.
Water	3 ounces.

Mix. Used in pruritus vaginæ. *Tilbury Fox.*

SODII IODIDUM—IODIDE OF SODIUM.

Prepared by decomposing a solution of iodide of iron with carbonate of sodium, filtering, evaporating, and crystallizing.

Iodide of sodium occurs in colorless crystals, of an acrid, saline taste, and freely soluble in water. Its effects are similar to those of iodide of potassium, for which it is occasionally substituted in the treatment of syphilis, etc.

Dose: 5 to 40 grains.

SATURATED SOLUTION OF IODIDE OF SODIUM.

Take of Iodide of sodium........................... 1 ounce.
Water ....to 1 ounce.

Dissolve. Dose: 5 to 10 minims, increased as desired, largely diluted with water. In syphilis. *E. L. Keyes.*

SODII PHOSPHAS (U. S. et al. Ph.)—PHOSPHATE OF SODIUM.

Take of Bone, calcined to whiteness.................120 ounces.
Sulphuric acid (by weight)............. .... 72 ounces.
Carbonate of sodium........................ sufficient.
Water...................................... sufficient.

Mix the bone and acid in an earthen vessel, add 8 pints of water, digest three days, stirring frequently, and adding water to replace that lost by evaporation. Then add 8 pints of boiling water, and strain, gradually adding more boiling water until it passes nearly tasteless. Allow it to settle, decant, and boil down to 8 pints. Heat in an iron vessel, neutralize with carbonate of sodium previously dissolved in hot water, filter, and set aside to crystallize.

Phosphate of sodium is in transparent crystals, which speedily effloresce when exposed to the air, have a cooling, saline taste, and are freely soluble in water.

It is used in scrofula, joint diseases, rickets, etc.

Dose: 20 to 40 grains.

SODII SALICYLAS—SALICYLATE OF SODIUM.

This salt may be prepared by carefully adding soda or carbonate of sodium to a saturated solution of salicylic acid in alcohol or diluted alcohol, and evaporating to dryness on a water-bath. Care should be taken not to exceed the point of saturation, for if there be an excess of alkali the solution will turn brown. It is safer to stop just short of saturation.

Charles Rice.

Dose: 5 to 30 grains. Commonly administered in aqueous solution, sweetened with syrup or glycerin.

Sodii Santonas (Ger.)—Santonate of Sodium.

Take of Santonin..................................100 parts.
Caustic soda, freshly prepared................13.2 parts.
Water ....450 parts.

Introduce the santonin into a flask, add the soda previously dissolved in the water, and heat on a water-bath until a clear solution results; then filter, evaporate till a pellicle forms, and set aside to crystallize. From the mother-water the retained santonin may be recovered by precipitating it with hydrochloric acid. *Charles Rice.*

Santonate of sodium is in colorless, transparent crystals, of a bitter, saline taste, and soluble in 3 parts of cold water; 120 parts of it correspond with 100 parts of santonin.

Dose: ½ to 3 grains, for children; 3 to 6 grains, for adults.

Elixir of Santonate of Sodium.

Take of Santonate of sodium......................307 grains.
Simple elixir.............................. 1 pint.

Dissolve. Each drachm represents 2 grains of santonin.

Charles Rice.

Dose: 15 minims to 1½ drachm, for children.

SODII SULPHIS (U. S.)—SULPHITE OF SODIUM.

Prepared by passing sulphurous acid gas through a solution of carbonate of sodium until it acquires an acid reaction, evaporating, and crystallizing.

Sulphite of sodium is in colorless crystals, of a cooling, sulphurous taste, and freely soluble in water. It is used in the same manner as hyposulphite of sodium, to arrest fermentation, etc.

Dose: 20 to 60 grains.

Mixture of Sulphite of Sodium.

Take of Sulphite of sodium .... 20 grains.
Infusion calumba........................ 1 ounce.

Mix. One dose. *Royal Chest Hospital.*

SODII SULPHOCARBOLAS—SULPHOCARBOLATE OF SODIUM.

Prepared by precipitating a solution of sulphocarbolate of barium with carbonate of sodium, filtering, evaporating, and crystallizing.

Sulphocarbolate of sodium is in colorless crystals, freely soluble in water. It has been used in low fevers, diphtheria, phthisis, etc.

Dose: 5 to 20 grains.

GARGLE OF SULPHOCARBOLATE OF SODIUM.

Take of	Sulphocarbolate of sodium	4 grains.
	Borax	18 grains.
	Glycerin	24 minims.
	Distilled water	to 1 ounce.

Mix, and dissolve. Antiseptic. *London Throat Hospital.*

MIXTURE OF SULPHOCARBOLATE OF SODIUM.

Take of	Sulphocarbolate of sodium	20 grains.
	Camphor water	1 ounce.

Mix. One dose. *Royal Chest Hospital.*

SPIGELIA (U. S.)—SPIGELIA—PINKROOT.

The root of Spigelia Marilandica, L. (*Nat. ord., Loganiaceæ*), an herbaceous perennial indigenous to the Southern States.

Pinkroot is an excellent anthelmintic for lumbrici. In very large doses it vomits and purges, and may act as a narcotic poison, producing vertigo, convulsions, and insensibility.

Dose: 1 to 2 drachms. It should be followed by a brisk cathartic.

PREPARATIONS.

Extractum Spigeliæ Fluidum (U. S.)—Fluid Extract of Spigelia.

Take of	Spigelia	16 ounces.
	Glycerin	½ pint.
	Alcohol	sufficient.
	Water	sufficient.

Mix 8 ounces of alcohol with 4 ounces each of glycerin and water, moisten the spigelia with 4 ounces of the mixture, and proceed according to the general formula, page 161. Continue the percolation with diluted alcohol until 26 ounces are obtained; reserve the first 10 ounces, add 4 ounces of glycerin to the remainder, evaporate to 6 ounces, and mix with the reserved portion.

Dose: 1 to 2 drachms.

Extractum Spigeliæ et Sennæ Fluidum (U. S.)—Fluid Extract of Spigelia and Senna.

Take of	Fluid extract of spigelia	10 ounces.
	Fluid extract of senna	6 ounces.
	Oil of anise	20 minims.
	Oil of caraway	20 minims.

Mix the extracts, and dissolve the oils in the mixture.

Dose: 1 to 4 drachms.

Infusum Spigeliæ (U. S.)—Infusion of Spigelia.

Take of	Spigelia	½ ounce.
	Boiling water	1 pint.

Macerate two hours in a covered vessel, and strain.

Dose: ½ to 1 ounce for children; 4 to 8 ounces for adults.

STILLINGIA (U. S.)—STILLINGIA—QUEEN'S DELIGHT.

The root of Stillingia sylvatica, L. (*Nat. ord., Euphorbiaceæ*), an herbaceous perennial indigenous to the United States, growing from Virginia southward. It is diaphoretic and alterative, and is used in scrofula, syphilis, chronic cutaneous diseases, chronic rheumatism, etc.

Dose: 15 to 40 grains.

PREPARATIONS.

Extractum Stillingiæ Fluidum (U. S.)—Fluid Extract of Stillingia.

Take of	Stillingia	16 ounces.
	Glycerin	4 ounces.
	Alcohol	sufficient.
	Water	sufficient.

Mix 12 ounces of alcohol, 3 of glycerin, and 1 of water, moisten the stillingia with 4 ounces of the mixture, and proceed according to the general formula, page 161. Finish the percolation with diluted alcohol, reserve 14 ounces, and add 1 ounce of glycerin to the remainder, before evaporation.

Dose: 15 to 40 minims.

COMPOUND SYRUP OF STILLINGIA.

Take of	Stillingia	6 parts.
	Turkey corn	6 parts.
	Blue flag	3 parts.
	Elder flowers	3 parts.
	Pipsissewa	3 parts.
	Coriander	2 parts.
	Prickly ash berries	2 parts.
	Sugar	55 parts.
	Alcohol	sufficient.
	Water	sufficient.

Having reduced the solid ingredients, except the sugar, to a coarse powder, moisten, pack, and percolate, with a menstruum of 1 part of alcohol and 3 parts of water. Reserve the first 35 parts of percolate, evaporate the next 25 parts to 7 parts, which are to be mixed with 3 parts of alcohol, and then with the reserved 35 parts. Lastly, dissolve the sugar in this liquid. *J. U. Lloyd.*

Dose: 1 to 4 drachms.

STRAMONIUM—THORN-APPLE.

STRAMONII FOLIA (U. S. et al. Ph.)—STRAMONIUM LEAVES.

STRAMONII SEMEN (U. S. et al. Ph.)—STRAMONIUM SEED.

The leaves and seed of Datura Stramonium, L. (*Nat. ord., Solanaceæ*) a large annual herb indigenous to Asia, but naturalized in all temperate regions, and found growing in waste places. It has a very strong, disagreeable odor, and a nauseous taste. Both the leaves and seed contain an alkaloid termed *daturia*, to which they owe their medicinal activity.

Stramonium is a narcotic, and in its action closely resembles belladonna. It is used in neuralgia, whooping-cough, epilepsy, asthma, etc. The dried leaves are often smoked for temporary relief in asthmatic paroxysms. In over-doses stramonium is a dangerous poison.

Dose: Of the leaves, 1 to 5 grains; of the seed, 1 to 3 grains.

PREPARATIONS.

Extractum Stramonii Foliorum (U. S., Fr.)—Extract of Stramonium Leaves.

Take of Stramonium leaves, recently dried	12 ounces.
Alcohol	1 pint.
Diluted alcohol	sufficient.

Moisten the stramonium with the alcohol, then percolate with diluted alcohol until 1 pint is obtained; allow this to evaporate spontaneously to 3 ounces. Continue the percolation until 2 pints more are obtained, or the stramonium is exhausted; evaporate this on a water-bath, at or below 160°, to the consistence of syrup, mix with the reserved portion, and continue the evaporation, at or below 120°, until it is reduced to a proper consistence.

Dose: ½ to 1 grain.

The French Codex and German Pharmacopœia prepare also an extract from the fresh leaves.

Extractum Stramonii Seminis (U. S., Br., Fr.)—Extract of Stramonium Seed.

Take of Stramonium seed, in powder	16 ounces.
Diluted alcohol	sufficient.

Macerate the powder in 1 pint of diluted alcohol for four days, then percolate until 2 pints are obtained. Distil off the alcohol, and evaporate on a water-bath until reduced to a proper consistence.

Dose: ¼ to ½ grain.

The British Pharmacopœia treats the powder first with ether, in order to deprive it of its oil, and then proceeds as above.

Tinctura Stramonii (U. S. et al. Ph.)—Tincture of Stramonium.

Take of Stramonium seed, in powder	4 ounces.
Diluted alcohol	sufficient.

Moisten, pack, and percolate to 2 pints.

Dose: 10 to 20 minims.

The French Codex prepares one tincture from fresh, another from dried leaves, but none from the seed.

Unguentum Stramonii (U. S.)—Stramonium Ointment.

Take of Extract of stramonium	60 grains.
Water	½ drachm.
Lard	420 grains.

Rub the extract with the water, then add the lard, and mix.

MIXTURE OF STRAMONIUM.

Take of Tincture of stramonium	½ drachm.
Ether	½ drachm.
Tincture of opium	5 minims.
Camphor water	to 1 ounce.

Mix. One dose. *Royal Chest Hospital.*

PILLS OF STRAMONIUM.

Take of Extract of stramonium	6 grains.
Extract of hyoscyamus	6 grains.
Extract of opium	6 grains.

Mix, and divide into 12 pills.

Dose: 1 pill every three to six hours. In dysmenorrhœa and neuralgia. *Bartholow.*

Take of Extract of stramonium	¼ grain.
Camphor	1 grain.
Squill	2 grains.
Syrup	sufficient.

Make 1 pill. *St. Mary's Hospital.*

STRYCHNIA (U. S. et al. Ph.)—STRYCHNIA—STRYCHNINE.

Take of Nux vomica, rasped	48 ounces.
Lime, in powder	6 ounces.
Hydrochloric acid (by weight)	3½ ounces.
Alcohol	sufficient.
Diluted alcohol	sufficient.
Diluted sulphuric acid	sufficient.
Water of ammonia	sufficient.
Purified animal charcoal	sufficient.
Water	sufficient.

Macerate the nux vomica twenty-four hours in 16 pints of water, acidulated with one-third of the hydrochloric acid; then boil two hours, express, and strain. Treat the residue twice successively in like manner, mix the decoctions, and evaporate to a thin syrup. Then add the lime, previously mixed with 1 pint of water, boil ten minutes, strain, wash, dry, and powder it. Treat the powder repeatedly with diluted alcohol to remove the brucia, until the washings are but faintly reddened by nitric acid. Then boil it with successive portions of alcohol until deprived of bitterness, mix the tinctures, distil off the alcohol, wash the residue, mix it with water, and add sufficient diluted sulphuric acid to dissolve the alcohol. Then add the charcoal, boil a few minutes, filter, evaporate, and crystallize. Dissolve the crystals in water, and add sufficient water of ammonia to precipitate the strychnine. Lastly, dry this on bibulous paper.

Strychnia is a white, or grayish-white powder, of an intensely bitter taste, nearly insoluble in water, slightly soluble in cold, and readily soluble in boiling alcohol.

Its effects do not differ materially from those of nux vomica, which see.

Dose: $\frac{1}{60}$ to $\frac{1}{12}$ grain.

PREPARATIONS.

Strychniæ Sulphas (U. S., Fr.)—Sulphate of Strych.

Take of	Strychnia	1 ounce.
	Diluted sulphuric acid	9 drachms, or sufficient.
	Distilled water	1 pint.

Mix the strychnia with the water, heat gently, and gradually add the acid until the alkaloid is dissolved. Filter, evaporate, and crystallize; drain the crystals, dry them rapidly on bibulous paper, and keep in a well-stopped bottle.

It is a white, crystalline, efflorescent salt, of an intensely bitter taste, and readily soluble in water. Owing to its solubility, it is frequently employed instead of strychnia.

Dose: $\frac{1}{60}$ to $\frac{1}{12}$ grain.

Hypodermic Injection of Strychnia.

Take of	Sulphate of strychnia	2 grains.
	Distilled water	1 ounce.

Dissolve with the aid of heat.

Five minims contain $\frac{1}{48}$ grain. *Bartholow.*

Mixtures of Strychnia.

Take of	Acetate of strychnia	1 grain.
	Compound tincture of cardamom	½ drachm.
	Alcohol	2½ drachms.
	Water	2½ drachms.
	Syrup	to 4 ounces.

Mix. Dose: 1 drachm. *Hospital Formulary.*

Take of Sulphate of strychnia	$\frac{1}{4}$ grain.
Diluted sulphuric acid	$\frac{1}{2}$ ounce.
Sulphate of morphia	2 grains.
Camphor water	$3\frac{1}{2}$ ounces.

Mix. Dose: 1 drachm every hour or two, in choleraic, and colliquative diarrhœa. *Bartholow.*

Take of Sulphate of strychnia	1 grain.
Pyrophosphate of iron	$\frac{1}{2}$ drachm.
Diluted phosphoric acid	$\frac{1}{2}$ ounce.
Syrup of ginger	$3\frac{1}{2}$ ounces.

Mix. Dose: 20 minims three times a day, for a child two years old. In infantile paralysis. *Hammond.*

Solution of Strychnia (Br.).

Take of Strychnia, in crystals	4 grains.
Diluted hydrochloric acid	6 minims.
Rectified spirit	2 drachms.
Distilled water	6 drachms.

Mix the acid with 4 drachms of the water, and dissolve the strychnia in the mixture by the aid of heat; then add the spirit and the remainder of the water.

Dose: 5 to 10 minims. It should be termed solution of hydrochlorate of strychnia.

SULPHUR (U. S. et al. Ph.)—SULPHUR.

Sulphur is found native in many volcanic countries, and is widely distributed in the mineral kingdom, in various combinations. It is a non-metallic element, crystallizable, of a yellow color, a feeble but peculiar odor and taste, insoluble in water, and volatilizable by heat.

It is employed internally as a mild laxative, especially in hemorrhoids, but its chief use is as a topical application in scabies, and other diseases of the skin.

Dose: 1 to 3 drachms.

PREPARATIONS.

Sulphur Sublimatum (U. S. et al. Ph.)—Sublimed Sulphur.

Prepared by subliming the crude article by heat. It is often strongly acid.

Sulphur Lotum (U. S. et al. Ph.)—Washed Sulphur.

Prepared by washing sublimed sulphur with water until entirely free from acid.

Sulphur Præcipitatum (U. S. et al. Ph.)—Precipitated Sulphur.

Take of		
Take of	Sublimed sulphur	12 ounces.
	Lime	18 ounces.
	Hydrochloric acid	sufficient.
	Water	sufficient.

Slake the lime with water, mix the sulphur with it, add 15 pints of water, boil two hours, adding water to preserve the measure, and filter. Dilute the filtered liquid with an equal bulk of water, and drop in the acid as long as a precipitate is produced. Lastly, wash and dry the precipitate.

Unguentum Sulphuris (U. S. et al. Ph.)—Sulphur Ointment.

Take of	Sublimed sulphur	1 ounce.
	Lard	2 ounces.

Mix.

The German preparation is identical with this; the British but half the strength. The French Codex directs: sulphur, 3 parts; oil of almonds, 2 parts; benzoated lard, 6 parts.

COMPOUND SULPHUR OINTMENT (Ger.).

Take of	Washed sulphur	1 part.
	Sulphate of zinc	1 part.
	Lard	8 parts.

Mix.

CONFECTION OF SULPHUR (Br.).

Take of	Sublimed sulphur	4 ounces.
	Bitartrate of potassium	1 ounce.
	Syrup of orange peel	4 ounces.

Rub well together.

Dose: 60 to 120 grains.

CONFECTION OF SULPHUR AND NUTMEG.

Take of	Sulphur	1 ounce.
	Nutmeg	2 drachms.
	Confection of rose	2 drachms.

Mix. Dose: ½ ounce at bedtime. In painful internal hemorrhoids.

Hospital for Ruptured and Crippled.

LOTIONS OF SULPHUR.

Take of	Sulphur	1 drachm.
	Glycerin	1 ounce.
	Rose water	10 ounces.

Mix. Apply two or three times daily. In acne of young women with disordered menses. *Ringer.*

Take of Sulphur	1 part.
Glycerin	1 part.
Rectified spirit	1 part.
Carbonate of potassium	1 part.
Sulphuric ether	1 part.

Mix. To be rubbed into the part affected with comedo.

Tilbury Fox.

Take of Lime	1 part.
Sulphur	2 parts.
Water	20 parts.

Slake the lime with some of the water, then add the remainder and the sulphur, boil to 12 parts, and filter. Used in scabies.

Vlemingkx's Solution.

Powder of Sulphur.

Take of Sublimed sulphur	1 ounce.
Bitartrate of potassium	1 ounce.
Sulphate of magnesium	1 ounce.
Carbonate of magnesium	1 ounce.

Mix thoroughly.

Dose: 1 to 4 drachms in a wineglass of sweetened water, before breakfast. For puerperal women predisposed to hemorrhoids.

Fordyce Barker.

Ointments of Sulphur.

Take of Sublimed sulphur	10 parts.
Tannic acid	5 parts.
Petroleum ointment	85 parts.

Mix. In acne and rosacea. *G. H. Fox.*

Take of Sulphur	6 parts.
Tar	6 parts.
Soft soap	16 parts.
Lard	16 parts.
Chalk	4 parts.

Mix. Used in scabies. *Hebra.*

SULPHURIS IODIDUM (U. S. et al. Ph.)—IODIDE OF SULPHUR.

Take of Iodine	4 ounces.
Sublimed sulphur	1 ounce.

Rub together, introduce the mixture into a flask, close the orifice loosely, and heat gently so as to darken the mass without melting it. Then increase the heat until the mass is melted, incline the flask in different directions, in order to return into the liquid any of the iodine which may have condensed on the inner surface of the vessel; cool, break the flask when the liquid has solidified, reduce the iodide to pieces, and keep in a well-stopped bottle.

Iodide of sulphur is a grayish black substance, having the odor of iodine, soluble in 60 parts of glycerin, but insoluble in water. Used externally in skin diseases.

PREPARATIONS.

Unguentum Sulphuris Iodidi (U. S., Br., Fr.)—Ointment of Iodide of Sulphur.

Take of Iodide of Sulphur.......................... 30 grains.
Lard .... 1 ounce.

Mix.

TABACUM (U. S. et al. Ph.)—TOBACCO.

The dried leaves of Nicotiana Tabacum, L. (*Nat. ord., Solanaceæ*), a plant indigenous to tropical America, but cultivated in most temperate, subtropical, and tropical countries.

Tobacco is narcotic and sedative, and, in full doses, emetic. In overdoses it is a dangerous poison. It is seldom used internally on account of the dangerous prostration which it is liable to produce. Externally, in the form of fomentation, it is often used in the treatment of orchitis. An infusion, or the smoke of burning tobacco, is occasionally introduced into the rectum for the relief of colic, ileus, strangulated hernia, etc. Tobacco has also been used with asserted benefit in traumatic tetanus.

PREPARATIONS.

Infusum Tabaci (U. S.)—Infusion of Tobacco.

Take of Tobacco.............................. 60 grains.
Boiling water............................ 1 pint.

Macerate an hour in a covered vessel, and strain.

Dose: As an enema, ½ to 4 ounces.

It should be employed with great care, since it is liable to induce excessive prostration.

Oleum Tabaci (U. S.)—Oil of Tobacco.

Take of Tobacco .... 12 ounces.

Put it into a retort of green glass, connected with a refrigerated receiver, to which is attached a tube for the escape of gases. Then, by means of a sand-bath, heat the retort to dull redness, and maintain that temperature until the empyreumatic oil ceases to come over. Lastly, separate the dark, oily liquid from the watery portion, and keep in a well-stopped bottle.

Employed externally.

Unguentum Tabaci (U. S.)—Tobacco Ointment.

Take of Tobacco	½ ounce.
Lard	8 ounces.
Water	sufficient.

Moisten, pack, and percolate the tobacco with the water until 4 ounces are obtained. Evaporate this to a soft extract, and mix it with the lard.

Vinum Tabaci (U. S.)—Wine of Tobacco.

Take of Tobacco	1 ounce.
Sherry wine	1 pint.

Macerate seven days, express and filter.

Dose: 10 to 30 minims.

This preparation, in nauseating doses, may be used in the paroxysms of spasmodic asthma, but lobelia is safer, and probably quite as efficient.

TARAXACUM (U. S. et al. Ph.)—DANDELION.

The root of Taraxacum Dens-leonis, Desf. (*Nat. ord., Compositæ*), a small herbaceous perennial indigenous to Europe, but naturalized and very common here.

Dandelion is tonic, diuretic, and slightly aperient. It is used chiefly in dyspepsia attended with hepatic derangement, and in dropsy dependent upon disease of the liver.

Dose: ½ to 2 drachms.

PREPARATIONS.

Extractum Taraxaci (U. S. et al. Ph.)—Extract of Dandelion.

Take of Dandelion, gathered in September .... 60 ounces.

Slice and bruise it, sprinkling on a little water, until reduced to a pulp. Then express and strain the juice, and evaporate it on a water-bath, to a proper consistence.

Dose: 20 to 60 grains.

The German preparation is made from the dried root.

Extractum Taraxici Fluidum (U. S.)—Fluid Extract of Dandelion.

Take of Dandelion	16 ounces.
Glycerin	4 ounces.
Alcohol	sufficient.
Water	sufficient.

Mix 8 ounces of alcohol, 3 of glycerin, and 5 of water, moisten the dandelion with 4 ounces of the mixture, and proceed according to the general formula, page 161. Finish the percolation with diluted alcohol, reserve 14 ounces, and add 1 ounce of glycerin to the remainder, before evaporation.

Dose : ½ to 2 drachms. Used chiefly as a vehicle.

Infusum Taraxaci (U. S.)—Infusion of Dandelion.

Take of Dandelion, bruised	1 ounce.
Boiling water	1 pint.

Macerate two hours in a covered vessel, and strain.

Dose : 1 to 2 ounces.

Succus Taraxaci (U. S., Br.)—Juice of Dandelion.

Take of Fresh dandelion	a convenient quantity.
Alcohol	sufficient.

Bruise the dandelion, express the juice, and to every 5 measures of juice add 1 of alcohol ; after seven days, filter.

Dose : 2 to 4 drachms.

The British Pharmacopœia adds 1 measure of rectified spirit to 3 of the juice.

COMPOUND ELIXIR OF TARAXACUM.

Take of Dandelion root	6 ounces.
Wild cherry bark	4 ounces.
Gentian root	1 ounce.
Orange peel	2 ounces.
Cinnamon	1 ounce.
Coriander	1 ounce.
Anise	¼ ounce.
Caraway	¼ ounce.
Cardamoms	¼ ounce.
Liquorice root	1 ounce.
Simple syrup	2½ pints.
Stronger alcohol	sufficient.
Water	sufficient.

Mix 1 volume of the alcohol with 3 of water, and, having reduced the solids to a powder, moisten, pack, and percolate with the diluted alcohol until 6½ pints of percolate have passed. Mix this with the syrup.

Dose : 1 to 2 drachms. Used as a vehicle for the administration of quinia, etc. *New Remedies.*

TEREBINTHINA (U. S. et al. Ph.)—TURPENTINE.

The concrete oleoresin obtained from Pinus palustris, Mill., and from other species of Pinus (*Nat. ord., Coniferæ*), and commonly known as *white turpentine.* As it exudes from the tree, it is a viscid liquid, of a yellowish color, a terebinthinate odor, and a pungent, bitterish taste. By exposure to the air, it loses some of its volatile principle (*oil of turpentine*), and concretes into solid masses, which are hard and brittle in winter, but become semi-liquid in summer.

Turpentine is seldom employed internally, oil of turpentine being commonly used instead. It is a constituent of numerous preparations for external use, as compound resin cerate, compound galbanum plaster, etc.

Dose: 20 to 60 grains.

TEREBINTHINA CANADENSIS (U. S., Br., Fr.)—CANADA TURPENTINE—CANADA BALSAM—BALSAM OF FIR.

The oleoresin obtained from Abies balsamea, Marshall (*Nat. ord., Coniferæ*), the balsam fir-tree of North America. It is a transparent, viscid liquid, of a yellowish color, and a peculiar, agreeable, balsamic odor and taste. Upon exposure to the air it slowly hardens into a brittle mass, owing to the loss of its volatile principle. Its chief use, like the preceding, is as a constituent of cerates, plasters, etc.

Several other turpentines are employed medicinally, but all agree essentially in their therapeutic effects. Quite recently the use of *Chian Turpentine* (*Terebinthina Chia*) the oleoresin obtained from Pistacia terebinthus, L. (*Nat. ord., Anacardiaceæ*), a small tree found in the basin of the Mediterranean, has been revived, and it has been much lauded as a remedy for cancer. Its claims, however, have not been substantiated, and it will soon take its place in the long list of discarded cancer specifics.

THUJA—ARBOR VITÆ—WHITE CEDAR.

The small branchlets with the scale-like leaves of Thuja occidentalis, L. (*Nat. ord., Coniferæ*), the common white cedar of the Northern States and Canada. It is cultivated in Europe, for ornamental purposes.

Thuja possesses properties somewhat resembling those of savin. Dr. J. R. Leaming, of New York, who has employed it for many years, finds it valuable in amenorrhœa from simple causes, in pulmonary hemorrhages, in malignant diseases, etc., and as a topical application to venereal and other warts, to cancerous ulcerations and tumors, etc.

Dose: Of the fresh leaves ½ to 1 drachm. It is not, however, administered in substance.

PREPARATIONS.

Tincture of Thuja (Ger.).

Take of Thuja (fresh)............................ 5 parts.
Alcohol.................................. 6 parts.

Having bruised the thuja, add the alcohol, macerate eight days, and filter.

Dose: ½ to 1 drachm.

The following preparations were prepared at the suggestion of Dr. J. R. Leaming, and have been largely employed by him, and by others upon his recommendation.

Elixir of Thuja.

Take of Thuja (fresh)............................ 12 ounces.
Alcohol, 1 part—Glycerin, 1 part............ sufficient.

Percolate to 12 ounces, and set aside. Continue the percolation to 12 ounces more, evaporate to 6 ounces, and mix with the reserved portion. Add an equal quantity (18 ounces) of simple elixir. Each drachm represents 20 grains of thuja.

Dose: 1 to 3 drachms. *W. H. Lawes.*

Fluid Extract of Thuja.

Take of Thuja.................................. 16 ounces.
Alcohol .... sufficient.

Percolate to 10 ounces, and set aside. Continue the percolation until the thuja is exhausted, evaporate to 6 ounces, and mix with the reserved portion.

Dose: ½ to 1 drachm. *W. H. Lawes.*

Glycerite of Thuja.

Take of Thuja .... 8 ounces.
Alcohol, 1 part—Glycerin, 1 part............ sufficient.

Percolate to 8 ounces, and evaporate to 4 ounces. *W. H. Lawes.*

May be made into suppositories, or mixed with the fluid extract, for application to the os uteri, upon a pessary of cotton. *J. R. Leaming.*

THYMOL—THYMOL.

A crystalline substance obtained from oil of thyme, oil of American horsemint, and from some other essential oils. By submitting oil of thyme to a low temperature for a few days, thymol separates in the form of nearly colorless crystals, of a pleasant odor, an aromatic, burning taste, and soluble in 1 part of alcohol, 120 parts of glycerin, and 1200 parts of water. It is a powerful antiseptic, like carbolic acid, but is in general less irritating. Used topically, in many different forms, as a surgical dressing, by inhalation in throat diseases, etc.

Inhalation of Thymol.

Take of Thymol	6 grains.
Rectified spirit	1 drachm.
Light carbonate of magnesium	3 grains.
Water	to 8 ounces.

Mix. 1 drachm in 20 ounces of water at 140° for each inhalation. A strong stimulant and disinfectant; very useful in pharyngitis and laryngitis when associated with exanthemata. *London Throat Hospital.*

Lotion of Thymol.

Take of Thymol	5 grains.
Rectified spirit	1 ounce.
Glycerin	1 ounce.
Water	to 8 ounces.

Dissolve. *Crocker.*

Ointment of Thymol.

Take of Thymol	8 grains.
Oil of almonds	1 drachm.
Cerate	5 drachms.

Mix. Used in eczema.

TRAGACANTHA (U. S. et al. Ph.)—TRAGACANTH.

A gummy exudation from several species of Astragalus (*Nat. ord., Leguminosæ*), shrubs indigenous to Persia.

Tragacanth occurs in hard and fragile flakes, or roundish masses, of a yellowish or brownish color, odorless, and almost tasteless.

It is insoluble in alcohol, and ether, but with water forms a jelly-like mucilage, which is nutritious and demulcent. Used in making pills, troches, etc.

PREPARATIONS.

Mucilago Tragacanthæ (U. S., Br., Fr.)—Mucilage of Tragacanth.

Take of Tragacanth	1 ounce.
Boiling water	1 pint.

Macerate twenty-four hours, stirring occasionally, then beat into a uniform consistence, and strain forcibly through muslin.

Used as a vehicle.

COMPOUND POWDER OF TRAGACANTH (Br.).

Take of Tragacanth	1 ounce.
Gum arabic	1 ounce.
Starch	1 ounce.
Sugar	3 ounces.

Rub them well together.

Dose: 20 to 60 grains.

ULMUS (U. S., Br., Fr.)—SLIPPERY-ELM BARK.

The inner bark of Ulmus fulva, Michx. (*Nat. ord.*, *Urticaceæ*), a medium sized tree indigenous to the United States.

Slippery-elm bark contains a large quantity of mucilaginous matter, and is a valuable demulcent in acute inflammatory diseases, as dysentery, and those of the genito-urinary organs.

PREPARATIONS.

Mucilago Ulmi (U. S.)—Mucilage of Slippery-elm Bark.

Take of Slippery-elm bark	1 ounce.
Boiling water	1 pint.

Macerate two hours in a covered vessel, and strain.

Taken *ad libitum*.

UVA URSI (U. S. et al. Ph.)—UVA URSI—BEARBERRY.

The leaves of Arctostaphylos Uva ursi, Spreng. (*Nat. ord.*, *Ericaceæ*), a very small shrub indigenous to the high latitudes of the Northern Hemisphere.

Uva ursi is tonic, astringent, and diuretic. It is used chiefly in catarrhal diseases of the genito-urinary organs.

Dose: 20 to 60 grains.

PREPARATIONS.

Decoctum Uvæ Ursi (U. S.)—Decoction of Uva Ursi.

Take of Uva Ursi	1 ounce.
Water	sufficient.

Boil for fifteen minutes, strain, and add sufficient water through the strainer to make 1 pint.

Dose: 1 to 2 ounces.

Extractum Uvæ Ursi Fluidum (U. S.)—Fluid Extract of Uva Ursi.

Take of Uva Ursi	16 ounces.
Glycerin	4 ounces.
Alcohol	sufficient.
Water	sufficient.

Mix 8 ounces of alcohol, 3 of glycerin, and 5 of water, moisten the uva ursi with ½ pint of the mixture, and proceed according to the general formula, page 161. Finish the percolation with diluted alcohol, reserve 14 ounces, and add 1 ounce of glycerin to the remainder, before evaporation.

Dose: 20 to 60 minims.

INFUSION OF BEARBERRY (Br.).

Take of Bearberry leaves	½ ounce.
Boiling distilled water	10 ounces.

Infuse in a covered vessel for two hours, and strain.

Dose: 1 to 2 ounces.

VALERIANA (U. S. et al. Ph.)—VALERIAN.

The root of Valeriana officinalis, L. (*Nat. ord., Valerianaceæ*) an herbaceous perennial indigenous to Europe and Asia, but cultivated to some extent in this country.

Valerian has a hot, pungent taste, and a very strong, disagreeable odor. It acts as a stimulant and antispasmodic, and is used chiefly in hysteria, though it is occasionally employed in other nervous diseases, as chorea, epilepsy, whooping-cough, etc.

Dose: 30 to 60 grains.

PREPARATIONS.

Extractum Valerianæ (U. S., Fr., Ger.)—Extract of Valerian.

Take of Valerian	12 ounces.
Alcohol	1 pint.
Diluted alcohol	sufficient.

Moisten the valerian with 4 ounces of the alcohol, pack it in a percolator, pour on the remainder, and continue the percolation with diluted alcohol until 1 pint of tincture is obtained. Allow this to evaporate spontaneously to 3 ounces. Continue the percolation until 2 pints more are obtained, evaporate this on a water-bath to the consistence of syrup, mix with the reserved portion, and evaporate, at or below 120°, to the proper consistence.

Dose: 10 to 30 grains.

Extractum Valerianæ Fluidum (U. S.)—Fluid Extract of Valerian.

Take of Valerian .... 16 troy ounces.
Stronger alcohol .... sufficient.

Moisten the valerian with 5 ounces of the alcohol, and proceed according to the general formula, page 161.

Dose: 30 to 60 minims.

Infusum Valerianæ (U. S., Br.)—Infusion of Valerian.

Take of Valerian .... ½ ounce.
Water .... sufficient.

Moisten, pack, and percolate until the filtered liquid measures 1 pint. Or, macerate the valerian in 1 pint of boiling water for two hours in a covered vessel, and strain.

Dose: 1 to 2 ounces.

Tinctura Valerianæ (U. S. et al. Ph.)—Tincture of Valerian.

Take of Valerian .... 4 ounces.
Diluted alcohol .... sufficient.

Moisten, pack, and percolate to 2 pints.

Dose: ½ to 2 drachms.

Tinctura Valerianæ Ammoniata (U. S., Br.)—Ammoniated Tincture of Valerian.

Take of Valerian.................................. 4 ounces.
Aromatic spirit of ammonia................ 2 pints.

Macerate seven days, express and filter.

Dose: ½ to 1 drachm.

Ethereal Tincture of Valerian (Ger., Fr.).

Take of Valerian .... 1 part.
Spirit of ether.......................... 5 parts.

Macerate eight days, express and filter.

Dose: ½ to 1 drachm.

Mixtures of Valerian.

Take of Ammoniated tincture of valerian............. 2 drachms.
Lacto-peptin.......................... 32 grains.
Bicarbonate of sodium.................... 12 grains.
Glycerin .... 2 drachms.
Water .... 6 drachms.
Orange flower water.......................... 1 ounce.

Mix. Dose: 20 to 60 minims in 1 or 2 drachms of warm water, as required. A carminative for infants, which contains no opiate.

F. A. Burrall.

Take of Fluid extract of valerian .... 30 parts.
Camphorated tincture of opium .... 90 parts.
Purified chloroform .... 1 part.

Mix. Dose: 1 drachm in a wineglass of cold water every hour. In colic and gastralgia. *Daniel Lewis.*

Pills of Valerian and Iron.

Take of Extract of valerian .... ½ drachm.
Dried sulphate of iron .... 7 grains.

Mix, and divide into 30 pills.

Dose: One pill three times a day. In obstinate constipation, and anasarca. *Hospital for Ruptured and Crippled.*

VERATRIA (U. S. et al. Ph.)—VERATRIA.

An alkaloid obtained from the seeds of Veratrum Sabadilla, Retzius (*Nat. ord., Melanthaceæ*) a bulbous plant indigenous to Mexico.

Its therapeutic effects are similar to those of veratrum viride, but it is seldom employed internally. Used externally in neuralgia, gout, rheumatism, etc.

Dose: $\frac{1}{30}$ to $\frac{1}{12}$ grain.

PREPARATION.

Unguentum Veratriæ (U. S., Br.)—Veratria Ointment.

Take of Veratria .... 20 grains.
Lard .... 1 ounce.

Rub the veratria with a little of the lard, then add the remainder, and mix.

British Pharmacopœia directs: veratria, 8 grains; prepared lard, 1 ounce; olive oil, ½ drachm.

VERATRUM VIRIDE (U. S., Br.)—AMERICAN HELLEBORE—GREEN HELLEBORE.

The rhizome of Veratrum viride, Ait. (*Nat. ord., Liliaceæ*), a tall, herbaceous perennial indigenous to North America, growing in moist situations, often associated with skunk's cabbage (*Symplocarpus fœtidus*). It should not be confounded with European green hellebore (*Helleborus viridis, L.*), a very different plant, nor yet with white hellebore (*Veratrum album, L.*), which, though officinal, is seldom or never used in this country, and is therefore omitted here.

The most important therapeutic action of veratrum is as a cardiac sedative. In large doses, or when too long continued, it induces vomiting and great prostration. It is used with great benefit in a variety of affections, but chiefly those of an acute inflammatory character, as in the earlier stages of pneumonia, pleurisy, acute rheumatism, puerperal inflammations, etc.

Dose: 1 to 4 grains.

PREPARATIONS.

Extractum Veratri Viridis Fluidum (U. S.)—Fluid Extract of American Hellebore.

Take of	American hellebore	16 ounces.
	Stronger alcohol	sufficient.

Moisten the hellebore with 5 ounces of the alcohol and proceed according to the general formula, page 161.

Dose: 1 to 4 minims.

This is the best form in which to administer veratrum, as the dose can be carefully regulated, and the effects noted.

Tinctura Veratri Viridis (U. S., Br.)—Tincture of American Hellebore.

Take of	American hellebore	16 ounces.
	Alcohol	sufficient.

Moisten, pack, and percolate to 2 pints.

Dose: 3 to 10 minims.

ZINCUM (U. S., Br., Fr.)—ZINC.

ZINCI ACETAS (U. S. et al. Ph.)—ACETATE OF ZINC.

Take of	Commercial oxide of zinc	2 ounces.
	Acetic acid	8½ ounces.
	Distilled water	5 ounces.

Mix the acid and water, digest the oxide in the mixture for half an hour, then heat to the boiling-point, filter while hot, and set aside to crystallize. Drain the crystals, and dry them on bibulous paper.

It is in white, micaceous, efflorescent crystals, which are freely soluble in water. Chiefly used as a topical astringent, though it may be employed internally as an antispasmodic and tonic, and, in large doses, as an emetic.

Dose: 1 to 2 grains, tonic; 10 to 20 grains, emetic.

Injections of Acetate of Zinc.

Take of	Acetate of zinc	12 grains.
	Water	4 ounces.

Dissolve. Glycerin may be substituted for ½ or 1 ounce of the water. Used in the later stages of gonorrhœa. *Bumstead.*

Take of Acetate of zinc .... 6 grains.
Acetate of morphia .... 1 grain.
Tannic acid .... 3 grains.
Orange flower water .... 1 drachm.
Water .... to 1 ounce.

Mix. The so-called *Matico Injection.* *Bumstead.*

OINTMENT OF ACETATE OF ZINC.

Take of Acetate of zinc .... 2 grains.
Rose water .... 1 drachm.
Cold cream .... 1 ounce.

Mix. Used in erythema and herpes. *Tilbury Fox.*

ZINCI BROMIDUM—BROMIDE OF ZINC.

Take of Bromine .... 10 parts.
Distilled water .... 20 parts.
Granulated zinc .... 5 parts.

Introduce the bromine and water into a flask, and gradually add the zinc. Warm to about 100°, preventing the escape of vapors by partially closing the flask with a small glass funnel. Digest a few hours, filter the colorless liquid through asbestos, and evaporate it in a porcelain capsule with a gentle heat, until a sample, removed on a cold glass rod, congeals. Then raise the heat for a short time, taking care that only aqueous vapors rise, and, finally, stir until a dry mass remains, which must be immediately transferred to small, glass-stopped vials.

It may also be prepared by dissolving precipitated carbonate of zinc in hydrobromic acid, filtering, and proceeding as above. *Charles Rice.*

It produces essentially the same effects as the other bromides.

Dose: ½ to 2 grains.

ZINCI CARBONAS PRÆCIPITATA (U. S., Br.)—PRECIPITATED CARBONATE OF ZINC.

Take of Sulphate of zinc .... 12 ounces.
Carbonate of sodium .... 12 ounces.
Water .... 8 pints.

Dissolve the salts separately, with the aid of heat, each in 4 pints of water, mix the solutions, and allow the precipitate to subside. Lastly, decant, wash the precipitate with hot water until the washings are nearly tasteless, and dry it with a gentle heat.

It is a soft, white powder, tasteless, and insoluble. Used externally as an astringent and desiccant. The impure native carbonate (*Calamine*) is used for the same purposes.

PREPARATIONS.

Ceratum Zinci Carbonatis (U. S.)—Cerate of Carbonate of Zinc.

Take of	Precipitated carbonate of zinc	2 ounces.
	Ointment	10 ounces.

Mix.

LOTION OF CARBONATE OF ZINC.

Take of	Levigated calamine	6 drachms.
	Oxide of zinc	6 drachms.
	Glycerin	2 ounces.
	Lime water	to 12 ounces.

Mix. *St. Mary's Hospital.*

POWDER OF CARBONATE OF ZINC.

Take of	Prepared calamine	½ ounce.
	Oxide of zinc	½ ounce.

Mix. As an absorbent powder. *British Skin Hospital.*

ZINCI CHLORIDUM (U. S. et al. Ph.)—CHLORIDE OF ZINC.

Take of solution of chloride of zinc a convenient quantity.

Evaporate to dryness, fuse, pour the liquid upon a flat stone, and when it has solidified, break in pieces, and keep in a well-stopped bottle.

It is a white, deliquescent salt, freely soluble in water, alcohol, and ether. Its chief use is as an escharotic, though it is occasionally employed internally.

Dose: ½ to 2 grains.

PREPARATIONS.

Liquor Zinci Chloridi (U. S., Br.)—Solution of Chloride of Zinc.

Take of	Zinc, in small pieces	6 ounces.
	Nitric acid	150 grains.
	Precipitated carbonate of zinc	150 grains.
	Hydrochloric acid	sufficient.
	Distilled water	sufficient.

Gradually add sufficient hydrochloric acid to the zinc to dissolve it, strain, add the nitric acid, and evaporate to dryness. Dissolve in 5 ounces of distilled water, add the carbonate, and agitate the mixture occasionally during twenty-four hours; then filter, adding enough distilled water through the filter to make 1 pint.

Used as a disinfectant, and, largely diluted, as an astringent and antiseptic.

CAUSTIC OF CHLORIDE OF ZINC.

Take of Chloride of zinc.......................... 1 part.
Oxide of zinc .... 1 part.

Mix. *British Skin Hospital.*

INHALATION OF CHLORIDE OF ZINC.

Take of Chloride of zinc.......................... 20 grains.
Water .... 8 ounces.

Dissolve. Use by means of a steam atomizer. Astringent and antiseptic. *G. M. Lefferts.*

INJECTION OF CHLORIDE OF ZINC.

Take of Chloride of zinc.......................... 1 grain.
Rose water.......................... 4 to 8 ounces.

Dissolve. Used in gonorrhœa; the injections should be frequently repeated. *Bartholow.*

ZINCI IODIDUM—IODIDE OF ZINC.

Prepared by digesting zinc in a mixture of iodine and water.

Iodide of zinc occurs as a white, deliquescent mass, of a metallic taste. It is used internally as a tonic and antispasmodic, and externally as a caustic, stimulant, etc.

Dose: ½ to 4 grains.

ZINCI OXIDUM (U. S. et al. Ph.)—OXIDE OF ZINC.

Take of Precipitated carbonate of zinc............... 12 ounces.

Expose it, in a shallow vessel, to a low red heat until the water and carbonic acid are expelled.

It is a yellowish-white powder, tasteless, and insoluble. Used internally as a tonic and antispasmodic, and topically in skin diseases.

Commercial oxide of zinc (*Zinci Oxidum Venale*) is also officinal. Used in preparations.

Dose: 1 to 8 grains.

PREPARATIONS.

Unguentum Zinci Oxidi (U. S. et al. Ph.)—Ointment of Oxide of Zinc.

Take of Oxide of zinc.......................... 80 grains.
Ointment of benzoin......................400 grains.

Rub thoroughly together.

British Pharmacopœia: oxide of zinc, 80 grains; benzoated lard, 1 ounce. French Codex: oxide of zinc, 1 part; benzoated lard, 9 parts. German Pharmacopœia: commercial oxide of zinc, 1 part; rose ointment 9 parts.

Lotion of Oxide of Zinc.

Take of Oxide of zinc .... 2 drachms.
Calamine .... ½ ounce.
Glycerin .... 2 drachms.
Rose water .... 8 ounces.

Mix. Used in eczema, generally where the surface is tender and red.
Tilbury Fox.

Ointments of Oxide of Zinc.

Take of Oxide of zinc .... 1 drachm.
Carbonate of lead .... 1 drachm.
Spermaceti .... 1 ounce.
Olive oil .... sufficient.

Make a soft ointment. Used in seborrhœa, when the skin is inflamed.
Neumann.

Take of Zinc ointment .... 1 ounce.
Carbolic acid .... 30 grains.

Mix. *Middlesex Hospital.*

Pills of Oxide of Zinc.

Take of Oxide of zinc .... 1 drachm.
Extract of belladonna .... 5 grains.

Mix, and divide into 20 pills.
Dose: 1 pill three times a day. In whooping-cough. *Bartholow.*

Take of Oxide of zinc .... 2½ grains.
Hydrochlorate of morphia .... ⅛ grain.
Extract of hops .... 2 grains.

Make 1 pill. Dose: 1 to 3 pills. *Consumption Hospital.*

Powders of Oxide of Zinc.

Take of Oxide of zinc .... 6 to 12 grains.
Saccharated pepsin .... 30 grains.
Subnitrate of bismuth .... 1 to 2 drachms.

Mix, and divide into 12 powders.
Dose: 1 powder every four to six hours. In summer diarrhœa of children. *Bartholow.*

Take of Oxide of zinc .... 1 ounce.
Calamine .... ½ ounce.
Corn meal .... 4 ounces.

Mix. Absorbent powder for excoriated surfaces. *Tilbury Fox.*

ZINCI PHOSPHIDUM—PHOSPHIDE OF ZINC.

May be prepared by adding phosphorus, in small pieces, to fused zinc in a covered crucible. It is a gray, crystalline, or friable mass. Used in nervous diseases.

Dose: $\frac{1}{10}$ to $\frac{1}{8}$ grain.

PILLS OF PHOSPHIDE OF ZINC AND NUX VOMICA.

Take of Phosphide of zinc	3 grains.
Extract of nux vomica	10 grains.
Confection of roses	sufficient.

Mix, and divide into 30 pills.

Dose: 1 pill after each meal, commencing ten days before the menstrual period. For the headaches which attend or precede menstruation, when due to defective innervation, and difficult ovulation.

When menstruation is about to commence, the patient should take two 5 grain capsules of *apiol* after each meal, continuing their use during the whole menstrual period.

A pill of phosphorus $\frac{1}{50}$ grain, and nux vomica $\frac{1}{8}$ grain, may be substituted for the pill of phosphide of zinc and nux vomica. *Fordyce Barker.*

ZINCI SULPHAS (U. S. et al. Ph.)—SULPHATE OF ZINC.

Prepared by dissolving zinc in diluted sulphuric acid, and crystallizing.

Sulphate of zinc is in colorless, transparent crystals, of a styptic, metallic taste, and readily soluble in water. In small doses it is tonic, antispasmodic, and astringent; in large doses, a prompt and efficient emetic. Externally it is used as a caustic, and, in solution, as an injection in gonorrhœa, leucorrhœa, etc.

Dose: 1 to 2 grains, tonic, etc.; 10 to 20 grains, emetic.

COLLYRIUM OF SULPHATE OF ZINC.

Take of Sulphate of zinc	1 to 2 grains.
Water	1 ounce.

Dissolve.

INHALATION OF SULPHATE OF ZINC.

Take of Sulphate of zinc	40 grains.
Water	8 ounces.

Dissolve. Used by means of steam atomizer. Astringent.

G. M. Lefferts.

Injections of Sulphate of Zinc.

Take of Sulphate of zinc .... 8 grains.
Acetate of lead .... 8 grains.
Chloride of ammonium .... 4 grains.
Alum .... 4 grains.
Rose water .... 1 ounce.

Mix. Used in gonorrhœa. *Bartholow.*

Take of Sulphate of zinc .... 30 grains.
Acetate of lead .... 30 grains.
Rose water .... 6 ounces.

Mix. Used in gonorrhœa. *Bumstead.*

Sulphate of Zinc Paint.

Take of Sulphate of zinc .... 1 drachm.
Glycerin .... ½ ounce.
Water .... ½ ounce.

Mix. Used with the laryngeal brush. Astringent.
G. M. Lefferts.

Take of Sulphate of zinc .... 60 grains.
Water .... 1 ounce.

Dissolve. Used like the preceding. *London Throat Hospital.*

Pills of Sulphate of Zinc.

Take of Sulphate of zinc .... 12 grains.
Opium .... 12 grains.
Ipecac .... 12 grains.

Mix, and make 12 pills.

Dose: 1 pill three or four times a day. In chronic diarrhœa and chronic dysentery. *Bartholow.*

ZINCI SULPHOCARBOLAS—SULPHOCARBOLATE OF ZINC.

Prepared by mixing two volumes of carbolic acid with one of sulphuric acid, heating to 290°, allowing to cool, diluting with water, and gently warming with a slight excess of oxide of zinc. On filtering and evaporating, crystals are obtained which should be dried on bibulous paper.

Used topically as a stimulant and antiseptic.

Inhalation of Sulphocarbolate of Zinc.

Take of Sulphocarbolate of zinc .... 5 grains.
Distilled water .... 1 ounce.

Dissolve. Used by means of a spray apparatus. In secondary syphilis of the pharynx and larynx. *London Throat Hospital.*

ZINCI VALERIANAS (U. S. et al. Ph.)—VALERIANATE OF ZINC.

Prepared by mixing hot solutions of valerianate of sodium and sulphate of zinc, cooling, and skimming off the crystals which form.

It occurs in white, pearly scales, having a faint odor of valerianic acid, a styptic taste, and is sparingly soluble in water. Used in hysteria, epilepsy, and other nervous affections.

Dose: 1 to 2 grains.

PILLS OF VALERIANATE OF ZINC.

Take of Valerianate of zinc	20 grains.
Extract of gentian	20 grains.
Extract of nux vomica	5 grains.

Mix, and divide into 20 pills.

Dose: 1 pill three or four times a day. In neuralgia dependent upon reflex irritation from the female pelvic organs. *Bartholow.*

Take of Valerianate of zinc	½ grain.
Sulphate of quinia	⅓ grain.
Compound rhubarb pill	1 grain.
Extract of gentian	2 grains.

Make 1 pill. *London Hospital.*

ZINGIBER (U. S. et al. Ph.)—GINGER.

The rhizome of Zingiber officinale, Roscoe (*Nat. ord., Zingiberaceæ*), a reed-like plant indigenous to Asia, but cultivated in most tropical countries.

Ginger is an aromatic stimulant and carminative. Applied externally it is a mild rubefacient. It is used in flatulent colic to relieve pain and expel flatus, in diarrhœa, etc., and is often combined with other medicines to correct their irritating properties.

Dose: 10 to 20 grains.

PREPARATIONS.

Extractum Zingiberis Fluidum (U. S.)—Fluid Extract of Ginger.

Take of Ginger	16 ounces.
Alcohol	sufficient.

Moisten the ginger with 4 ounces of alcohol, and proceed according to the general formula, page 161.

Dose: 10 to 20 minims.

Infusum Zingiberis (U. S.)—Infusion of Ginger.

Take of Ginger .... ½ ounce.
Boiling water.................................. 1 pint.

Macerate for two hours in a covered vessel, and strain.

Dose: 1 to 2 ounces.

Oleoresina Zingiberis (U. S.)—Oleoresin of Ginger.

Take of Ginger.................................... 12 ounces.
Stronger ether.................................. 12 ounces.
Alcohol.. sufficient.

Moisten the ginger with the ether, then percolate with alcohol until 12 ounces of liquid have passed. Distill off most of the ether, then expose the residue in a capsule until the volatile part has evaporated. Preserve in a well-stopped bottle.

Dose: ⅛ to 1 minim, or grain.

Syrupus Zingiberis (U. S., Br.)—Syrup of Ginger.

Take of Fluid extract of ginger...................... 1 ounce.
Carbonate of magnesium...................160 grains.
Sugar .... 72 ounces.
Water.. 42 ounces.

Rub the extract first with the carbonate and 2 ounces of sugar, then with the water added gradually, and filter. To the filtered liquid add the remainder of the sugar, dissolve with a gentle heat, and strain while hot.

Dose: 1 to 2 drachms.

Tinctura Zingiberis (U. S. et al. Ph.)—Tincture of Ginger.

Take of Ginger.................................... 8 ounces.
Alcohol.. sufficient.

Moisten, pack, and percolate to 2 pints.

Dose: 10 to 30 minims.

STRONGER TINCTURE OF GINGER (Br.).

Take of Ginger .... 10 ounces.
Rectified spirit.............................. sufficient.

Moisten, pack, and percolate to 20 ounces.

Dose: 5 to 20 minims.

INDEX.

www.ingramcontent.com/pod-product-compliance
Lightning Source LLC
Chambersburg PA
CBHW021349210326
41599CB00011B/814

* 9 7 8 3 3 3 7 1 8 7 1 7 0 *